# Introduction to Finite Element Analysis

S. Unnikrishnan Nair · S. Somanath
Editors

# Introduction to Finite Element Analysis

## A Textbook for Engineering Students

With Application of “FEAST$^{SMT}$” through worked out problems

Springer

*Editors*
S. Unnikrishnan Nair
VSSC
Indian Space Research Organisation
Thiruvananthapuram, India

S. Somanath
Indian Space Research Organisation
Bengaluru, India

ISBN 978-981-19-7988-0 ISBN 978-981-19-7989-7 (eBook)
https://doi.org/10.1007/978-981-19-7989-7



This Springer imprint is published by the registered company Springer Nature Singapore Pte Ltd.
The registered company address is: 152 Beach Road, #21-01/04 Gateway East, Singapore 189721, Singapore

# Foreword

The Finite Element Method (FEM), which originated as an offshoot of Matrix Analysis of Structures six decades back, has today evolved into the most powerful tool for numerical analysis of complex problems in various disciplines of engineering. Although the bulk of current applications of FEM is still in the field of Structural and Solid Mechanics, engineers working in other areas like Fluid Mechanics and Heat Transfer also extensively use it to obtain approximate numerical solutions. In the early years of its development, the "stiffness matrices" associated with various structural elements like beams and plates were derived using elementary concepts in Strength of Materials. Subsequently, the variational basis for FEM was developed which significantly amplified its scope of application and also established it firmly as a numerical technique, along with the Finite Difference Method (FDM), to solve boundary value problems in engineering science. In recent years, FEM has overtaken FDM due to its ease of modeling complicated domains and boundary conditions as well as governing differential equations. FEM differs from FDM in the sense that the governing differential formulation for a boundary value problem is first converted into a variation formulation (dealing with minimization of a suitably derived functional), which is then solved approximately in the spirit of the Rayleigh–Ritz method. However, unlike the latter, which requires an approximation for the fundamental unknown variable (the displacement field in the case of Solid Mechanics) to be introduced over the entire domain, and is therefore not a trivial task, FEM first discretizes the domain into elements and introduces the above approximation using simple polynomial functions over the realm of individual elements. The treatment of traction boundary condition is simple in FEM since the variational principle (like minimum potential energy theorem) naturally takes care of it. Other tools like the Galerkin and penalty/Lagrange multiplier methods are also available to cast governing differential equations and boundary/constraint conditions into a finite element framework.

The present book, authored by experienced engineers and scientists, gives a comprehensive introduction to FEM for final year undergraduate or first year graduate students. It begins with a crisp summary of basic concepts in Linear Algebra and Calculus of Variations and then develops the FEM procedure for static analysis in Solid/Structural Mechanics. Also, the students are introduced briefly to dynamic and buckling analysis as well as to heat transfer applications using FEM. The main strength of this book is the exposition to the companion FEM package, $\text{FEAST}^{\text{SMT}}$, developed indigenously, in the last four chapters to solve the various problems detailed above. Overall, this book should serve to establish the fundamentals of FEM and motivate students to apply it to solve practical engineering as well as research problems. Indeed, FEM combined with computer-aided design and optimization procedures is now an essential tool in industries for robust design of engineering components. This book, along with its companion $\text{FEAST}^{\text{SMT}}$ student version computer code, should prepare students to take up such challenging tasks in their career. Also, it should provide the prerequisites to students to learn further advanced finite element concepts in nonlinear Structural Analysis with applications in areas like metal forming and impact/crashworthiness design.

Prof. R. Narasimhan
Indian Institute of Science
Bengaluru, India

# Preface

Finite element method is one of the most popular numerical techniques for solving partial differential equations describing physical processes. The subject of finite element is rooted in concepts which are, to some extent, esoteric to many students and practicing engineers and hence is an impediment to many beginners. There are several textbooks on finite element method and it is very difficult to make the right choice on textbooks, especially for the undergraduates, to have a reasonable entry into the subject for a fair exposure, covering all contemporaneous aspects of finite element method.

Indian Spaec Research Organisation (ISRO) recognized the need to make indigenous products for aerospace applications and has been developing general-purpose finite element software FEAST$^{SMT}$ (acronym for Finite Element Analysis of Structure) for structural engineering applications over the last three decades. This software is completely an indigenous effort complying with industry standard practices and has embraced many of the user-friendly features of commercially available finite element software products. The software is being continuously updated and enhanced in its capability by the team and many of our collaborators across many institutions. Appropriate quality and reliability practices have been adopted during the continuous development process of FEAST$^{SMT}$ to ensure accuracy of the results and robustness of the software. Presently, FEAST$^{SMT}$ is available commercially for industries, educational institutions, and students.

To popularize the usage of FEAST$^{SMT}$ as true Indian Finite Element Analysis software, increase awareness and attract researchers working in the domains closely associated with finite element methods, many outreach programs are conducted annually across India. Feedback from the user communities of FEAST$^{SMT}$ comprising of academia and industries was the motivation for conceiving a standard university-level textbook with a link to the powers of FEAST$^{SMT}$ for undergraduate students and reference for practicing engineers. The book is authored by experienced scientists and engineers who have immensely contributed to the growth of this domain within our country and having exposure of using this software for the space program.

My heartiest appreciations to the team of contributing authors lead by editors, who have put in coordinated efforts and displayed immense patience to improve the overall presentation of the subject and applications. Professor R. Narasimhan of Indian Institute of Science, Bengaluru, an iconic figure in the realm of finite element analysis, research, and teaching, has found time to go through the book, offered many valuable suggestions, and penned the foreword. I thank him profusely for this kind gesture.

I am sure that the students, practicing engineers, and curious individuals interested to know about finite element method from its theory to software applications, can richly benefit from this textbook. The introductory format of the book will evince interest in readers to pursue profound facets of the subject. Further, they will gain confidence in using a general-purpose finite element software FEAST$^{SMT}$, which will augment their skill-set while they move into industry-level operations in their professions.

I congratulate the authors and editors for their efforts to come out with this textbook banking on their core strengths in the domain of structural mechanics with an aim to popularize FEAST$^{SMT}$ software which is a true symbol of *Atmanirbaratha*.

S. Somanath
Chairman
Indian Space Research Organisation
Bengaluru, India

# Acknowledgements

The idea of writing a textbook in finite elements has stemmed from the need to have a student-friendly textbook for the undergraduate students of engineering. The content of the book has been arrived at after series of brainstorming sessions with the authors of various chapters and with a focus to make this book a self-teaching text for undergraduate students.

We are extremely thankful to Prof. (Dr.) R. Narasimhan, IISc, an excellent teacher of finite element analysis and an expert in this domain, who has found time to go through this text book, given many suggestions and has written the foreword for the book.

We thank Springer Nature, a globally renowned publisher, for spontaneously agreeing to publish this work.

We acknowledge with profound thanks all those who have directly and indirectly contributed to improving the quality and content of the book.

S. Unnikrishnan Nair
S. Somanath

# Editorial Note

The idea of writing a textbook on finite element analysis for undergraduate students has been in vogue for some time. The editorial board had many brainstorming sessions and finalized various attributes of the book based on discussions and feedback from students of finite element analysis. Efforts have been put by all the contributing authors to introduce the subject to students in a lucid way with examples from everyday life.

The book is written with the objective of introducing the undergraduate students to the world of finite element analysis. Enough care is taken to ensure that the basics are explained in a simple language to appreciate the fundamentals of various domains of structural engineering like statics, dynamics, thermal, buckling, and how finite element method can be applied to solve problems in these domains. The book familiarizes the students about the usage of commercial FEAST$^{SMT}$ software and contains problems that will enable them to carry out finite element modeling and analysis for many practical engineering problems, specifically in the domain of aerospace systems.

Finite element software package called FEAST$^{SMT}$ was developed by ISRO along with many academic and institutional collaborators and now being marketed as a commercial software for industry, academia and students. Over a period of time, FEAST$^{SMT}$ evolved into a general-purpose finite element analysis package with many user-friendly features and capabilities that are being used by many in aerospace, industries, and academia. The software gets updated every year based on the feedback from user community and new features are constantly being added by designers. FEAST$^{SMT}$ is a true example of "*Atmanirbharatha*" and collaborative developments in high-end engineering software.

We are extremely happy to dedicate the book to the student community and they are encouraged to give their invaluable suggestions and opinions on the contents of the book for further improving it in subsequent editions.

S. Unnikrishnan Nair
S. Somanath

# Contents

# Editors and Contributors

## About the Editors

**Dr. S. Unnikrishnan Nair** is the Founding Director of Human Space Flight Centre, Indian Space Research Organisation, Bengaluru, India. He obtained his bachelor's degree from MACE, Kerala University; master's degree (M.E.) from Indian Institute of Science, Bengaluru; and Ph.D. (Acoustics) from Indian Institute of Technology Madras, Chennai. He is the Founding Director of Human Space Flight Centre where Gaganyaan project activities are being carried out. He is currently the Director of Vikram Sarabhai Space Centre, Thiruvananthapuram.

**Dr. S. Somanath** is currently Chairman of Indian Space Research Organisation and Secretary Department of Space (DOS). He obtained his bachelor's degree from TKM Engineering College, Kerala University and M.E. from Indian Institute of Science (IISc), Bengaluru. He is an Expert in launch vehicle systems and has served as Director of LPSC and later as Director of VSSC, Thiruvananthapuram, before becoming Chairman of ISRO/Secretary DOS.

## Contributors

**A. K. Asraff** DD, MDA/LPSC, Thiruvananthapuram, Kerala, India

**Philip George** ADAD/AHTG/AERO, Veli, VSSC, Thiruvananthapuram, Kerala, India

**K. Jayakumar** PSLV/VSSC, Thiruvananthapuram, Kerala, India

**P. V. Anil Kumar** Former DH, FSDD/SDAG/STR, Thiruvananthapuram, Kerala, India

**R. Marimuthu** DH, FSDD/SDAG/STR, Veli, VSSC, Thiruvananthapuram, Kerala, India

**S. Unnikrishnan Nair** Director, VSSC, Indian Space Research Organisation, Thiruvananthapuram, India

**P. Raveendranath** Department of Aerospace Engineering, Indian Institute of Space Science and Technonlogy, Valiamala, Thiruvananthapuram, Kerala, India

**K. Renganathan** Former DH, VAD/SDEG/STRVSSC, Thiruvananthapuram, Kerala, India

**T. J. Raj Thilak** FSDD/SDAG/STR, Veli, VSSC, Thiruvananthapuram, Kerala, India

# Chapter 1
# Introduction

**T. J. Raj Thilak and S. Unnikrishnan Nair**

*Nature is pleased with simplicity—Sir Isaac Newton (1642–1727)*

## 1.1 History of Finite Element Method

The exponential growth in computing power has led to remarkable changes in our daily life, especially in engineering domains. The availability of such high performance machines has enabled engineers to design sophisticated products incorporating multidisciplinary fields. To arrive at a suitable design, engineers have to convert physics of real-life system into an equivalent mathematical system represented by differential or integral equations. For simple systems, these equations can be solved using variable separable methods, differential operator methods, etc., to obtain exact solutions for the given boundary conditions. However, real-life problems are always complex, and it is very difficult to get closed form solutions or analytical solutions for such complex systems, hence we resort to numerical methods for solving these

T. J. R. Thilak (✉)
FSDD/SDAG/STR, Veli, VSSC, Thiruvananthapuram 695022, India
e-mail: tj_rajthilak@vssc.gov.in

S. U. Nair
Director, VSSC, Indian Space Research Organisation, Thiruvananthapuram, India

 
S. U. Nair and S. Somanath (eds.), *Introduction to Finite Element Analysis*,
https://doi.org/10.1007/978-981-19-7989-7_1

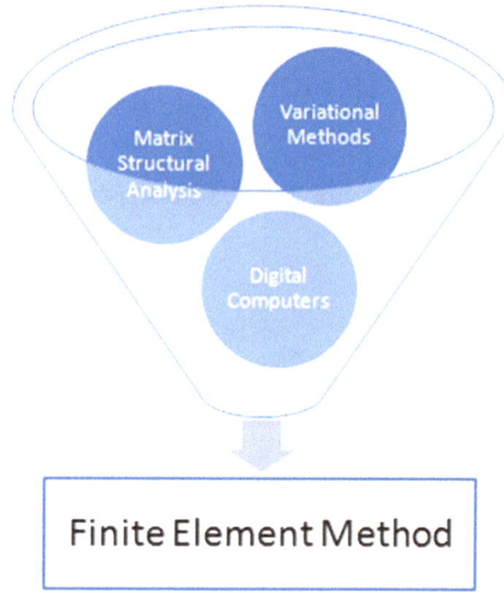


**Fig. 1.1** Finite element method—a confluence of three different domains

problems. Numerical methods provide approximate but acceptable solutions. The number of mathematical operations increases in proportion to the complexity of the system and the solution of the resulting mathematical equations becomes a Herculean task by hand or a calculator, and that is where the advancements in computing power have helped us solve increasingly complex real-life problems.

For generating mathematical equations, equivalent mathematical models have to be formulated and represented in a form that can be converted as a computational code, which can be fed to the computer. The mathematical models of real-life problems formulated in the form of differential equations are converted to simultaneous algebraic equations of unknown variables called field variables. The Finite Difference Method (FDM) and the Finite Element Method (FEM) are the most popular numerical tools among several methods to derive these simultaneous algebraic equations. The present day state-of-the-art analysis or simulation is a technology that has evolved over many years of research and development. It is the confluence of three different domains as depicted in Fig. 1.1 that were developed independently over the years by many scientists.

The three domains in chronological order are Variational Approximation Theory, Matrix Structural Analysis, and Digital Computers. The crude form of approximation theory can be dated back to third century BC when Archimedes proposed straight lines of known length to measure the perimeter of a circle, triangles of known area to calculate the area of a circle, and tetrahedral of known volume to compute the volume of a cylinder, as depicted in Fig. 1.2. He used simple geometries, whose dimensions can be measured easily, to represent complex geometries. Archimedes' work formed the basis for the concept of discretization. In finite element parlance, the basic geometry is called an *element* and the vertices of the elements, where they are joined together, are called *nodes*.

If the elements are represented by simple functions and connected at nodes where the functions or their derivatives or integrals are continuous, and if the functions can be mapped onto a scalar variable, then a *functional* can be defined for the system. In

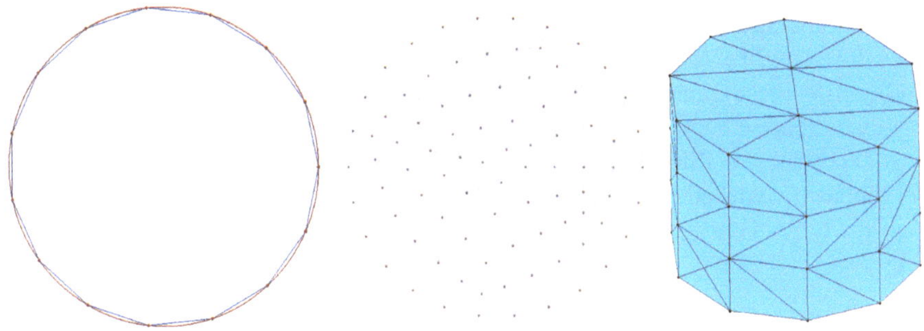

**Fig. 1.2** Archimedes' concept of discretization with known geometries

structural mechanics, functional is the total potential of the system and the functions are parameters of stress and strain which in turn are functions of displacements (field variable in structural mechanics). The variational approximation theory deals with extremizing these functionals with respect to the field variables to obtain the solution.

Variational approximation theory uses the concept of calculus which was invented independently by two great scientists, Sir Isaac Newton and Gottfried Wilhelm Leibniz. As geometries were used to study the shapes, calculus was used to study the changes in the dependent component with respect to an independent component. It can be used to generalize the evolution or the propagation of a system. Leonhard Euler was the first to use the concept of discretization in differential equations in which he discretized the system using piecewise linear functions to calculate the solution at the interface between elements. It was further explored in the engineering domain by Alexander Hrennikoff who used 'lattice analogy' to analyze flat plates in plane stress and bending conditions. He replaced the continuum model of plates with lattices of equilateral triangles formed using bars as their edges. Hrennikoff's model worked correctly only for specific geometries with specific material properties. However, his research was a pioneering work in the field of finite element analysis (FEA) even before the name was coined. Richard Courant is considered to be the first scientist who discretized the cross section of a circular hollow shaft into finite triangular subdomains to calculate its torsional stiffness.

Although many scientists have contributed to the development of finite element method (FEM), the credit for inventing the modern-day FEM can be attributed to M. Jonathan Turner who generalized the derivation of the stiffness matrix of individual elements and assembled them together to form a global stiffness matrix for the entire model. In 1956, Turner along with Clough, Martin, and Topp published a revolutionary paper on *"Stiffness and deflection analysis of complex structures" in Journal of Aeronautical Sciences*. It was Clough who coined the name Finite Element Method in 1960 in his paper titled *"Finite Element Method in Plane Stress Analysis" for the 2nd ACSE Conference on Electronic Computation*. From then onwards, huge research was carried out in FEM. The first general purpose FEM for plane stress analysis was programmed by E. Wilson and the first textbook on FEM was written by O. C. Zienkiewicz. NASA developed the first general purpose FEA package called

NASTRAN, and many other companies followed the commercial exploitation of FEA software like ANSYS and ABAQUS, to name a few. FEAST$^{SMT©}$ is the latest addition to these general purpose FEA packages which is developed by Vikram Sarabhai Space Centre, Indian Space Research Organisation (ISRO) drawing on the vast experience of ISRO in the domain of aerospace structural analysis and design for the last five decades.

## 1.2 Finite Element Method

Finite Element Method is a numerical tool for solving differential equations of a boundary value problem to obtain an approximate solution for the field variables at discrete points. A boundary value problem is a differential equation with a set of constraints applied on boundary, called boundary conditions.

FEM uses variational approaches that are based on the principle of energy minimization. The principle of energy minimization states that when a boundary condition is applied to a body (such as force or displacement), the body takes only that configuration where the total energy is minimum, as depicted in Fig. 1.3. Here, the tree subjected to a wind load deforms to a configuration for which its total energy is minimum. To develop finite element formulation, the partial differential equations must be stated in an integral form called *weak form*. The name originates from the fact that the solution to the weak form will be accurate at discrete node points, whereas, the solution obtained using differential equations otherwise known as strong form, is accurate at all the points in the continuum. Also, the highest order derivative that appears in the integral form should be first-order which is an essential requirement for commercial FEM model constructions.

The finite element method involves splitting the whole domain into a set of finite subdomains called elements. Then solve the problem in subdomains with a set of

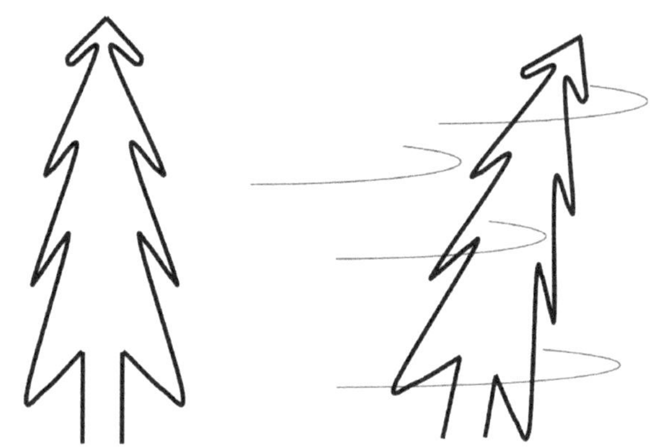

**Fig. 1.3** Deformed configuration of tree under wind load is the minimum energy configuration

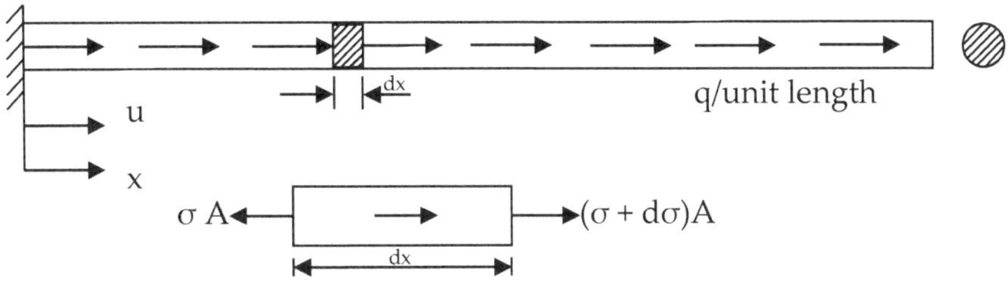


**Fig. 1.4** Bar under uniform traction

elemental equations and systematically combine elemental equations into a global system of equations to obtain final solution. The subdomains are connected to each other at points called nodes, and each node has a set of *degrees of freedom* (DOFs) assigned to it. The degrees of freedom could be temperature, displacement, etc. called the field variable depending upon the problem being analyzed. These nodes connecting the elements together define mathematical interactions of degrees of freedom in the domain. The interaction among the degrees of freedom is estimated by a numerical integration over the element. The FE formulation finally results in a set of algebraic equations, which can be solved numerically for the unknown field variables. As mentioned, the unknown field variables in FEM can be temperature, displacements, pressure, etc. However generally, in structural analysis, displacements are used as the unknown field variables for ease of computational implementation.

As the number of elements increases, the finite element model approaches perfect representation of system. This is analogous to the situation in which the polygon approaches the shape of a circle as the number of sides of polygon approaches infinity. It is not practical to split the domain into infinite number of elements, however a sufficiently large number of elements will yield a solution that is an approximation to the problem being solved.

The concept of finite element analysis and its comparison with closed form and analytical solutions is illustrated with the following example. Consider a steel bar of length '*L*' with Young's modulus '*E*' and area of cross section '*A*' fixed at one end and a uniform traction of '*q/unit length*' applied throughout the length of the bar, as depicted in Fig. 1.4. Different approaches are used to solve this problem.

### *1.2.1 Derive the Differential Equation and Obtain the Closed Form Solution*

Consider an element of length $dx$.

$\sigma$ Stress on structure due to the applied load.
$\varepsilon$ Strain on structure due to the applied load $-\ \frac{du}{dx}$.

The force equilibrium equation for the element $dx$ gives

$$\sigma A = qdx + (\sigma + d\sigma)A$$

$$\sigma A = qdx + \sigma A + d\sigma A$$

$$\frac{d\sigma}{dx} = -\frac{q}{A}.$$

Substituting $\sigma = E\varepsilon$,

$$\frac{Ed\varepsilon}{dx} = -\frac{q}{A}.$$

Substituting $\varepsilon = \frac{du}{dx}$,

$$\frac{d^2u}{dx^2} = -\frac{q}{AE}.$$

The above equation is the governing differential equation of the problem defined for $0 \leq x \leq L$. Exact solution can be obtained by integrating the governing differential equation twice and applying boundary conditions,

$$\frac{du}{dx} = -\frac{q}{AE}x + C_1$$

$$u = -\frac{q}{2AE}x^2 + C_1x + C_2. \tag{1.1}$$

Applying boundary conditions, $u = 0$ at $x = 0$ (displacement is 0 at fixed end), in Eq. 1.1 gives

$$0 = 0 + C_2$$

$$C_2 = 0.$$

At $x = L$, $\frac{du}{dx} = 0$ (strain is zero at free end); applying in equation for $\frac{du}{dx}$ gives

$$0 = -\frac{q}{AE}L + C_1$$

$$C_1 = \frac{qL}{AE}.$$

Substituting the values of $C_1$ and $C_2$ in Eq. 1.1 gives

$$u = -\frac{q}{2AE}x^2 + \frac{qL}{AE}x$$

$$u = \frac{qx}{AE}\left(L - \frac{x}{2}\right). \tag{1.2}$$

Equation 1.2 is the solution for the governing differential equation for $0 \leq x \leq L$. For a bar of length 2 m, area of cross section 0.01 m$^2$, Young's modulus 70 GPa, acted upon by a uniformly distributed load of 10,000 N/m, the displacement at the free end is

$$u = \frac{qL}{AE}\left(L - \frac{L}{2}\right)$$

$$u = \frac{qL^2}{2AE}. \tag{1.3}$$

substituting the numerical values in Eq. (1.3), the displacement at the free end is

$$u = 2.857 \times 10^{-5}\ \text{m}.$$

### 1.2.2 Analytical Solution

The analytical solution for the same problem can be obtained by assuming a trial solution and substituting it in the governing differential equation. We have the governing differential equation from the previous section,

$$\frac{d^2u}{dx^2} = -\frac{q}{AE}.$$

Assume a linear polynomial as a trial solution,

$$u = a_0 + a_1 x$$

where $a_0$ and $a_1$ are constants to be determined.

At $x = 0$, $u = 0$ gives

$$a_0 = 0.$$

Differentiating $u$ with respect to $x$ gives

$$\frac{du}{dx} = a_1.$$

Second differentiation of $u$ with respect to $x$ gives

$$\frac{d^2u}{dx^2} = 0,$$

i.e.

$$-\frac{q}{AE} = 0$$

which cannot be true for $q \neq 0$.

Therefore, using a linear trial function we cannot obtain the solution for a second order differential equation.

Assume a second order polynomial as a trial solution

$$u = a_0 + a_1 x + a_2 x^2.$$

To obtain solution for the governing differential equation,

$$\frac{d^2u}{dx^2} = -\frac{q}{AE}.$$

At $x = 0$, $u = 0$ gives

$$a_0 = 0.$$

Differentiating $u$ with respect to $x$ gives

$$\frac{du}{dx} = a_1 + 2a_2 x.$$

Second differentiation of $u$ with respect to $x$ gives

$$\frac{d^2u}{dx^2} = 2a_2 = -\frac{q}{AE}$$

$$\therefore a_2 = -\frac{q}{2AE}$$

$$u = a_1 x - \frac{q}{2AE} x^2.$$

At $x = L$, $\frac{du}{dx} = 0$,

$$0 = a_1 - \frac{2qL}{2AE}$$

$$a_1 = \frac{qL}{AE}$$

$$u = \frac{qL}{AE}x - \frac{q}{2AE}x^2$$

$$u = \frac{qx}{AE}\left(L - \frac{x}{2}\right).$$

At $x = L$,

$$u = \frac{qL^2}{AE} - \frac{qL^2}{2AE}$$

$$u = \frac{qL^2}{2AE}.$$

For a bar of length 2 m, area of cross section 0.01 m$^2$, Young's modulus 70 GPa, acted upon by a uniformly distributed load of 10,000 N/m, the deformation at free end will be given by the above equation as $u = 2.857 \times 10^{-5}$m, which is the same as that in Sect. 1.2.1.

Assume a third order polynomial as a trial function,

$$u = a_0 + a_1x + a_2x^2 + a_3x^3$$

for obtaining the solution for the governing differential equation,

$$\frac{d^2u}{dx^2} = -\frac{q}{AE}.$$

$$\text{At } x = 0,\quad u = 0 \text{ gives}$$

$$a_0 = 0.$$

Differentiating $u$ with respect to $x$ gives

$$\frac{du}{dx} = a_1 + 2a_2x + 3a_3x^2$$

$$\text{At, } x = L\ \frac{du}{dx} = 0,$$

$$a_1 + 2a_2L + 3a_3L^2 = 0.$$

Second differentiation of $u$ with respect to $x$ gives

$$\frac{d^2u}{dx^2} = 2a_2 + 6a_3x = -\frac{q}{AE}$$

$$\text{At } x = 0, \frac{d^2u}{dx^2} = -\frac{q}{AE},$$

$$-\frac{q}{AE} = 2a_2$$

$$a_2 = -\frac{q}{2AE}$$

$$\text{At } x = L, \frac{d^2u}{dx^2} = -\frac{q}{AE},$$

$$-\frac{q}{AE} = 2\left(-\frac{q}{2AE}\right) + 6a_3L$$

$$6a_3L = 0$$

$$a_3 = 0.$$

$$\text{Now, at } x = L, \frac{du}{dx} = 0,$$

$$a_1 + 2a_2L + 3a_3L^2 = 0,$$

i.e.

$$a_1 + 2a_2L = 0$$

$$a_1 = -2a_2L$$

$$= -2\left(-\frac{q}{2AE}\right)L.$$

i.e.

$$a_1 = \frac{qL}{AE}.$$

Therefore, the solution of the equation is

$$u = a_1x + a_2x^2$$

$$= \frac{qL}{AE}x - \frac{q}{2AE}x^2,$$

i.e.

$$u = \frac{qx}{AE}\left(L - \frac{x}{2}\right)$$

which is same as the solution obtained with second order trial solution. It can be observed that even with a higher order trial solution, the coefficients for all the higher order terms become zero and the solution converges to the closed form solution.

### 1.2.3 Finite Element Solution

Let the bar of length $L$ be idealized by a one-dimensional *truss element* (finite element having only axial degree of freedom at each of the two nodes at its ends). Even though the bar is idealized using a one-dimensional truss element, the area of cross section is given as an input to the truss element. Let $F_1$ and $F_2$ be the forces acting at nodes and $u_1$ and $u_2$ be the corresponding displacements at the respective nodes at each end, as shown in Fig. 1.5.

Stiffness for the truss element can be derived from Hooke's law as follows.

From Hooke's law,

$$\text{stress } \alpha \text{ stain}$$
$$\sigma \;\alpha\; \varepsilon$$
$$\sigma = E\varepsilon.$$

Substituting for stress and strain,

$$\frac{F}{A} = E\frac{\delta l}{L}$$
$$\frac{F}{\delta l} = \frac{AE}{L}.$$

But $\frac{F}{\delta l}$ is defined as the stiffness of the structure.

Let $k = \frac{F}{\delta l} = \frac{AE}{L}$.

Force balancing at node 1 gives $k(u_1 - u_2) = F_1 + R_1$.

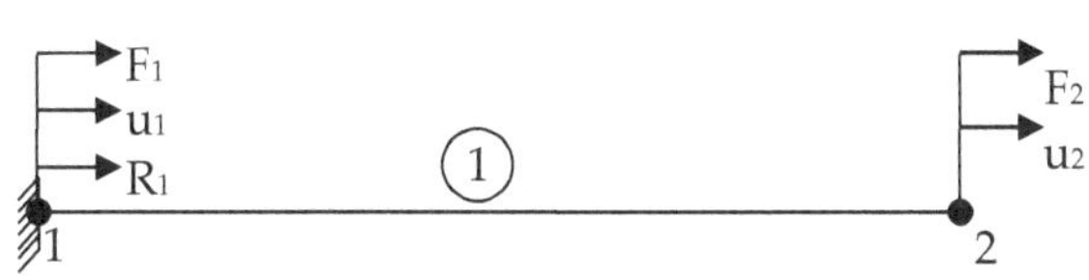


**Fig. 1.5** Bar discretized with single truss element

Similarly, at node 2, $\quad k(u_2 - u_1) = F_2.$

where $R_1$ is the reaction force at the fixed end, since the other end is free reaction force is zero at that end.

In matrix form, the above equations can be written as

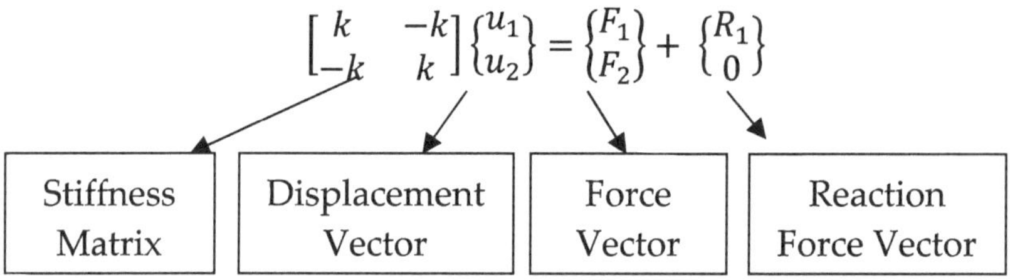


From this, the stiffness matrix for the bar can be rewritten as (substituting $k = \frac{AE}{L}$)

$$k = \frac{AE}{L}\begin{bmatrix} 1 & -1 \\ -1 & 1 \end{bmatrix}.$$

Writing the force equilibrium equations,

$$\frac{AE}{L}\begin{bmatrix} 1 & -1 \\ -1 & 1 \end{bmatrix}\begin{Bmatrix} u_1 \\ u_2 \end{Bmatrix} = \begin{Bmatrix} F_1 \\ F_2 \end{Bmatrix} + \begin{Bmatrix} R_1 \\ 0 \end{Bmatrix}. \tag{1.4}$$

This can be written as

$$\frac{AE}{L}u_1 - \frac{AE}{L}u_2 = F_1 + R_1$$

$$-\frac{AE}{L}u_1 + \frac{AE}{L}u_2 = F_2.$$

Since it is a one-element truss, it can be assumed that the uniformly distributed load acting on the element is shared equally between the nodes. The total load on the structure is $qL$. The total load is redistributed to the nodes equally by applying $\frac{qL}{2}$ at each node.

Therefore, the load vector is

$$\begin{Bmatrix} F_1 \\ F_2 \end{Bmatrix} = \begin{Bmatrix} \frac{qL}{2} \\ \frac{qL}{2} \end{Bmatrix}.$$

At $x = 0$, $u_1 = 0$. Substituting these values in Eq. 1.4, we get

$$-\frac{AE}{L}u_2 = F_1 + R_1$$

$$F_2 = \frac{AE}{L}u_2.$$

But $F_1 = F_2 = \frac{qL}{2}$. Substituting these values in the above equations, we get

$$\frac{qL}{2} = \frac{AE}{L}u_2$$

$$u_2 = \frac{qL^2}{2AE}.$$

Therefore,

$$-\frac{qL}{2} = \frac{qL}{2} + R_1,$$

i.e.

$$R_1 = -qL.$$

For the truss problem discussed, the deformation at the free end will be

$$u_2 = \frac{qL^2}{2AE}$$

$$u_2 = 2.857 \times 10^{-5}\ \text{m}.$$

If the bar is discretized with two truss elements as shown in Fig. 1.6, the load vector and the stiffness matrix have to be calculated for each element and it has to be added at respective nodes or assembled to get the global system load vector and global system stiffness matrix. In the present case, the length of each element is $\frac{L}{2}$ as node 2 is at the center of the truss.

From each element, the element load vector and element stiffness matrix are given as

$$\{F\} = \begin{Bmatrix} \frac{qL}{4} \\ \frac{qL}{4} \end{Bmatrix}; [K] = \frac{2AE}{L}\begin{bmatrix} 1 & -1 \\ -1 & 1 \end{bmatrix}.$$

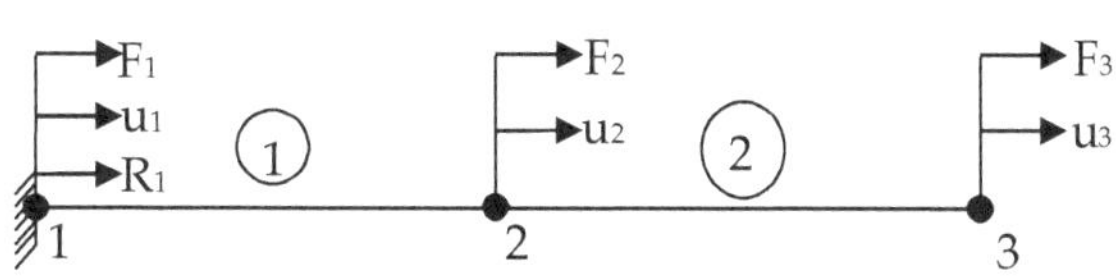

**Fig. 1.6** Truss discretized with two elements

For element no. 1, there is no contribution from node 3, therefore the third row of the force vector and the stiffness matrix and the third column of the stiffness matrix will be zero. Similarly, for second element, there will not be any contribution from node 1, and the corresponding first row of the force vector and the stiffness matrix and the first column of the stiffness matrix will be zero. Therefore, the force vector or load vector and stiffness matrix for both the elements can be written as

$$\{F^1\} = \begin{Bmatrix} \frac{qL}{4} \\ \frac{qL}{4} \\ 0 \end{Bmatrix} [K^1] = \frac{2AE}{L}\begin{bmatrix} 1 & -1 & 0 \\ -1 & 1 & 0 \\ 0 & 0 & 0 \end{bmatrix}$$

and

$$\{F^2\} = \begin{Bmatrix} 0 \\ \frac{qL}{4} \\ \frac{qL}{4} \end{Bmatrix} [K^2] = \frac{2AE}{L}\begin{bmatrix} 0 & 0 & 0 \\ 0 & 1 & -1 \\ 0 & -1 & 1 \end{bmatrix}.$$

By adding both the element force vector and stiffness matrix, we get global load vector and stiffness matrix,

$$\{F\} = \{F^1\} + \{F^2\} = \begin{Bmatrix} \frac{qL}{4} \\ \frac{qL}{2} \\ \frac{qL}{4} \end{Bmatrix}; \qquad [K] = [K^1] + [K^2] = \frac{2AE}{L}\begin{bmatrix} 1 & -1 & 0 \\ -1 & 2 & -1 \\ 0 & -1 & 1 \end{bmatrix}$$

$$\frac{2AE}{L}\begin{bmatrix} 1 & -1 & 0 \\ -1 & 2 & -1 \\ 0 & -1 & 1 \end{bmatrix}\begin{Bmatrix} u_1 \\ u_2 \\ u_3 \end{Bmatrix} = \begin{Bmatrix} \frac{qL}{4} \\ \frac{qL}{2} \\ \frac{qL}{4} \end{Bmatrix} + \begin{Bmatrix} R_1 \\ 0 \\ 0 \end{Bmatrix}. \tag{1.5}$$

Expanding it into a system of linear equations,

$$\frac{2AE}{L}u_1 - \frac{2AE}{L}u_2 = \frac{qL}{4} + R_1$$

$$-\frac{2AE}{L}u_1 + \frac{4AE}{L}u_2 - \frac{2AE}{L}u_3 = \frac{qL}{2}$$

$$-\frac{2AE}{L}u_2 + \frac{2AE}{L}u_3 = \frac{qL}{4}.$$

Substituting $u_1 = 0$, the above equations become

$$-\frac{2AE}{L}u_2 = \frac{qL}{4} + R_1$$

$$\frac{4AE}{L}u_2 - \frac{2AE}{L}u_3 = \frac{qL}{2}$$

$$-\frac{2AE}{L}u_2 + \frac{2AE}{L}u_3 = \frac{qL}{4}.$$

Adding the last two equations,

$$\frac{2AE}{L}u_2 = \frac{3qL}{4}$$

$$u_2 = \frac{3qL^2}{8AE},$$

i.e.

$$-\frac{2AE}{L} \times \frac{3qL^2}{8AE} + \frac{2AE}{L}u_3 = \frac{qL}{4}$$

$$-\frac{3qL}{4} + \frac{2AE}{L}u_3 = \frac{qL}{4}$$

$$\frac{2AE}{L}u_3 = \frac{4qL}{4}$$

$$\frac{2AE}{L}u_3 = qL;\ u_3 = \frac{qL^2}{2AE}.$$

With the values of $u_1$, $u_2$, and $u_3$ known, the reaction force $R_1$ can be found out as follows:

$$\frac{2AE}{L} \times 0 - \frac{2AE}{L} \times \frac{3qL^2}{8AE} = \frac{qL}{4} + R_1$$

$$-\frac{3qL}{4} = \frac{qL}{4} + R_1$$

$$-\frac{3qL}{4} = \frac{qL}{4} + R_1$$

$$R_1 = -qL.$$

For the problem stated earlier, the values of $u_1$, $u_2$, and $u_3$ are as follows:

$$u_1 = 0; \quad u_2 = \frac{3qL^2}{8AE} = 2.142 \times 10^{-5}\ \text{m}$$

$$u_3 = \frac{qL^2}{2AE} = 2.857 \times 10^{-5}\ \text{m}.$$

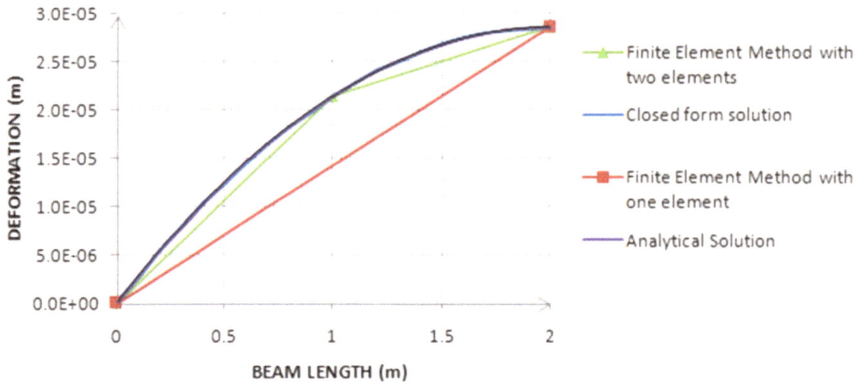


**Fig. 1.7** Variation of $u$ along the length of beam for different methods

All three methods have yielded same solution at the free end. Figure 1.7 shows the variation of displacement along the length for various methods. The following points can be inferred from the plot:

1. All three methods give the same answer at both ends in this case. However, the displacement is different at intermediate points when comparing FE and closed form solutions.
2. With two elements, the solution is approaching the closed form solution. With more number of elements, finite element solution approaches the closed form solution.

In structural analysis, typically the unknown field variable will be displacements at the nodes. From the nodal displacement values, the displacement within the element will be interpolated using assumed polynomials or trial functions. These trial functions need not be spanned for the entire domain or structure but rather on the individual subdomains or elements. Usually, the order of the trial function will be the same for all the elements used in the structure.

## 1.3 Applications of FEA

FEM has its origin in aerospace industries for the obvious reason that aerospace structures, often made of thin shells stiffened with beams and spars, need a thorough analysis for optimizing weight and yet should perform under harsh conditions to which they are exposed. However, the underlying differential equations of FEM were common for different domains, and its application was expanded to almost all the engineering domains. In the entire finite element analysis, user is interested in finding the distribution of field variables and its derivatives. The following sections give the field variables and its derivatives for various disciplines of engineering.

### 1.3.1 Solid Mechanics

Figure 1.4 shows a bar fixed at one end with a uniform traction applied throughout the bar. The governing differential equation for the bar with Young's modulus '$E$', area of cross section '$A$', fixed at one end, and a uniform traction '*q/unit length*' applied throughout the bar is

$$\frac{d}{dx}\left(AE\frac{du}{dx}\right)+q=0 \tag{1.6}$$

where

$u$ Displacement (field variable).

In this case, stress and strain are derived variables.

### 1.3.2 Heat Conduction

Figure 1.8 shows a bar with $T = T_0$ at one end and $T = T_x$ at the other end. The governing differential equation for a bar with thermal conductivity '$k$' and area of cross section '$A$', maintained at temperature $T_0$ at one end and $T_x$ at the other end having a heat generation of $Q$ W/m, is

$$\frac{d}{dx}\left(kA\frac{dT}{dx}\right)+Q=0 \tag{1.7}$$

where

$T$ Temperature (field variable).

Heat Flux is the derived variable in this case.

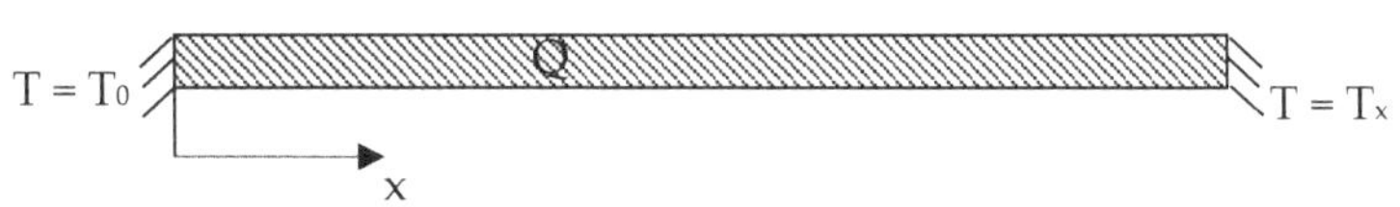


**Fig. 1.8** Finite element method applied to heat transfer

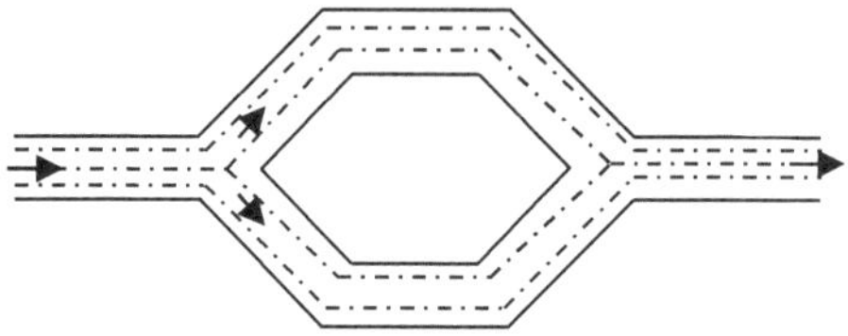

Fig. 1.9 Finite element method applied to fluid flow

### 1.3.3 Fluid Flow Through Pipes

Figure 1.9 shows a pipe network with total flow rate $Q$ m$^3$/s, diameter $D$, and dynamic viscosity $\mu$. The governing differential equation for this network is given as

$$\frac{d}{dx}\left(\frac{\pi D^4}{128\mu}\frac{dp}{dx}\right) + Q = 0 \tag{1.8}$$

where

$p$ Pressure (field variable)
$Q$ Flow rate, derived variable.

### 1.3.4 Electrostatics

The governing differential for a nonmagnetic medium, as shown in Fig. 1.10, with dielectric constant $\varepsilon_r$ between two infinite parallel plates separated at a distance '$d$' with one plate maintained at fixed potential $V_0$ while the other plate is grounded, i.e. $V = 0$, is

$$\frac{d}{dx}\left(\varepsilon_r\frac{dv}{dx}\right) + \frac{\rho_v}{\varepsilon_0} = 0 \tag{1.9}$$

where

$v$ electric potential (field variable)
$\varepsilon_r$ dielectric constant of the medium
$\varepsilon_0$ reference dielectric constant
$\rho_v$ electron volume charge density.

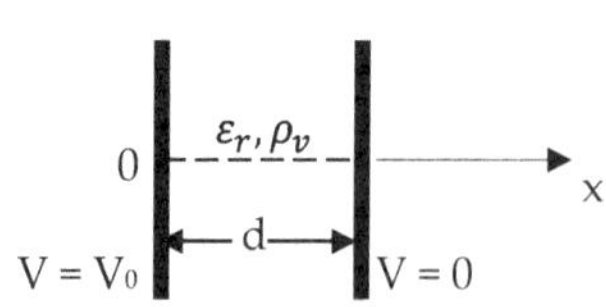


Fig. 1.10 Finite element method applied to electrostatics

**Table 1.1** Field variables and derived variables for different domains

| Domain | Field variable | Derived variable |
|---|---|---|
| Solid mechanics | Displacement | Strain, stress |
| Fluid flow | Pressure | Flow rate |
| Heat transfer | Temperature | Heat flux |
| Electrostatics | Electric potential | Electric flux |

Electric flux is the derived variable here.

Table 1.1 shows the summary of the finite element method applied to different engineering domains, their field variables, and derived variables.

## 1.4 Different Numerical Methods

Although finite element method is a popular method in structural and heat transfer analysis domains, there are other methods, which are equally popular in other domains. Generally, numerical methods can be divided into two broad categories, one that requires mesh connectivity or 'mesh methods' and the other that does not require mesh connectivity data or 'mesh-free methods'. Some numerical methods fall under both the categories, for example, eXtended Finite Element Method (XFEM). Table 1.2 lists various numerical methods that falls under mesh methods and mesh free methods.

### *1.4.1 Mesh Methods*

#### Finite Difference Method (FDM)

FDM differs from FEM in how the governing equations are discretized; while FDM discretizes the governing 'differential equation', FEM discretizes the 'integral form'

**Table 1.2** Different types of numerical methods

| Mesh methods | Mesh-free methods |
|---|---|
| Finite Element Method (FEM) | Smoothed Particle Hydrodynamics (SPH) |
| Finite Difference Method (FDM) | Element-Free Galerkin Method (EFGM) |
| Finite Volume Method (FVM) | Material Point Method (MPM) |
| Boundary Element Method (BEM) | Meshless Local Petrov–Galerkin (MLPG) |
| eXtended Finite Element Method (XFEM) | |

of equations. In this method, the governing differential equation is expanded using Taylor's series and after neglecting the higher order terms, finite difference method is used to form the algebraic equations. FDM is generally used to solve fluid flow problems and heat transfer analysis problems.

**Finite Volume Method (FVM)**

In this method, the field variable is evaluated per element by finding the integral over each element and dividing the integral by the volume of the element. Contrary to elements that are connected at node points in FEM, the mesh can be unstructured, i.e. there is no concept of node in FVM and hence the adjacent elements need not share a common edge in FVM. Due to this advantage, FVM is quite popular in Computational Fluid Dynamics (CFD).

**Boundary Element Method (BEM)**

In this method, the 'volume integral' form of equations has to be converted to 'surface integral' form of equations and calculations are done on surfaces containing the volume. Therefore, only the surfaces have to be meshed thus reducing one dimension of the problem. BEM is appropriate for representing infinite and semi-infinite domains and is particularly appropriate for solving linear problems. The main disadvantage of this method is that the resulting characteristic matrix is a full unsymmetrical matrix making it time-consuming to solve, whereas in other methods the characteristic matrix is a sparse symmetric matrix.

### *1.4.2 Mesh Free Methods*

**Smoothed Particle Hydrodynamics (SPH)**

SPH is based on the Lagrangian representation of the domain and was initially developed for analyzing astrophysical problems, but is widely used for simulating fluid flows and high-speed impact studies. In SPH, the domain is discretized into a set of discrete elements called particles. The properties of a specific particle interact with other neighboring particles within a spatial range controlled by weight function or a smoothing function.

**Element-Free Galerkin Method (EFGM)**

In EFGM, the discretization is in the form of scattering a set of nodes on the domain including its boundaries. Moving Least Square approximation (MLS) procedure is used to approximate the field variable at a node point using the nodal parameters of field variables that fall into a domain called support domain. This method is widely used in structural and heat transfer problems.

**Material Point Method (MPM)**

MPM is widely used for multibody and multi physics simulations because of the ease in detecting contact without interpenetration. MPM can be coupled with FEM for simulating impact and metal deformation studies. In this method, at the end of each step an inner step called convective step occurs in which the mesh is reset to the original position, whereas the material points remain in their current positions.

**Meshless Local Petrov–Galerkin (MLPG)**

MLPG is very similar to EFGM, the difference being that EFGM requires a background element which satisfies the integral form of weighted residual method, and this has to be satisfied for the entire domain. In MLPG, the background element is confined to a very small local domain around the node. This alleviates the stability issues that arise in EFGM. Similar to EFGM, this method is also widely used in structural and heat transfer analysis domains.

### *1.4.3 Mixed Method*

**eXtended Finite Element Method (XFEM)**

XFEM is a numerical method that locally enriches the field variable approximations with local partitions of unity enrichment functions. This method extends the FEM by enriching the solution space by adding enrichment function such as Heaviside functions. This method is widely used in fracture mechanics to track the crack path without refining the mesh.

### *1.4.4 Advantages of Using FEM Over Other Methods*

- It is a proven technique for a wide variety of field problems: for example, stress analysis, heat transfer analysis, computational fluid dynamics, fracture analysis, etc.
- Any complex structure with complex loads and boundary conditions can be modeled resembling the actual structure.
- Combination of structures with different material and physical properties can be easily simulated.
- There is no restriction on the size of the elements used for the analysis.
- The important advantage is that FEM is supported by a lot of commercial packages when compared to other methods.

## 1.5 FEAST$^{SMT©}$

ISRO has been actively pursuing finite element analysis for many decades in the area of aerospace and space vehicle structures. The experience accumulated over time is effectively translated into a user-friendly structural analysis package called FEAST$^{SMT}$. PreWin is the Graphical User Interface (GUI)-based pre/post-processor of FEAST$^{SMT}$. It provides industry-standard interactive graphical tools for geometric modeling, mesh generation, model editing, and result visualization. FEAST$^{SMT}$ solver is seamlessly integrated with PreWin as a single application. Substructured and multithreaded implementation of the solver ensures high performance by exploiting the multi-core architecture of modern computing platforms. With advanced solution algorithms, the solver is able to handle large-order problems of structural engineering. The current version of FEAST$^{SMT}$ is capable of solving linear analysis problems with metallic and composite models. The analysis capabilities of FEAST$^{SMT}$ include

- Static
- Free Vibration
- Buckling
- Transient response
- Frequency response
- Random Response
- Inertia relief method
- Visco-elastic
- Thermo-elasticity
- Heat Transfer
- Electrostatic
- Magneto static.

The graphical pre/post processor of FEAST$^{SMT}$ PreWin provides in addition to the common functions, multi-port views controllable individually, import/export of models from similar proprietary finite element software, and unique selection options for complicated models. The software supports a rich set of element libraries, linear and eigen value solvers, load types, constraints including single point constraints, multipoint constraints, tying and rigid link, import and export of Craig Bampton reduced matrices, variable data specification using expressions and data tables.

FEAST$^{SMT}$ is being continuously updated based on users' feedback and experience gained in ISRO. Both commercial and academic versions of FEAST$^{SMT}$ are available. Every year, ISRO conducts an annual users meeting to converse with the users' community to get feedback based on which updated versions of the software are released periodically.

## 1.6 Exercise Problems

1.6.1 What is Finite Element Method?
1.6.2 What is a degree of freedom?
1.6.3 What are the criteria for selecting a trial function?
1.6.4 Assume a trigonometric trial function and get the analytical solution for the equation that was derived for the problem shown in Sect. 1.2.1

$$\frac{d^2u}{dx^2} = -\frac{q}{AE}.$$

**Hint:** The assumed function should satisfy the boundary conditions of the domain.
1.6.5 For Fig. 1.7, if both the ends are fixed and the load applied is a thermal load of temperature $\Delta T = 100$ °C throughout, what is the reaction at the end? Assume the coefficient of thermal expansion as $1 \times 10^{-06}$.
1.6.6 What are the limitations of FEM?
1.6.7 What is a field variable? Give examples from different domains.
1.6.8 What is a boundary value problem? Give an example.

## Bibliography

Clough RW (1960) The finite element method in plane stress analysis. In: Proceedings of the 2nd ACSE conference on electronic computation, Pittsburgh, PA, pp 345–378
Hrennikoff A (1941) Solution of problems in elasticity by the framework method. J Appl Mech 8:169–175
Leveque RJ (2004) Finite-volume methods for hyperbolic problems. Cambridge University Press. ISBN 0-521-00924-3
Lewis RW, Nithiarasu P, Seetharamu KN (2004) Fundamentals of the finite element method for heat and fluid flow. Wiley. ISBN 9780470847886
Liu GR (2003) Mesh free methods—moving beyond the finite element method. CRC press. ISBN 0-8493-1238-8
Macneal RH (1994) Finite elements: their design and performance. Marcel Dekker, Inc. ISBN 0-8247-9162-2
Turner MJ, Clough RW, Martin HC, Topp LJ (1956) Stiffness and deflection analysis of complex structures. J Aeronaut Sci 23(9):803–823
Wunderlich W, Pilkey WD (2003) Mechanics of structures variational and computational methods, 2nd edn. CRC Press. ISBN 0-8493-0700-7
Zienkiewicz OC, Taylor RL (2000) The finite element method. The basis, vol 1, 5th edn. Butterworth–Heinemann. ISBN 0 7506 5049 4

# Chapter 2
# Linear Algebra

**K. Renganathan**

*An equation has no meaning for me unless it expresses a thought of God.*
*—Srinivasa Ramanujan*

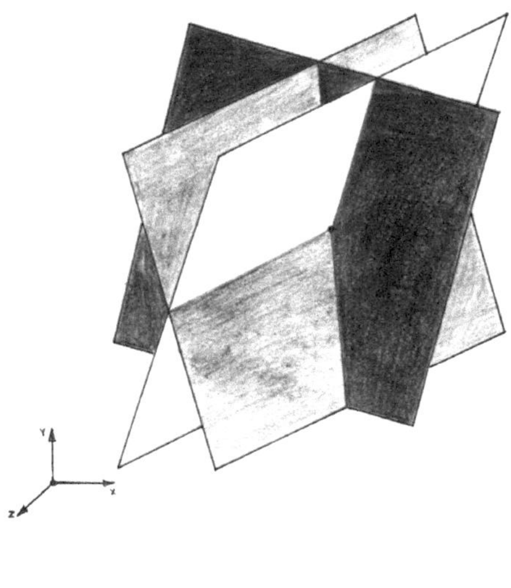

## 2.1 Introduction

The practical implementation of finite element analysis usually involves extensive use of matrix algebra. The evaluation of integrals is an important task in finite element analysis and involves many operations on systems of equations such as addition,

K. Renganathan (✉)
Former DH, VAD/SDEG/STRVSSC, Thiruvananthapuram, Kerala 695035, India
e-mail: drrengavssc@gmail.com

 
S. U. Nair and S. Somanath (eds.), *Introduction to Finite Element Analysis*,
https://doi.org/10.1007/978-981-19-7989-7_2

subtraction, inner products, matrix inversion, matrix decomposition, etc. These are best handled by matrices and hence sound knowledge of linear algebra is essential for developing efficient finite element codes. It is assumed that the reader is familiar with the elementary properties of matrices and determinants. Here, we present the definition of some special matrices and some results for reference. However, we are refraining from giving any proof, which is extensively covered in standard textbooks.

Linear Algebra comprises the theory and application of linear systems of equations, linear transformations, and eigenvalue problems. Arthur Cayley (1821–1895), who was a professor at Cambridge University, developed the theory of Matrices in the year 1860. Matrix is a powerful tool in modern mathematics due to its wide application in every branch of science and engineering. With the advancement of computers, the implementation of matrix methods has been greatly facilitated.

## 2.2 Matrices

A *matrix* is a rectangular array of numbers (real or complex) or functions arranged along horizontal and vertical lines, which are known as *rows and columns*. Generally, a matrix is denoted by **A, B, C**, etc., whereas the elements of the matrices are denoted by *a, b, c*, etc. If a matrix has '*m*' rows and '*n*' columns then it is said to be of *order or size* '$m \times n$' and pronounced as '*m by n*'. In general, the element $a_{ij}$ denotes the $i^{\text{th}}$ row and $j^{\text{th}}$ column element of the matrix. Thus, an '$n \times n$' matrix is of the form

$$\mathbf{A} = \begin{bmatrix} a_{11} & a_{12} \cdots & a_{1n} \\ \vdots & \ddots & \vdots \\ a_{n1} & a_{n2} \cdots & a_{nn} \end{bmatrix}.$$

## 2.3 Special Matrices

*Null matrix*: A matrix in which all elements are zero, e.g. $\mathbf{A} = \begin{pmatrix} 0 & \cdots & 0 \\ \vdots & \ddots & \vdots \\ 0 & \cdots & 0 \end{pmatrix}$.

*Row matrix*: A matrix which has only one row and it is a matrix of order $1 \times n$, e.g. $\mathbf{A} = [1\ 2\ 3\ 4]$.

*Column matrix*: A matrix that has only one column and has order $m \times 1$,

$$\text{e.g.} \quad A = \begin{bmatrix} 1 \\ 2 \\ 3 \end{bmatrix}$$

*Square matrix*: A matrix in which the number of rows is equal to the number of columns (i.e. $m = n$). For only such a matrix, the leading or principal diagonal and determinant exist. Here, *Trace* (**A**) = Sum of the leading diagonal elements of **A**.

$$\text{i.e.} \quad Trace\,(A) = a_{11} + a_{22} + a_{33} + \cdots + a_{nn}.$$

*Singular*: A square matrix is said to be *singular* if its determinant is zero; otherwise *non- singular.*

*Diagonal matrix*: A square matrix in which all the elements except the diagonal are zero,

$$\text{e.g.} \quad A = \begin{bmatrix} 3 & 0 & 0 \\ 0 & 11 & 0 \\ 0 & 0 & 2 \end{bmatrix}.$$

*Scalar matrix*: A diagonal matrix in which the diagonal elements are the same,

$$\text{e.g.} \quad A = \begin{bmatrix} 3 & 0 & 0 \\ 0 & 3 & 0 \\ 0 & 0 & 3 \end{bmatrix}.$$

*Unit matrix*: A scalar matrix in which all the diagonal elements are 1 and is denoted as $\mathbf{I_n}$.

*Transpose*: A matrix obtained by writing the rows as columns and is denoted by $\mathbf{A^T}$

*Triangular matrix*: A matrix in which all the elements below the diagonal are zero is called an *upper triangular matrix,* and a square matrix in which all the elements above the diagonal are zero is called a *lower triangular matrix.* A matrix is said to be *triangular* if it is either upper or lower triangular. A *unit matrix* is a scalar, upper, and lower triangular matrix.

*Symmetric matrix*: A square matrix is said to be *symmetric* if $a_{ij} = a_{ji}$, $\forall i, j$, i.e. $\mathbf{A^T} = \mathbf{A}$,

$$\text{e.g.} \quad A = \begin{bmatrix} 1 & -1 & -2 \\ -1 & 2 & 3 \\ -2 & 3 & 4 \end{bmatrix}.$$

*Skew symmetric matrix*: A square matrix is said to be *skew symmetric* if $a_{ij} = -a_{ji}$, $\forall i, j$,

i.e. $\mathbf{A^T = -A}$. Here, the principal diagonal elements of skew-symmetric matrix must be all zero, e.g. $A = \begin{bmatrix} 0 & -1 & -2 \\ 1 & 0 & 3 \\ 2 & -3 & 0 \end{bmatrix}$.

$$\text{Also,}\quad A = \frac{1}{2}\left(A + A^T\right) + \frac{1}{2}\left(A - A^T\right).$$

Here, the Symmetric part of $\mathbf{A} = \frac{1}{2}(\mathbf{A} + \mathbf{A^T})$; Skew Symmetric part of $\mathbf{A} = \frac{1}{2}\,(\mathbf{A} - \mathbf{A^T})$.

## 2.4 Matrix Algebra

*Equality of matrices*: *Two* matrices **A** and **B** are said to be equal if and only if (a) they are of the same order (b) the corresponding elements of **A** and **B** are equal.

*Addition/Subtraction of matrices*: If **A** and **B** are two matrices of the same order, then their sum (**A + B**) is of the same order and is obtained by adding the corresponding elements of **A** and **B**. Similarly, the difference (**A − B**) is obtained by subtracting each element of **B** from the corresponding elements of **A**.

*Multiplication of matrix by a scalar*: Let **A** be any matrix and let 'k' be any scalar.

Then, k [A] = [kA]; Multiplying each element of **A** by 'k'.

*Multiplication of matrices* **A** *and* **B**: *Let* **A** *be any matrix of order* $m \times n$ and **B** be of any matrix of order $n \times p$. Then, the product matrix [**AB**] is of order $m \times p$.

*Minor*: of an element $a_{ij}$ in **A** is the determinant of the square matrix that remains after deleting the $i^{th}$ row and $j^{th}$ column of **A** and is denoted as $\mathbf{M_{ij}}$.

*Cofactor*: Let **A** be a square matrix and let $\mathbf{M_{ij}}$ be the minor of an element be $a_{ij}$.

Then, cofactor of $a_{ij}$ is $\mathbf{A_{ij}} = (-1)^{\mathbf{i+j}}\mathbf{M_{ij}}$.

*Inverse of a matrix* **A**: *Let* **A** *be any matrix. Then, a matrix* **B** *is said to be the inverse of* **A** *if* **AB = BA = I,** i.e. $\mathbf{B = A^{-1}}$. *Thus,* $\mathbf{AA^{-1} = A^{-1}A = I}$.

*The matrix* $\mathbf{A^{-1}}$ *is obtained as*: $\mathbf{A}^{-1} = \frac{(\textbf{cofactor of A})^T}{|A|} = \frac{adj A}{|A|}$.

*Properties of matrices*: For any matrices **A, B, C,** and a scalar k, the following properties of matrices can be easily verified:

- In general, $\mathbf{AB} \neq BA$, $\mathbf{AB = 0}$ need not imply either $\mathbf{A = 0}$ or $\mathbf{B = 0}$ and **AB = AC** need not imply **B = C**.

- In general, $\mathbf{A} + (\mathbf{B} + \mathbf{C}) = (\mathbf{A} + \mathbf{B}) + \mathbf{C}$, $\mathbf{A}(\mathbf{BC}) = (\mathbf{AB})\mathbf{C}$ **and** $k(\mathbf{A} + \mathbf{B}) = k\mathbf{A} + k\mathbf{B}$.
- We have, $(\mathbf{A}^T)^T = \mathbf{A}$, $(k\mathbf{A})^T = k\mathbf{A}^T$, $(\mathbf{AB})^T = \mathbf{B}^T\mathbf{A}^T$, $(\mathbf{AB})^{-1} = \mathbf{B}^{-1}\mathbf{A}^{-1}$, $(\mathbf{A}^{-1})^{-1} = \mathbf{A}$, $(\mathbf{A}^{-1})^T = (\mathbf{A}^T)^{-1}$ and $(\mathbf{A} \pm B)^T = \mathbf{A}^T \pm B^T$.
- A square matrix with real elements is said to be orthogonal if $\mathbf{A}^{-1} = \mathbf{A}^T$, i.e. $\mathbf{AA}^T = \mathbf{A}^T\mathbf{A} = \mathbf{I}$. Here, (i) If **A** and **B** are orthogonal, then **AB** is also orthogonal (ii) If **A** is orthogonal then $\mathbf{A}^{-1}$ and $\mathbf{A}^T$ are also orthogonal (iii) If **A** is orthogonal then $|A| = \pm 1$.
- If $\mathbf{A}^2 = \mathbf{A}$, then **A** is called the Idempotent matrix.
- If $\mathbf{A}^2 = \mathbf{I}$, then **A** is called an involutory matrix.
- If **A** and **B** are symmetric matrices, then $\mathbf{A} \pm \mathbf{B}$, $\mathbf{AB} + \mathbf{BA}$ are symmetric whereas **AB** and **BA** are not symmetric and $\mathbf{AB} - \mathbf{BA}$ is skew symmetric.

## 2.5 Linear Dependence and Rank of a Matrix

*Vector*: Any quantity having $n$ components is a vector of order $n$.

*Linear combination*: is of the form $\alpha_1 x_1 + \alpha_2 x_2 + \cdots + \alpha_n x_n = 0$,

where $\alpha_1, \alpha_2, \ldots, \alpha_n$ are scalars and $x_1, x_2, \ldots, x_n$ are members of vector **x**.

*Linearly independent*: The linear combination $\alpha_1 x_1 + \alpha_2 x_2 + \cdots + \alpha_n x_n = 0$ is said to be *linearly independent* if $\alpha_1, \alpha_2, \ldots, \alpha_n$ are all zero, i.e. we cannot find a linear combination among the given vectors. If **A** is a non-singular matrix then all the rows/columns of **A** are linearly independent.

*Linearly dependent*: The linear combination $\alpha_1 x_1 + \alpha_2 x_2 + \cdots + \alpha_n x_n = 0$ is said to be linearly dependent if all $\alpha_1, \alpha_2, \ldots, \alpha_n$ are not zero, i.e. it is possible to find a linear combination among the given vectors.

*Rank of a matrix*: The order of the maximum square submatrix whose determinant is not equal to zero. It is also the number of independent rows or columns of **A** and the number of non-zero independent rows or columns in the upper triangular matrix.

### *2.5.1 Properties of Rank*

The following properties of ranks can be easily verified:

- Rank of a null matrix is zero; The rank of an identity matrix $\mathbf{I}_n = n$.
- Rank $(\mathbf{AB}) \leq \min\{\text{rank}(\mathbf{A}), \text{rank}(\mathbf{B})\}$; Rank $(\mathbf{A} + \mathbf{B}) \leq \text{rank}(\mathbf{A}) + \text{rank}(\mathbf{B})$.
- Rank $(\mathbf{A} - \mathbf{B}) \geq \text{rank}(\mathbf{A}) - \text{rank}(\mathbf{B})$.

## 2.6 Solution of System of Linear Simultaneous Equations

Consider a system of linear simultaneous equations of the form

$$a_{11}x + a_{12}y + a_{13}z = b_1;\ a_{21}x + a_{22}y + a_{23}z = b_2;\ a_{31}x + a_{32}y + a_{33}z = b_3.$$

The above system in matrix form can be written AX = B, where

$$\mathbf{A} = \text{coefficient matrix} = \begin{bmatrix} a_{11} & a_{12} & a_{13} \\ a_{21} & a_{22} & a_{123} \\ a_{31} & a_{32} & a_{33} \end{bmatrix};\ \mathbf{X} = \text{Unknown vector} = \begin{pmatrix} x \\ y \\ z \end{pmatrix};$$

$$\text{B} = \text{Constant vector} = \begin{pmatrix} b_1 \\ b_2 \\ b_3 \end{pmatrix};\ \mathbf{AB} = \text{Augmented matrix} = \begin{bmatrix} a_{11} & a_{12} & a_{13} & b_1 \\ a_{21} & a_{22} & a_{23} & b_2 \\ a_{31} & a_{32} & a_{33} & b_3 \end{bmatrix}.$$

### 2.6.1 *Rouche's Theorem*

Consider a non-homogeneous system of linear simultaneous equations of the form. AX = B. Then:

- If Rank (**A**) $\neq$ Rank (**AB**), then the given system is *inconsistent* and hence it has no solution.
- If Rank (**A**) = Rank (**AB**) < the number of unknowns; then the given system is consistent and has an infinite number of non-zero solutions.
- If Rank (**A**) = Rank (**AB**) = the number of unknowns; then the given system is *consistent* and has a unique solution.

### 2.6.2 *Homogeneous System*

A homogeneous system is of the form **AX** = **0**. Then:

- If Rank (**A**) = number of unknowns, then the given system will have only a *trivial solution,* i.e. all values of unknowns are zero.
- If Rank (**A**) < number of unknowns, then the given system will have infinitely many non-zero solutions.

## 2.7 Eigenvalues and Eigenvectors

The word eigen is originated from the German word 'eigen' which means 'proper' or 'characteristic'. Eigenvalues and eigenvectors are very common in many disciplines and have a wide range of applications in vibration analysis, stability analysis, etc.

Eigenvalue problem is one in which it has discrete solutions and they are called eigenvalues. Their matching vectors are called eigenvectors. For example, in vibration analysis, eigenvalues represent the natural frequencies of the structure and the eigenvectors are the corresponding mode shapes or the pattern of vibration of the structure. Similarly, in structural buckling analysis, eigenvalues are the load at which buckling takes place and the corresponding eigenvectors are the buckled shapes of the structure.

Let **A** be a square matrix of order n and let λ be any scalar. Then:

- $|\boldsymbol{A} - \lambda \boldsymbol{I}| = a_0 + a_1\lambda + a_2\lambda^2 + a_3\lambda^3 + \cdots + a_n\lambda^n$ is the *characteristic polynomial.*
- $|\boldsymbol{A} - \lambda \boldsymbol{I}| = a_0 + a_1\lambda + a_2\lambda^2 + a_3\lambda^3 + \cdots + a_n\lambda^n = 0$ is the *characteristic equation.*
- The roots of the characteristic equation $|\boldsymbol{A} - \lambda \boldsymbol{I}| = 0$ are the characteristic roots or eigenvalues and are denoted as $\lambda_1, \lambda_2, \lambda_3, \ldots, \lambda_n$.
- In the matrix equation $\boldsymbol{AX} = \lambda \boldsymbol{X}$, the vector **X** is the characteristic vector or eigenvector corresponding to the eigenvalue λ.

### *2.7.1 Properties of Eigenvalues*

- The eigenvalues of **A** and $\mathbf{A}^T$ are the same, as the determinant can be expanded along rows or columns.
- The eigenvalues of an upper triangular or a lower triangular or diagonal matrices are the diagonal elements only, as the determinant value of such matrices is the product of diagonal elements.
- Sum of the eigenvalues = Sum of the principal diagonal elements of **A** = trace of **A**, *i.e.* $\lambda_1 + \lambda_2 + \lambda_3 + \cdots + \lambda_n = a_{11} + a_{22} + a_{33} + \cdots + a_{nn}$.
- Product of the eigenvalues = The value of the determinant A, i.e. $\lambda_1\lambda_2\lambda_3 \cdots \lambda_n = |A|$.
- If λ is an eigenvalue of a matrix **A**, then the eigenvalue of $\mathbf{A}^m$ is $\lambda^m$, the eigenvalue of $\mathbf{A}^{-1}$ is $\lambda^{-1}$, the eigenvalue of **kA** is **k** λ, the eigenvalue of k**A + 5I** is k λ **+** 5, etc.
- The eigenvalues of a real symmetric matrix are always real, whereas eigenvalues of skew-symmetric matrices are either zero or purely imaginary.
- If λ is an eigenvalue of an orthogonal matrix **A**, then $\frac{1}{\lambda}$ is also an eigenvalue of **A**.
- If λ is an eigenvalue of **A**, then the eigenvalue of the matrix adj **A** = $\frac{|A|}{\lambda}$
- If λ is an eigen value of **A**, then the eigenvalue of the matrices $a_0\boldsymbol{A}^2 + a_1\boldsymbol{A} + a_2\boldsymbol{I}$ is $a_0\lambda^2 + a_1\lambda + a_2$.

### *2.7.2 Properties of Eigenvectors*

- Eigenvector **X** of a matrix **A** is not unique, i.e. if $X_i$ is an eigenvector of a matrix, then $CX_i$ is also an eigenvector, where C is any scalar.

- If two or more eigenvalues are equal, the eigenvector may or may not be linearly independent.
- Two eigenvectors are called orthogonal if $X_1^T \cdot X_2 = 0$, where $X_1, X_2$ are column vectors.
- Eigenvectors of a symmetric matrix corresponding to different eigenvalues are orthogonal.
- The eigenvectors of $\mathbf{A}, \mathbf{A^{-1}}, \mathbf{A^T}, \mathbf{A^m}$ are the same.
- The eigenvectors corresponding to distinct eigenvalues of matrix **A** are linearly independent.

## 2.8 Diagonalization of a Matrix

### *2.8.1 Diagonalization by Similarity Transformation*

Matrices can be easily handled if they can be diagonalized. The mathematical operations become simplified if the matrix is in diagonal form.

Two matrices **A** and **B** are said to be similar if there exists a non-singular matrix **P** such that $\mathbf{B} = \mathbf{P^{-1}AP}$. Here **B** is said to be obtained from **A** by a *similarity transformation.*

Two similar matrices have the same eigenvalues.

The process of finding a matrix **M** such that $\mathbf{M^{-1}AM = D}$, where **D** is a diagonal matrix, is called *diagonalization of the matrix.* As **A** and **D** are similar matrices it has the same eigenvalues and the diagonal matrix **D** has eigenvalues as the diagonal elements. Here, the transformation $\mathbf{M^{-1}AM = D}$ is called a similarity transformation.

### *2.8.2 Diagonalization of a Real Symmetric Matrix by Orthogonal Reduction*

Let **A** be a real symmetric matrix. Thus, eigenvectors corresponding to distinct eigenvalues of **A** are pairwise orthogonal (i.e.$X_1 X_2^T = I$). Let **N** be the matrix whose columns are the normalized eigenvectors of **A**. Then, **N** is orthogonal, and hence $\mathbf{N^{-1}} = \mathbf{N^T}$. Hence, we have $\mathbf{D = N^{-1}AN = N^TAN}$, which is an *orthogonal transformation or orthogonal reduction.* Hence, the symmetric matrix **A** can be diagonalized by an orthogonal reduction $\mathbf{N^TAN = D}$.

## 2.9 Quadratic Forms

A homogeneous polynomial of the second degree in any number of variables is called a quadratic form.

*Nature of Quadratic form*: Let $\mathbf{Q} = \mathbf{X}^T\mathbf{A}\mathbf{X}$ be a quadratic form of n variables $X_1$, $X_2$, $X_3$, ..., $X_n$. If the rank of **A** is *r*, then the canonical form of **Q** consists of only *r* square terms. The number of positive square terms in the canonical form is called the *index (s)* of the quadratic form. The difference in the number of positive and negative square terms $s - (r - s) = 2s - r$, is called the *signature* of the quadratic form.

The quadratic form $\mathbf{Q} = \mathbf{X}^T\mathbf{A}\mathbf{X}$ in **n** variables is said to be:

- ***Positive definite***: if $r = n$ and $s = n$ or if all the eigenvalues of A are positive or if the principal minors $\mathbf{D_1}, \mathbf{D_2}, \mathbf{D_3}, \ldots, \mathbf{D_n}$ are all positive, i.e. $\mathbf{D_n} > 0$ for all *n* or if $\mathbf{X}^T\mathbf{A}\mathbf{X} > \mathbf{0}$ for all $\mathbf{X} \neq \mathbf{0}$.
- ***Negative definite***: if $r = n$ and $s = 0$ or if all the eigenvalues of **A** are negative or if the principal minors $\mathbf{D_1}, \mathbf{D_3}, \mathbf{D_3}$, etc. are all negative and $\mathbf{D_2}, \mathbf{D_4}, \mathbf{D_6}$, etc. are all positive, i.e. $(-1)^n\,\mathbf{D}_n > 0$ for all *n* or if $\mathbf{X}^T\mathbf{A}\mathbf{X} < 0$ for all $\mathbf{X} \neq 0$.
- ***Positive semi definite***: if $r < n$ and $s = r$ or if all the eigenvalues of $\mathbf{A} \geq 0$ *and at least* one eigenvalue is zero or the $\mathbf{D_n} \geq 0$ and at least one $\mathbf{D_i} = 0$ or if $\mathbf{X}^T\mathbf{A}\mathbf{X} \geq 0$ for all $\mathbf{X} \neq 0$.
- ***Negative semi definite***: if $r < n$ and $s = 0$ or if all the eigenvalues of $\mathbf{A} \leq 0$ and atleast one eigen value is zero or $(-1)^n\,\mathbf{D}_n \geq 0$ and at least one $\mathbf{D}_i = 0$ or if $\mathbf{X}^T\mathbf{A}\mathbf{X} \leq 0$ for all $\mathbf{X} \neq 0$.
- ***Indefinite*** in all other cases or if **A** has both positive and negative eigenvalues if $\mathbf{X}^T\mathbf{A}\mathbf{X} > 0$ for some **X** and $< 0$ for some other **X**.

### Examples

The above unconstrained system with two masses connected with springs will have eigenvalues zero and positive numbers, and hence it is positive semi-definite (Fig. 2.1).

The above system when **constrained** will have only positive eigenvalues, and hence it is **positive definite** (Fig. 2.2).

In buckling analysis, if the structure is under tension, then the system will have a negative eigenvalue, and hence it is **negative definite** (Fig. 2.3).

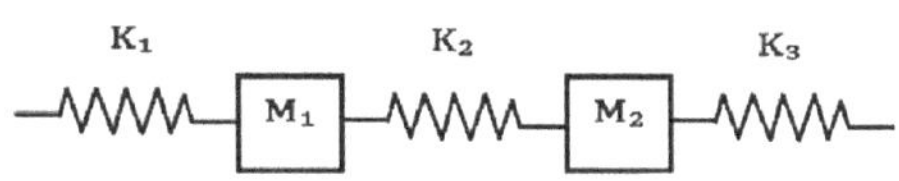


**Fig. 2.1.** Free spring mass system

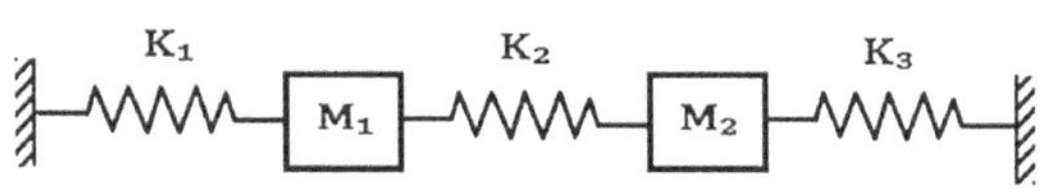


**Fig. 2.2** Constrained spring mass system

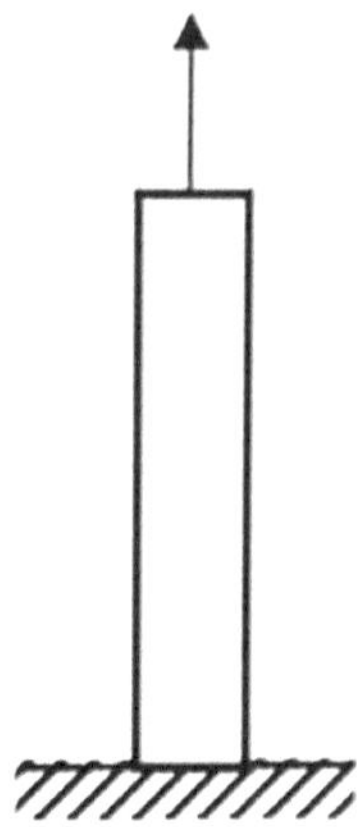

**Fig. 2.3** Column subjected to tensile load

## 2.10 Solution Schemes of a System of Linear Simultaneous Equations

Simultaneous equations are equations where the unknown variables satisfy all the equations. Simultaneous linear algebraic equations are very much useful for numerically solving the mathematical models of problems in Engineering and Mathematical Physics. Stress analysis, heat transfer, vibration analysis, etc. are some typical engineering problems for which the finite element formulation is adopted and the solution of which typically involves solving simultaneous linear equations.

If the equations are written in matrix form, the solution can be easily obtained by performing some matrix operations. For example,

Let **[A]{X} = {B}**, where **[A]** and **{B}** are known parameters and **{X}** is the vector of unknowns.

Multiplying both sides by $\mathbf{[A]^{-1}}$, the inverse of **[A]**, we get $[A]^{-1}$ **[A]{X}** = $[A]^{-1}$**{B}**, but $[A]^{-1}$ **[A] = I**, is an identity matrix.

So, it becomes **{X}** = $[A]^{-1}$**{B}**.

The solutions vector**{X}** is obtained by taking the inverse of matrix [**A**] and multiplying it with vector **{B}**.

There are two approaches to solve a system of linear simultaneous equations in matrix form; the (a) Direct method and (b) Iterative method.

### *2.10.1 Direct Methods*

(a) Method of determinants (Crammer's rule)

Let $\mathbf{AX} = \mathbf{B}$ be the given system of equations where $D = |\mathbf{A}| \neq 0$.

Then, the above system of equations will have a unique solution and is given by

$$x_1 = \frac{D_1}{D}; x_2 = \frac{D_2}{D}; x_3 = \frac{D_3}{D},$$

where $D_1$ is the determinant obtained by replacing the first column **A** with the vector **B**. Similarly, for $D_2$ and $D_3$.

Crammer's rule is suitable for small matrices. However, for a large system of matrices, it is very difficult to compute the determinant of the matrix, and hence it is not suitable for computational implementation.

(b) **Matrix inversion method**

Let **AX = B** be the given system of the equation where **A** is the non-singular matrix.

Multiplying by $\mathbf{A^{-1}}$ both sides, $A^{-1}$ **(AX)** $= A^{-1}$**B; (**$\mathbf{A^{-1}}$ **A) X =** $\mathbf{A^{-1}B}$**; IX =** $\mathbf{A^{-1}B}$,

i.e. $\mathbf{X} = \mathbf{A^{-1}B} = \frac{adjA}{|A|}$ B, which is the solution.

(c) **LU Decomposition Method**

The LU (Lower-Upper) decomposition is a matrix decomposition, which writes the matrix as a product of a lower and upper triangular matrix. The LU decomposition is used to solve a system of linear equations and also find the inverse of a matrix.

Let **AX** = **B** be any given system of linear equations, where **A** is a square matrix of order n.

Let **A** = **LU**, where **L** is a lower triangular and **U** is an upper triangular matrix.

Then, we have $$\boldsymbol{LUX = B} \tag{2.1}$$

Let $$\boldsymbol{UX = Y} \tag{2.2}$$

Then Eq. (2.1) becomes $$\boldsymbol{LY = B}. \tag{2.3}$$

From Eq. (2.3), **Y** can be found and from Eq. (2.2), **X** can be found.

(d) **Gaussian Elimination Method**

Gaussian elimination method is a direct method that consists of transforming the given system of simultaneous equations to an equivalent upper triangular system. From this system, the required solution can be obtained by the method of back substitution.

Consider three linear equations in three unknowns of the form

$$a_{11}x + a_{12}y + a_{13}z = b_1; a_{21}x + a_{22}y + a_{23}z = b_2; a_{31}x + a_{32}y + a_{33}z = b_3.$$

Here we can represent the 'augmented matrix', $\boldsymbol{AB} = \begin{bmatrix} a_{11} & a_{12} & a_{13} & b_1 \\ a_{21} & a_{22} & a_{23} & b_2 \\ a_{31} & a_{32} & a_{33} & b_3 \end{bmatrix}$

$$\sim \begin{bmatrix} a_{11} & a_{12} & a_{13} & b_1 \\ 0 & a'_{22} & a'_{23} & b'_2 \\ 0 & a'_{32} & a'_{33} & b'_3 \end{bmatrix} R_2 \to R_2 + \left(\frac{-a_{21}}{a_{11}}\right) R_1, \quad R_3 \to R_3 + \left(\frac{-a_{31}}{a_{11}}\right) R_1$$

$$\sim \begin{bmatrix} a_{11} & a_{12} & a_{13} & b_1 \\ 0 & a'_{22} & a'_{23} & b'_2 \\ 0 & 0 & a''_{33} & b''_3 \end{bmatrix} R_3 \to R_3 + \left(\frac{-a'_{32}}{a'_{22}}\right) R_2.$$

Which is the required upper triangular matrix.
We have, **AX = B**,

$$\text{i.e.,} \quad a_{11}x + a_{12}y + a_{13}z = b_1;$$
$$a'_{22}y + a'_{23}z = b'_2;$$
$$a''_{33}z = b''_3.$$

By back substitution, we get the solution in the order $z$, $y$, and $x$.

Here, the elements $a_{11}$, $a_{22}$, etc. are called pivotal elements. In the elimination process, if any one of the pivot elements vanishes (or) becomes very small compared to other elements in that column, then we attempt to rearrange the remaining rows so as to obtain a non-vanishing pivot (or) to avoid the multiplication by a large number. This strategy is called *pivoting* and there are two types. They are (a) Partial pivoting (b) Complete pivoting.

(e) **Gauss–Jordan method**

Here, augmented matrix, $\boldsymbol{AB} = \begin{bmatrix} a_{11} & a_{12} & a_{13} & b_1 \\ a_{21} & a_{22} & a_{23} & b_2 \\ a_{31} & a_{32} & a_{33} & b_3 \end{bmatrix}$

$$\sim \begin{bmatrix} a_{11} & a_{12} & a_{13} & b_1 \\ 0 & a'_{22} & a'_{23} & b'_2 \\ 0 & a'_{32} & a_{33} & b'_3 \end{bmatrix} R_2 \to R_2 + \left(\frac{-a_{21}}{a_{11}}\right) R_1; \quad R_3 \to R_3 + \left(\frac{-a_{31}}{a_{11}}\right) R_1$$

$$\sim \begin{bmatrix} a_{11} & 0 & a'_{13} & b'_1 \\ 0 & a'_{22} & a'_{23} & b'_2 \\ 0 & 0 & a''_{33} & b''_{33} \end{bmatrix} R_3 \to R_3 + \left(\frac{-a'_{32}}{a'_{22}}\right) R_2; \quad R_1 \to R_1 + \left(\frac{-a_{12}}{a'_{22}}\right) R_1.$$

Instead of eliminating 'y' only from the third equation, we have eliminated it from the first equation also. The linear equation **AX = B** becomes

$$a_{11}x + a'_{13}z = b'_1; \tag{2.4}$$

$$a'_{22}y + a'_{23}z = b'_2; \tag{2.5}$$

$$a''_{33}z = b''_3. \tag{2.6}$$

From the 3rd equation, '$z$' can be found, which can be simultaneously substituted in the first and second equations to get '$y$' and '$x$'.

The round-off error is the quantity R which must be added to the finite representation of computed numbers in order to make it the true representation of that number. Gaussian elimination is preferred for a large system of equations and it requires less number of multiplications compared to Gauss–Jordan method. It fails when any one of the pivots is zero or if it is a very small number, as the elimination progresses.

Both the above methods are efficient numerical procedures and can be implemented on high-speed digital computers.

### 2.10.2 Iterative Methods

(a) **Gauss–Jacobi iterative Method**

Consider three linear equations in three unknowns of the form

$$a_{11}x + a_{12}y + a_{13}z = b_1;\ a_{21}x + a_{22}y + a_{23}z = b_2;\ a_{31}x + a_{32}y + a_{33}z = b_3;$$

From the above three equations, we have

$$x = \frac{1}{a_{11}}(b_1 - a_{12}y - a_{13}z); \tag{2.7}$$

$$y = \frac{1}{a_{22}}(b_2 - a_{21}x - a_{23}z); \tag{2.8}$$

$$z = \frac{1}{a_{33}}(b_3 - a_{31}x - a_{32}y). \tag{2.9}$$

Let $x^{(0)} = 0$, $y^{(0)} = 0$, and $z^{(0)} = 0$ be the initial approximation. Substituting in the RHS of Eqs. (2.7), (2.8), and (2.9), we get the first approximation $x^{(1)}$, $y^{(1)}$, and $z^{(1)}$. Substituting the first iterative value again in the RHS of Eqs. (2.7), (2.8), and (2.9), we get the second approximation, and so on. Hence, the (n + 1)th approximation is given by

$$\begin{aligned} x^{(n+1)} &= \frac{1}{a}\left(b_1 - a_{12}y^{(n)} - a_{13}z^{(n)}\right); \\ y^{(n+1)} &= \frac{1}{a_{22}}\left(b_2 - a_{21}x^{(n)} - a_{23}z^{(n)}\right); \\ z^{(n+1)} &= \frac{1}{a_{33}}\left(b_3 - a_{31}x^{(n)} - a_{32}y^{(n)}\right). \end{aligned}$$

The above procedure should be continued till two successive approximations converge to the desired accuracy.

(b) **Gauss–Seidal iterative method**

Consider the three linear equations in three unknowns of the form

$$a_{11}x + a_{12}y + a_{13}z = b_1; \quad a_{21}x + a_{22}y + a_{23}z = b_2; \quad a_{31}x + a_{32}y + a_{33}z = b_3.$$

From the above three equations, we have

$$x = \frac{1}{a_{11}}(b_1 - a_{12}y - a_{13}z); \tag{2.10}$$

$$y = \frac{1}{a_{22}}(b_2 - a_{21}x - a_{23}Z); \tag{2.11}$$

$$z = \frac{1}{a_{33}}(b_3 - a_{31}x - a_{32}y). \tag{2.12}$$

Let $x^{(0)} = 0$, $y^{(0)} = 0$, and $z^{(0)} = 0$ be the initial approximation. Substituting $y^{(0)} = 0$, $z^{(0)} = 0$ in the RHS of Eq. (2.10), we get the first approximation $x^{(1)}$. Now, substituting $x^{(1)}$, $z^{(0)}$ in Eq. (2.11), we get $y^{(1)}$. Substituting $x^{(1)}$, $y^{(1)}$ in Eq. (2.12), we get $z^{(1)}$.Thus, the first approximation for $x^{(1)}$, $y^{(1)}$, and $z^{(1)}$ is obtained. In this method, whenever, an approximation is found out for one variable, it is immediately used in the subsequent equation. Hence, the (n + 1)th approximation is given by

$$\begin{aligned} x^{(n+1)} &= \frac{1}{a_{11}}\left(b_1 - a_{12}y^{(n)} - a_{13}z^{(n)}\right) \\ y^{(n+1)} &= \frac{1}{a_{22}}\left(b_2 - a_{21}x^{(n+1)} - a_{23}z^{(n)}\right) \\ z^{(n+1)} &= \frac{1}{a_{33}}\left(b_3 - a_{31}x^{(n+1)} - a_{32}y^{(n+1)}\right). \end{aligned}$$

The above procedure should be continued till two successive approximations converge to the desired accuracy.

The conditions for convergence of both schemes are $|a_{11}| > |a_{12}| + |a_{13}|$, $|a_{22}| > |a_{21}| + |a_{23}|$, and $|a_{33}| > |a_{31}| + |a_{32}|$. The convergence of the Gauss–Seidal method is two times faster than Gauss–Jacobi method.

## 2.11 Worked Out Examples

2.11.1 If $\mathbf{A}+\mathbf{B}=\begin{bmatrix}1 & -1\\ 3 & 0\end{bmatrix}$ and $\mathbf{A}-\mathbf{B}=\begin{bmatrix}3 & 1\\ 1 & 4\end{bmatrix}$ calculate the product AB

*Solution*:

$$\text{Given,}\quad \boldsymbol{A}+\boldsymbol{B}=\begin{bmatrix}1 & -1\\ 3 & 0\end{bmatrix} \tag{2.13}$$

$$\text{And}\quad \boldsymbol{A}-\boldsymbol{B}=\begin{bmatrix}3 & 1\\ 1 & 4\end{bmatrix}. \tag{2.14}$$

Adding and subtracting Eqs. (2.13) and (2.14), we get

$$2\boldsymbol{A}=\begin{bmatrix}4 & 0\\ 4 & 4\end{bmatrix}\text{ and }2\boldsymbol{B}=\begin{bmatrix}-2 & -2\\ 2 & -4\end{bmatrix}.$$

Hence, we have, $\boldsymbol{A}=\begin{bmatrix}2 & 0\\ 2 & 2\end{bmatrix}$ and $\boldsymbol{B}=\begin{bmatrix}-1 & -1\\ 1 & -2\end{bmatrix}$,

$$\therefore \boldsymbol{AB}=\begin{bmatrix}2 & 0\\ 2 & 2\end{bmatrix}\begin{bmatrix}-1 & -1\\ 1 & -2\end{bmatrix}=\begin{bmatrix}-2 & -2\\ 0 & -6\end{bmatrix}=-2\begin{bmatrix}1 & 1\\ 0 & 3\end{bmatrix}.$$

2.11.2 Show that $\mathbf{A}=\begin{bmatrix}\boldsymbol{cos\theta} & -\boldsymbol{sin\theta}\\ \boldsymbol{sin\theta} & \boldsymbol{cos\theta}\end{bmatrix}$ is orthogonal, and hence find $\mathbf{A}^{-1}$

*Solution*:

We have, $\boldsymbol{AA}^T=\begin{bmatrix}cos\theta & -sin\theta\\ sin\theta & cos\theta\end{bmatrix}\begin{bmatrix}cos\theta & sin\theta\\ -sin\theta & cos\theta\end{bmatrix}$

$$=\begin{bmatrix}cos^2\theta+sin^2\theta & 0\\ 0 & sin^2\theta+cos^2\theta\end{bmatrix}=\begin{bmatrix}1 & 0\\ 0 & 1\end{bmatrix}=I.$$

Similarly, we can prove , $\boldsymbol{A}^T\boldsymbol{A}=\boldsymbol{I}$.
Hence, we have , $\boldsymbol{AA}^T=\boldsymbol{A}^T\boldsymbol{A}=\boldsymbol{I}$; $\therefore$ the matrix **A** is orthogonal.
For the orthogonal matrix, we have $\mathbf{A^{-1}}=\mathbf{A^{T}}$ .*Hence*, $A^{-1}=\begin{bmatrix}cos\theta & sin\theta\\ -sin\theta & cos\theta\end{bmatrix}$.

2.11.3 Express $\mathbf{A} = \begin{bmatrix} 3 & 5 & 7 \\ -9 & -11 & 4 \\ 12 & -16 & 6 \end{bmatrix}$ as the sum of unit lower triangular and upper triangular matrix

*Solution*:

Let $\boldsymbol{A} = \boldsymbol{LU} = \begin{bmatrix} 1 & 0 & 0 \\ l_{21} & 1 & 0 \\ l_{31} & l_{32} & 1 \end{bmatrix} \begin{bmatrix} u_{11} & u_{12} & u_{13} \\ 0 & u_{22} & u_{23} \\ 0 & 0 & u_{33} \end{bmatrix}$,

$$\text{i.e.} \quad \begin{bmatrix} 3 & 5 & 7 \\ -9 & -11 & 4 \\ 12 & -16 & 6 \end{bmatrix} = \begin{bmatrix} u_{11} & u_{12} & u_{13} \\ l_{21}u_{11} & l_{21}u_{12} + u_{22} & l_{21}u_{13} + u_{23} \\ l_{31}u_{11} & l_{31}u_{12} + l_{32}u_{22} & l_{31}u_{13} + l_{32}u_{23} + u_{33} \end{bmatrix}.$$

Comparing, we get

$$u_{11} = 3; \quad u_{12} = 5; \quad u_{13} = 7; \quad l_{21}u_{11} = -9, \qquad \therefore l_{21} = -3;$$

$$l_{21}u_{12} + u_{22} = -11, \quad \therefore (-3)(5) + u_{22} = -11, \quad \therefore u_{22} = 4;$$

$$l_{21}u_{13} + u_{23} = 4, \quad \therefore (-3)(7) + u_{23} = 4, \qquad \therefore u_{23} = 25;$$

$$l_{31}u_{11} = 12, \quad \therefore l_{31} = 4;$$

$$l_{31}u_{12} + l_{32}u_{22} = -16, \quad \therefore (4)(5) + l_{32}(4) = -16; \qquad \therefore l_{32} = -9;$$

$$l_{31}u_{13} + l_{32}u_{23} + u_{33} = 6, \quad \text{i.e. } (4)(7) + (-9)(25) + u_{33} = 6; \quad \therefore u_{33} = 203.$$

Hence, **L** and **U** can be found.

2.11.4 Find the symmetric and skew-symmetric part of the matrix $\mathbf{A} = \begin{bmatrix} 1 & 2 & -1 \\ 3 & 5 & 4 \\ -2 & 3 & 1 \end{bmatrix}$

*Solution*:

We have, symmetric part $= \left(\frac{1}{2}\right)\left(\boldsymbol{A} + \boldsymbol{A}^T\right)$

$$= \left(\frac{1}{2}\right)\left(\begin{bmatrix} 1 & 2 & -1 \\ 3 & 5 & 4 \\ -2 & 3 & 1 \end{bmatrix} + \begin{bmatrix} 1 & 3 & -2 \\ 2 & 5 & 3 \\ -1 & 4 & 1 \end{bmatrix}\right)$$

$$= \left(\frac{1}{2}\right)\begin{bmatrix} 2 & 5 & -3 \\ 5 & 10 & 7 \\ -3 & 7 & 2 \end{bmatrix} = \begin{bmatrix} 1 & \frac{5}{2} & -\frac{3}{2} \\ \frac{5}{2} & 5 & \frac{7}{2} \\ \frac{-3}{2} & \frac{7}{2} & 1 \end{bmatrix}.$$

Skew symmetric part $= \left(\frac{1}{2}\right)\left(\boldsymbol{A} - \boldsymbol{A}^T\right)$

$$= \frac{1}{2}\begin{bmatrix} 0 & -1 & 1 \\ 1 & 0 & 1 \\ -1 & -1 & 0 \end{bmatrix} = \begin{bmatrix} 0 & -\frac{1}{2} & \frac{1}{2} \\ \frac{1}{2} & 0 & \frac{1}{2} \\ -\frac{1}{2} & -\frac{1}{2} & 0 \end{bmatrix}.$$

2.11.5 Find the rank of the following matrices

(a) $\boldsymbol{A} = \begin{bmatrix} 1 & 2 & 3 \\ 2 & 3 & 4 \\ 0 & 2 & 2 \end{bmatrix}$; (b) $\boldsymbol{A} = \begin{bmatrix} 1 & -1 & 2 \\ 2 & 2 & 2 \\ 3 & -3 & 6 \end{bmatrix}$; (c) $\boldsymbol{A} = \begin{bmatrix} 1 & 2 & 1 & 2 \\ 1 & 3 & 2 & 2 \\ 2 & 4 & 3 & 4 \\ 3 & 7 & 4 & 6 \end{bmatrix}$;

(d) Given the rank of the matrix $\boldsymbol{A} = \begin{bmatrix} 1 & -2 & 3 & 4 \\ 3 & 1 & 0 & 3 \\ 5 & 4 & -3 & k \end{bmatrix}$ is 2. Find $k$ ?

*Solution*:

(a) Here $|\boldsymbol{A}| = \begin{vmatrix} 1 & 2 & 3 \\ 2 & 3 & 4 \\ 0 & 2 & 2 \end{vmatrix} = 1(6-8) - 2(4-0) + 3(4-0) = -2 - 8 + 12 \neq 0,$

$\therefore$ Rank of $\mathbf{A} = 3$.

(b) Here, $|\boldsymbol{A}| = \begin{vmatrix} 1 & -1 & 2 \\ 2 & 2 & 2 \\ 3 & -3 & 6 \end{vmatrix} = 1(12+6) + 1(12-6) + 2(-6-6)$

$$= 18 + 6 - 24 = 0,$$

$\therefore$ Rank of $(\boldsymbol{A}) \neq 3$; But $\begin{vmatrix} 1 & -1 \\ 2 & 2 \end{vmatrix} = 2 + 2 = 4 \neq 0$; $\therefore$ Rank of $\mathbf{A} = 2$.

(c) We have, $\boldsymbol{A} = \begin{bmatrix} 1 & 2 & 1 & 2 \\ 1 & 3 & 2 & 2 \\ 2 & 4 & 3 & 4 \\ 3 & 7 & 4 & 6 \end{bmatrix}$

$$\sim \begin{bmatrix} 1 & 2 & 1 & 2 \\ 0 & 1 & 1 & 0 \\ 0 & 0 & 1 & 0 \\ 0 & 1 & 1 & 0 \end{bmatrix} R_2 \to R_2 + (-1)R_1;\ R_3 \to R_3 + (-2)R_1;\ R_4 \to R_4 + (-3)R_1$$

$$\sim \begin{bmatrix} 1 & 2 & 1 & 2 \\ 0 & 1 & 1 & 0 \\ 0 & 0 & 1 & 0 \\ 0 & 0 & 0 & 0 \end{bmatrix} R_4 \to R_4 + (-1)R_2,$$

$\therefore$ Rank of $\boldsymbol{A}$ = Number of non − zero rows in the upper triangular matrix = 3.

(d) Given, $A = \begin{bmatrix} 1 & -2 & 3 & 4 \\ 3 & 1 & 0 & 3 \\ 5 & 4 & -3 & k \end{bmatrix}$

$$\sim \begin{bmatrix} 1 & -2 & 3 & 4 \\ 0 & 7 & -9 & -9 \\ 0 & 14 & -18 & (k-20) \end{bmatrix} R_2 \to R_2 + (-3)R_1;\ R_3 \to R_3 + (-5)R_1$$

$$\sim \begin{bmatrix} 1 & -2 & 3 & 4 \\ 0 & 7 & -9 & -9 \\ 0 & 0 & 0 & (k-2) \end{bmatrix} R_3 \to R_3 + (-2)R_2.$$

Given, rank (**A**) =2 => Number of non-zero rows = 2,
$\therefore$ We have, $k = 2$.

2.11.6 Investigate the linear dependence or independence of the set of vector (1, 3, 4, 2), (3, −5, 2, 2), and (2, −1, 3, 2)

*Solution*:

Let
$$x_1 = (1, 3, 4, 2); \tag{2.15}$$
$$x_2 = (3, -5, 2, 2); \tag{2.16}$$
$$x_3 = (2, -1, 3, 2). \tag{2.17}$$

Now, find a linear combination of the relations (2.15), (2.16), and (2.17) so that the first component will be zero.

Therefore, we have,
$$x_2 - 3x_1 = (0, -14, -10, -4); \tag{2.18}$$
$$x_3 - 2x_1 = (0, -7, -5, -2). \tag{2.19}$$

Now, find a linear combination of the relations (2.18) and (2.19) so that the second component will be zero. Hence we have

$$2(x_3 - 2x_1) - (x_2 - 3x_1) = (0, 0, 0, 0); \quad \text{i.e.,} \quad 2x_3 - x_2 - x_1 = 0.$$

Hence, the given vectors are linearly dependent and the relation between them is

$$2x_3 - x_2 - x_1 = 0.$$

2.11.7 Investigate, for what value of $\boldsymbol{\lambda}$ ***and*** $\boldsymbol{\mu}$ the simultaneous equations, $\boldsymbol{x}+\boldsymbol{y}+z=6$

$x+2y+3z=10$, $x+2y+\lambda z=\mu$ have (i) no solution, (ii) a unique solution, and (iii) an infinite number of solutions.

*Solution*:

Given system of equation can be written in matrix form as **AX = B**

Here, we have $[\boldsymbol{AB}] = \begin{bmatrix} 1 & 1 & 1 & 6 \\ 1 & 2 & 3 & 10 \\ 1 & 2 & \lambda & \mu \end{bmatrix}$

$$\sim \begin{bmatrix} 1 & 1 & 1 & 6 \\ 0 & 1 & 2 & 4 \\ 0 & 1 & (\lambda-1) & (\mu-6) \end{bmatrix} R_2 \to R_2 + (-1)R_1;\ R_3 \to R_3 + (-1)R_1$$

$$\sim \begin{bmatrix} 1 & 1 & 1 & 6 \\ 0 & 1 & 2 & 4 \\ 0 & 0 & (\lambda-3) & (\mu-10) \end{bmatrix} R_3 \to R_3 + (-1)R_2.$$

Here, we have

(i) *When* $\lambda = 3$ *and* $\mu = 10$
R (**AB**) = 2, R (**A**) = 2. Hence, consistent and infinite number of solutions.
(ii) *When* $\lambda = 3$ *and* $\mu \neq 10$
Here, R (**AB**) = 3, R (**A**) = 2. Hence, inconsistent and no solution.
(iii) *When* $\lambda \neq 3$ *and* $\mu =$ *any value*
Here, R (**AB**) = 3, R (**A**) = 3, and hence consistent, has a unique solution.

2.11.8 Test for consistency to the system $\boldsymbol{x}+2\boldsymbol{y}+3z=0$, $2\boldsymbol{x}+\boldsymbol{y}+4z=0$, and $5x-6y+2z=0$, and solve if it is consistent.

*Solution*:

The given system is of the form **AX** = **0**.

Here, $[A] = \begin{bmatrix} 1 & 2 & 3 \\ 2 & 1 & 4 \\ 5 & -6 & 2 \end{bmatrix}$

$$\sim \begin{bmatrix} 1 & 2 & 3 \\ 0 & -3 & -2 \\ 0 & -16 & -13 \end{bmatrix} R_2 \to R_2 + (-2)R_1;\ R_3 \to R_3 + (-5)R_1$$

$$\sim \begin{bmatrix} 1 & 2 & 3 \\ 0 & -3 & -2 \\ 0 & 0 & {}^{-7}/_{3} \end{bmatrix} R_3 \to R_3 + \left(\frac{16}{-3}\right)R_2.$$

Here, we see that R (**A**) = 3 = number of variables

Hence, the given system will have only trivial solution $x=0, y=0, z=0$.

2.11.9 Find the values of a and b for which the equations $\boldsymbol{x}+\boldsymbol{y}+z = 3$, $\boldsymbol{x}+2\boldsymbol{y}+2z = 6$

$x+ay+3z = b$ have (i) no solution, (ii) unique solution, and (iii) many solutions.

*Solution*:

$$\text{Here, we have} \quad [\boldsymbol{AB}] = \begin{bmatrix} 1 & 1 & 1 & 3 \\ 1 & 2 & 2 & 6 \\ 1 & a & 3 & b \end{bmatrix}$$

$$\sim \begin{bmatrix} 1 & 1 & 1 & 3 \\ 0 & 1 & 1 & 3 \\ 0 & (a-1) & 2 & (b-3) \end{bmatrix} R_2 \to R_2 + (-1)R_1;\ R_3 \to R_3 + (-1)R_1$$

$$\sim \begin{bmatrix} 1 & 1 & 1 & 3 \\ 0 & 1 & 1 & 3 \\ 0 & 0 & (3-a) & (b-3a) \end{bmatrix} R_3 \to R_3 + (-(a-1))R_2.$$

(i) If $a \neq 3$ and b is any value, R (**A**) = 3, R (**AB**) = 3.
Hence, in this case, consistent and the given system has a unique solution.

(ii) If a = 3 and b = 9, we have R(**A**) = 2, R(**AB**) = 2.
Hence, consistent and has an infinite number of solutions.

(iii) If a = 3 and $b \neq 9$, we have, R(**A**) = 2, R(**AB**) = 3.
Hence, the given system is inconsistent and it has no solution.

2.11.10 Solve, completely, the system of equation

$x + y - 2z + 3w = 0$, $x - 2y + z - \omega = 0$, $4x + y - 5z + 8w = 0$, and $5x - 7y + 2z - w = 0$

*Solution*:

The given system is of the form, $\mathbf{AX = 0}$.

$$\text{Here, we have} \quad [A] = \begin{bmatrix} 1 & 1 & -2 & 3 \\ 1 & -2 & 1 & -1 \\ 4 & 1 & -5 & 8 \\ 5 & -7 & 2 & -1 \end{bmatrix}$$

$$\sim \begin{bmatrix} 1 & 1 & -2 & 3 \\ 0 & -3 & 3 & -4 \\ 0 & -3 & 3 & -4 \\ 0 & -12 & 12 & -16 \end{bmatrix} R_2 \to R_2 + (-1)R_1;\ R_3 \to R_3 + (-4)R_1;\ R_4 \to R_4 + (-5)R_1$$

$$\sim \begin{bmatrix} 1 & 1 & -2 & 3 \\ 0 & -3 & 3 & -4 \\ 0 & 0 & 0 & 0 \\ 0 & 0 & 0 & 0 \end{bmatrix} R_3 \to R_3 + (-1)R_2;\ R_4 \to R_4 + (-4)R_2.$$

Here, we have, R (**A**) = Number of non-zero rows = 2 < unknowns.

Hence, the given system will have infinite number of non-zero solutions.

We have, $\mathbf{AX} = \mathbf{0}$; i.e., $\begin{bmatrix} 1 & 1 & -2 & 3 \\ 0 & -3 & 3 & -4 \\ 0 & 0 & 0 & 0 \\ 0 & 0 & 0 & 0 \end{bmatrix} \begin{Bmatrix} x \\ y \\ z \\ w \end{Bmatrix} = \begin{Bmatrix} 0 \\ 0 \\ 0 \\ 0 \end{Bmatrix}$.

Hence, we have $x + y - 2z + 3w = 0$; $-3y + 3z - 4w = 0$; $0 = 0$; $0 = 0$.

Let $w = a$, $z = b$, $\therefore y = \frac{4a-3b}{-3}$; and $x = -3a + 2b + \frac{(4a-3b)}{3} = \frac{1}{3}(-5a + 3b)$,

$\therefore$ The solution is $x = \frac{1}{3}(-5a + 3b)$, $y = \frac{1}{3}(3b - 4a)$, and $z = b$, $w = a$, where *a and b* are arbitrary.

2.11.11 Determine the value of $\boldsymbol{\lambda}$ for which the following equations may possess non-trivial solution, $3\boldsymbol{x} + \boldsymbol{y} - \boldsymbol{\lambda z} = 0$, $4\boldsymbol{x} - 2\boldsymbol{y} - 3\boldsymbol{z} = 0$, and $2\boldsymbol{\lambda x} + 4\boldsymbol{y} + \boldsymbol{\lambda z} = 0$

*Solution*:

The coefficient matrix $[\boldsymbol{A}] = \begin{bmatrix} 3 & 1 & -\lambda \\ 4 & -2 & -3 \\ 2\lambda & 4 & \lambda \end{bmatrix}$.

If the given system, $\mathbf{AX} = \mathbf{0}$, has a non-trivial solution, then $|\boldsymbol{A}| = 0$,

$$\text{i.e.} \begin{vmatrix} 3 & 1 & -\lambda \\ 4 & -2 & -3 \\ 2\lambda & 4 & \lambda \end{vmatrix} = 0; \quad \text{i.e.,} \quad 3(-2\lambda + 12) - 1(4\lambda + 6\lambda) - \lambda(16 + 4\lambda) = 0,$$

i.e. $4\lambda^2 + 32\lambda - 36 = 0$; i.e. $4(\lambda - 1)(\lambda + 9) = 0$; $\therefore$ we have $\lambda = 1, -9$.

2.11.12 Find the eigenvalues and eigenvectors of the matrix $\mathbf{A} = \begin{bmatrix} -2 & 2 \\ 2 & 1 \end{bmatrix}$

*Solution*:

The characteristic equation of **A** is $|\boldsymbol{A} - \boldsymbol{\lambda I}| = 0$,

$$\text{i.e.,} \begin{vmatrix} -2-\lambda & 2 \\ 2 & 1-\lambda \end{vmatrix} = 0; \quad \text{i.e.} \quad (-2-\lambda)(1-\lambda) - 4 = 0,$$

i.e., $\lambda^2 + \lambda - 6 = 0$; i.e., $(\lambda + 3)(\lambda - 2) = 0$; $\therefore \lambda = -3, 2$ are the eigenvalues

*When* $\lambda = -3$.

Consider the equation $\boldsymbol{AX} = \boldsymbol{\lambda X}$; ***i.e.,*** $(\boldsymbol{A} - \boldsymbol{\lambda I})\boldsymbol{X} = 0$.

Hence, we have, $x + 2y = 0$ and $2x + 4y = 0$,

i.e., $x + 2y = 0$; Let y = 1; $\therefore x = -2$,

$$\therefore \text{The eigen vector is } \boldsymbol{X}_1 = \begin{Bmatrix} x \\ y \end{Bmatrix} = \begin{Bmatrix} -2 \\ 1 \end{Bmatrix}.$$

*When* $\lambda = 2$

We have $-4x + 2y = 0$ and $2x - y = 0$,
i.e., $2x - y = 0$; Let y = 1; $\therefore x = \frac{1}{2}$,

$$\therefore \text{The eigen vector is } \boldsymbol{X}_2 = \begin{Bmatrix} x \\ y \end{Bmatrix} = \begin{Bmatrix} \frac{1}{2} \\ 1 \end{Bmatrix} = \begin{Bmatrix} 1 \\ 2 \end{Bmatrix}.$$

2.11.13 Find the eigenvalues and eigenvectors of the matrix $\mathbf{A} = \begin{bmatrix} 1 & 2 & 2 \\ 2 & 1 & 2 \\ 2 & 2 & 1 \end{bmatrix}$

*Solution*:

The characteristic equation of **A** is $|\boldsymbol{A} - \boldsymbol{\lambda I}| = 0$; *i.e.* $\lambda^3 - 3\lambda^2 - 9\lambda - 5 = 0$,

$$\text{i.e. } (\lambda + 1)\left(\lambda^2 - 4\lambda - 5\right) = 0; \text{ i.e. } (\lambda + 1)(\lambda + 1)(\lambda - 5) = 0,$$
$$\therefore \text{The eigenvalues of } \boldsymbol{A} \text{ are } \lambda = -1, \ -1, 5.$$

*When* $\lambda = -1$.
Consider, the equation $\boldsymbol{AX} = \boldsymbol{\lambda X}$; i.e., $(\boldsymbol{A} - \boldsymbol{\lambda I})\boldsymbol{X} = 0$;

$\therefore$ we have , $2x + 2y + 2z = 0$; $2x + 2y + 2z = 0$; and $2x + 2y + 2z = 0$,
i. e. $x + y + z = 0$;

Let z = 0; $\therefore x + y = 0$,

$$\therefore x = -1 \text{ when } y = 1.$$

Similarly, *Let* $y = 0$; $\therefore x + z = 0$; $\therefore z = -1$, when $x = 1$.

$$\therefore \text{The eigen vectors are } \begin{Bmatrix} -1 \\ 1 \\ 0 \end{Bmatrix} \text{ and } \begin{Bmatrix} 1 \\ 0 \\ -1 \end{Bmatrix}.$$

*when* $\lambda = 5$

Here, we have

$$2x - y - z = 0; \tag{2.20}$$

$$x - 2y + z = 0; \tag{2.21}$$

$$x + y - 2z = 0. \tag{2.22}$$

Eliminating z from (2.20) and (2.21) and then from (2.21) and (2.22), we get.

$3x - 3y = 0$ and $3x - 3y = 0$; *i.e.* $x - y = 0$; when $y = 1$, $x = 1$, and $z = 1$.

$\therefore$ The eigenvector is $\begin{Bmatrix} 1 \\ 1 \\ 1 \end{Bmatrix}$.

2.11.14 If 3 and 15 are two eigenvalues of $\mathbf{A} = \begin{bmatrix} 8 & -6 & 2 \\ -6 & 7 & -4 \\ 2 & -4 & 3 \end{bmatrix}$, find the eigenvalues of $\mathbf{A}^2$, $6\mathbf{A} + 8\mathbf{I}$ and $\mathbf{A}^{-1}$

*Solution*:

We have, sum of the eigenvalues = sum of the principal diagonal elements,

$$\text{i.e. } \lambda + 3 + 15 = 8 + 7 + 3; \quad i.e. \quad \lambda = 18 - 18 = 0.$$

Hence, the eigenvalues of **A** are 0, 3, and 15.
Therefore, the eigenvalues of $\mathbf{A^2}$ are $0^2$, $3^2$, and $15^2$; *i.e.* 0, 9, and 225.
The eigenvalues of $(6A + 8I)$ *are* $6(0) + 8, 6(3) + 8, 6(3) + 15$ i.e., 8, 26, 33.
The eigenvalues of $\mathbf{A^{-1}}$ are $0^{-1}$, $3^{-1}$, and $15^{-1}$, ***(i.e.)*** $\infty, \frac{1}{3}, \frac{1}{15}$.

2.11.15 Diagonalize the matrix $\mathbf{A} = \begin{bmatrix} 4 & 1 \\ 2 & 3 \end{bmatrix}$

*Solution*:

The characteristic equation is given by $|A - \lambda I| = 0$,

i.e. $\begin{vmatrix} 4-\lambda & 1 \\ 2 & 3-\lambda \end{vmatrix} = 0$; i.e. $(4-\lambda)(3-\lambda) - 2 = 0$; i.e. $\lambda^2 - 7\lambda + 10 = 0$,
i.e. $(\lambda - 2)(\lambda - 5) = 0$; $\therefore \lambda = 2, 5$ are the eigenvalues
when $\lambda = 2$
we have $(A - \lambda I)X = 0$,
i.e., $2x + y = 0$ and $2x + y = 0$; when $y = 1, x = -\frac{1}{2}$

$\therefore X_1 = \begin{Bmatrix} -\frac{1}{2} \\ 1 \end{Bmatrix} = \begin{Bmatrix} -1 \\ 2 \end{Bmatrix}$.

when $\lambda = 5$
we have, $-x + y = 0, 2x - 2y = 0$; i.e., $x - y = 0$, i.e., when $y = 1$, we get $x = 1$,

$\therefore X_2 = \begin{Bmatrix} 1 \\ 1 \end{Bmatrix}$;

The modal matrix, $B = \begin{bmatrix} -1 & 1 \\ 2 & 1 \end{bmatrix}$.

We have, $|\boldsymbol{B}| = -3$, $adj\,\boldsymbol{B} = \begin{bmatrix} 1 & -1 \\ -2 & -1 \end{bmatrix}$; $\therefore \boldsymbol{B}^{-1} = \frac{adj\,\boldsymbol{B}}{|\boldsymbol{B}|} = \frac{1}{3}\begin{bmatrix} -1 & 1 \\ 2 & 1 \end{bmatrix}$.

$$Consider, \quad \boldsymbol{B}^{-1}\boldsymbol{A}\boldsymbol{B} = \frac{1}{3}\begin{bmatrix} -1 & 1 \\ 2 & 1 \end{bmatrix}\begin{bmatrix} 4 & 1 \\ 2 & 3 \end{bmatrix}\begin{bmatrix} -1 & 1 \\ 2 & 1 \end{bmatrix} = \frac{1}{3}\begin{bmatrix} 6 & 0 \\ 0 & 15 \end{bmatrix} = \begin{bmatrix} 2 & 0 \\ 0 & 5 \end{bmatrix} = \mathbf{D}.$$

2.11.16 Find $A^4$, for $\mathbf{A} = \begin{bmatrix} 1 & 1 & 1 \\ 0 & 2 & 1 \\ -4 & 4 & 3 \end{bmatrix}$ by diagonalizing

*Solution*:

The characteristic equation of $\mathbf{A}$ is $|\boldsymbol{A} - \lambda \boldsymbol{I}| = \begin{bmatrix} 1-\lambda & 1 & 1 \\ 0 & 2-\lambda & 1 \\ -4 & 4 & 3-\lambda \end{bmatrix} = 0.$

Expanding $(1-\lambda)(6 - 5\lambda + \lambda^2) = 0$; i.e., $(1-\lambda)(\lambda - 2)(\lambda - 3) = 0$,
$\therefore$ We have, $\lambda = 1, 2, 3$

$$When \;\; \lambda = 1, \quad \boldsymbol{X}_1 = \begin{Bmatrix} 1 \\ 2 \\ -2 \end{Bmatrix}; \;\; When \;\; \lambda = 2,$$

$$\mathbf{X}_2 = \begin{Bmatrix} 1 \\ 1 \\ 0 \end{Bmatrix}; \;\; When \;\; \lambda = 3, \quad \mathbf{X}_3 = \begin{Bmatrix} 1 \\ 1 \\ 1 \end{Bmatrix}.$$

Hence, the modal matrix, $\boldsymbol{B} = \begin{bmatrix} 1 & 1 & 1 \\ 2 & 1 & 1 \\ -2 & 0 & 1 \end{bmatrix}$.

Here, we can find, $\boldsymbol{B}^{-1} = \begin{bmatrix} -1 & 1 & 0 \\ 4 & -3 & -1 \\ -2 & 2 & 1 \end{bmatrix}$.

We know that $\boldsymbol{B}^{-1}\boldsymbol{A}\boldsymbol{B} = \boldsymbol{D} =$ a diagonal matrix $= \begin{bmatrix} 1 & 0 & 0 \\ 0 & 2 & 0 \\ 0 & 0 & 3 \end{bmatrix}$.

Also, $\boldsymbol{B}^{-1}\boldsymbol{A}^4\boldsymbol{B} = \boldsymbol{D}^4 = \begin{bmatrix} 1^4 & 0 & 0 \\ 0 & 2^4 & 0 \\ 0 & 0 & 3^4 \end{bmatrix} = \begin{bmatrix} 1 & 0 & 0 \\ 0 & 16 & 0 \\ 0 & 0 & 81 \end{bmatrix}$.

Hence, $\boldsymbol{A}^4 = \boldsymbol{B}\boldsymbol{D}^4\boldsymbol{B}^{-1} = \begin{bmatrix} 1 & 1 & 1 \\ 2 & 1 & 1 \\ -2 & 0 & 1 \end{bmatrix}\begin{bmatrix} 1 & 0 & 0 \\ 0 & 16 & 0 \\ 0 & 0 & 81 \end{bmatrix}\begin{bmatrix} -1 & 1 & 0 \\ 4 & -3 & -1 \\ -2 & 2 & 1 \end{bmatrix}$

$$= \begin{bmatrix} 1 & 1 & 1 \\ 2 & 1 & 1 \\ -2 & 0 & 1 \end{bmatrix} \begin{bmatrix} -1 & 1 & 0 \\ 64 & -48 & -16 \\ -162 & 162 & 81 \end{bmatrix} = \begin{bmatrix} -99 & 115 & 65 \\ -100 & 116 & 65 \\ -160 & 160 & 81 \end{bmatrix}.$$

2.11.17 Diagonalize the symmetric matrix, $\mathbf{A} = \begin{bmatrix} 8 & -6 & 2 \\ -6 & 7 & -4 \\ 2 & -4 & 3 \end{bmatrix}$

*Solution*:

The characteristic equation is $|\boldsymbol{A} - \lambda \boldsymbol{I}| = \begin{bmatrix} 8-\lambda & -6 & 2 \\ -6 & 7-\lambda & -4 \\ 2 & -4 & 3-\lambda \end{bmatrix} = 0.$

Expanding, $\lambda^3 - 18\lambda^2 + 45\lambda = 0$; i.e. $\lambda(\lambda^2 - 18\lambda + 45) = 0$,

i.e., $\lambda(\lambda - 3)(\lambda - 15) = 0$; $\therefore \lambda = 0, 3, 15$ are the eigen values

when $\lambda = 0$:

$|A - \lambda I|X = 0$ gives

$$8x - 6y + 2z = 0; \tag{2.23}$$

$$-6x + 7y - 4z = 0; \tag{2.24}$$

$$2x - 4y + 3z = 0. \tag{2.25}$$

From Eqs. (2.23) and (2.24), we get $\frac{x}{1} = \frac{y}{2} = \frac{z}{2}$ .

$$\therefore \text{The eigen vector is } \boldsymbol{X}_1 = \begin{Bmatrix} x \\ y \\ z \end{Bmatrix} = \begin{Bmatrix} 1 \\ 2 \\ 2 \end{Bmatrix}.$$

Similarly, when $\lambda = 3$, $\boldsymbol{X}_2 = \begin{Bmatrix} 2 \\ 1 \\ -2 \end{Bmatrix}$ and when $\lambda = 15$, $\boldsymbol{X}_3 = \begin{Bmatrix} 2 \\ -2 \\ 1 \end{Bmatrix}$ .

Hence, we have, $\boldsymbol{Q}$ = Matrix formed by normalising eigen vectors

$$= \begin{bmatrix} \frac{1}{3} & \frac{2}{3} & \frac{2}{3} \\ \frac{2}{3} & \frac{1}{3} & -\frac{2}{3} \\ \frac{2}{3} & -\frac{2}{3} & \frac{1}{3} \end{bmatrix} = \frac{1}{3}\begin{bmatrix} 1 & 2 & 2 \\ 2 & 1 & -2 \\ 2 & -2 & 1 \end{bmatrix}.$$

Then, $$\boldsymbol{Q}^T\boldsymbol{A}\boldsymbol{Q} = \begin{bmatrix} 0 & 0 & 0 \\ 0 & 3 & 0 \\ 0 & 0 & 15 \end{bmatrix} = \boldsymbol{D}.$$

2.11.18 Write down the quadratic form corresponding to the matrix $\mathbf{A} = \begin{bmatrix} 11 & 5 & -1 \\ 5 & 2 & 6 \\ -1 & 6 & 3 \end{bmatrix}$

*Solution*:

Quadratic form, $Q = 11x^2 + 2y^2 + 3z^2 + 10xy - 2xz + 12yz$.

2.11.19 Reduce $8\boldsymbol{x}^2 + 7\boldsymbol{y}^2 + 3\boldsymbol{z}^2 - 12\boldsymbol{xy} - 8\boldsymbol{yz} + 4\boldsymbol{xz}$ into canonical form by orthogonal reduction. Test for definiteness

*Solution*:

The matrix of the quadratic form is $\boldsymbol{A} = \begin{bmatrix} 8 & -6 & 2 \\ -6 & 7 & -4 \\ 2 & -4 & 3 \end{bmatrix}$.

Here, the eigenvalues are $\lambda = 0, 3,$ and 15.

The corresponding eigenvectors are $\begin{Bmatrix} 1 \\ 2 \\ 2 \end{Bmatrix}$, $\begin{Bmatrix} 2 \\ 1 \\ -2 \end{Bmatrix}$, and $\begin{Bmatrix} 2 \\ -2 \\ 1 \end{Bmatrix}$.

Hence, we have $\boldsymbol{N} = \frac{1}{3}\begin{bmatrix} 1 & 2 & 2 \\ 2 & 1 & -2 \\ 2 & -2 & 1 \end{bmatrix}$; then, $\boldsymbol{N}^T\boldsymbol{A}\boldsymbol{N} = \begin{bmatrix} 0 & 0 & 0 \\ 0 & 3 & 0 \\ 0 & 0 & 15 \end{bmatrix}$.

Consider the orthogonal transformation $\boldsymbol{X} = \boldsymbol{NY}$. This will reduce the given quadratic form to $Q = \lambda_1 y_1^2 + \lambda_2 y_2^2 + \lambda_3 y_3^2 = 0y_1^2 + 3y_2^2 + 15y_3^2 = 3y_2^2 + 15y_3^2 . Q = \lambda_1 y_1^2 + \lambda_2 y_2^2 + \lambda_3 y_3^2 = 0y_1^2 + 3y_2^2 + 15y_3^2 = 3y_2^2 + 15y_3^2$.

As the eigenvalues are $\lambda = 0, 3, \textit{and}$ 15, the quadratic system is semi positive definite.

2.11.20 Reduce the quadratic form $\boldsymbol{Q} = 3\boldsymbol{x}^2 + 5\boldsymbol{y}^2 + 3\boldsymbol{z}^2 - 2\boldsymbol{yz} + 2\boldsymbol{zx} - 2\boldsymbol{xy}$ to the canonical form. Find the rank, index, and signature

*Solution*:

The matrix associated with the given quadratic form is $\boldsymbol{A} = \begin{bmatrix} 3 & -1 & 1 \\ -1 & 5 & -1 \\ 1 & -1 & 3 \end{bmatrix}$.

Here, we have $\lambda^3 - 11\lambda^2 + 36\lambda - 36 = 0$,

i.e. $(\lambda - 2)(\lambda - 3)(\lambda - 6) = 0$. $\therefore$ The eigen values are $\lambda = 2, 3$ and 6.

$\therefore$ The canonical form is $Q = \lambda_1 y_1^2 + \lambda_2 y_2^2 + \lambda_3 y_3^2 = 2y_1^2 + 3y_2^2 + 6y_3^2$.

Here, Rank = Number of terms in the canonical form = 3.
Index = Number of positive term = 3.
Signature = Number of positive term − Number of negative term = 3 − 0 = 3.
Nature of the Quadratic form is positive definite.

2.11.21 Show that the form $x^2 + 2y^2 + 3z^2 + 2yz + 2xy - 2xz$ is indefinite

*Solution*:

The matrix associated with the quadratic form is $A = \begin{bmatrix} 1 & 1 & -1 \\ 1 & 2 & 1 \\ -1 & 1 & 3 \end{bmatrix}$.

Here, the principal minors are given by

$$D_1 = |1| = 1 > 0;\ D_2 = \begin{vmatrix} 1 & 1 \\ 1 & 2 \end{vmatrix} = 2 - 1 = 1 > 0;$$

$$D_3 = \begin{vmatrix} 1 & 1 & -1 \\ 1 & 2 & 1 \\ -1 & 1 & 3 \end{vmatrix} = 1(6-1) - 1(3+1) - 1(1+2) = 5 - 4 - 3 = -2 < 0.$$

As, $\mathbf{D_1 > 0, D_{2>} > 0,}$ **and** $\mathbf{D_3 < 0}$, the given quadratic form is indefinite.

2.11.22 Solve the equations $3x + y + 2z = 3$, $2x - 3y - z = -3$, $x + 2y + z = 4$ by

(i) Method of determinants (ii) Method of matrices

*Solution*:

(i) By determinants

The given system can be written in the matrix form as $\mathbf{AX = B}$, where

$$A = \begin{bmatrix} 3 & 1 & 2 \\ 2 & -3 & -1 \\ 1 & 2 & 1 \end{bmatrix}.$$

$$\text{Here, } D = \begin{vmatrix} 3 & 1 & 2 \\ 2 & -3 & -1 \\ 1 & 2 & 1 \end{vmatrix} = 3(-3+2) - 1(2+1) + 2(4+3) = -3 - 3 + 14 = 8.$$

Similarly $D_1 = \begin{vmatrix} 3 & 1 & 2 \\ -3 & -3 & -1 \\ 4 & 2 & 1 \end{vmatrix} = 8$; $D_2 = \begin{vmatrix} 3 & 3 & 2 \\ 2 & -3 & -1 \\ 1 & 4 & 1 \end{vmatrix} = 16$; $D_3 = \begin{vmatrix} 3 & 1 & 3 \\ 2 & -3 & -3 \\ 1 & 2 & 4 \end{vmatrix} = -8$.

The solution is given by

$$x = \frac{D_1}{D} = \frac{8}{8} = 1;\ y = \frac{D_2}{D} = \frac{16}{8} = 2;\ z = \frac{D_3}{D} = -\frac{8}{8} = -1.$$

Hence, the solution is $x = 1, y = 2, z = -1$.

(ii) By matrices

We have, $\boldsymbol{A} = \begin{bmatrix} 3 & 1 & 2 \\ 2 & -3 & -1 \\ 1 & 2 & 1 \end{bmatrix}$ $and\ |\boldsymbol{A}| = 8$.

Also, cofactor of $\boldsymbol{A} = \begin{bmatrix} -1 & -3 & 7 \\ 3 & 1 & 5 \\ 5 & 7 & -11 \end{bmatrix}$.

$$\therefore \boldsymbol{A}^{-1} = \frac{(Cofactor)^T}{|\boldsymbol{A}|} = \frac{1}{8}\begin{bmatrix} -1 & 3 & 5 \\ -3 & 1 & 7 \\ 7 & 5 & -11 \end{bmatrix}.$$

$$\therefore \boldsymbol{X} = \boldsymbol{A}^{-1}\boldsymbol{B} = \frac{1}{8}\begin{bmatrix} -1 & 3 & 5 \\ -3 & 1 & 7 \\ 7 & 5 & -11 \end{bmatrix}\begin{Bmatrix} 3 \\ -3 \\ 4 \end{Bmatrix} = \begin{Bmatrix} 1 \\ 2 \\ -1 \end{Bmatrix}.$$

Hence, the solution is $x = 1, y = 2, z = -1$.

2.11.23 Solve: $\boldsymbol{x} + 3\boldsymbol{y} + 3\boldsymbol{z} = 16, \boldsymbol{x} + 4\boldsymbol{y} + 3\boldsymbol{z} = 18, \boldsymbol{x} + 3\boldsymbol{y} + 4\boldsymbol{z} = 19$ by

(a) Gaussian elimination method (b) Gauss–Jordan method

*Solution*:

Here we have $\boldsymbol{A} = \begin{bmatrix} 1 & 3 & 3 \\ 1 & 4 & 3 \\ 1 & 3 & 4 \end{bmatrix}$, $\boldsymbol{X} = \begin{Bmatrix} x \\ y \\ z \end{Bmatrix}$, and $\boldsymbol{B} = \begin{Bmatrix} 16 \\ 18 \\ 19 \end{Bmatrix}$.

(a) Gaussian elimination method

$$\text{Augmented matrix} = [\boldsymbol{AB}] = \begin{bmatrix} 1 & 3 & 3 & 16 \\ 1 & 4 & 3 & 18 \\ 1 & 3 & 4 & 19 \end{bmatrix}$$

$$\sim \begin{bmatrix} 1 & 3 & 3 & 16 \\ 0 & 1 & 0 & 2 \\ 0 & 0 & 1 & 3 \end{bmatrix} \begin{matrix} R_2 \rightarrow R_2 + (-1)R_1 \\ R_3 \rightarrow R_3 + (-1)R_1 \end{matrix}.$$

Which is the required upper triangular matrix.

We have, $\mathbf{AX} = \mathbf{B}$; i.e. $\begin{bmatrix} 1 & 3 & 3 \\ 0 & 1 & 0 \\ 0 & 0 & 1 \end{bmatrix} \begin{Bmatrix} x \\ y \\ z \end{Bmatrix} = \begin{Bmatrix} 16 \\ 2 \\ 3 \end{Bmatrix}$,

i.e. $x + 3y + 3z = 16$, $y = 2$, and $z = 3$. Hence, the solution is $x = 1$, $y = 2$, $z = 3$.

(b) Gauss–Jordan method

We have $[\boldsymbol{AB}] = \begin{bmatrix} 1 & 3 & 3 & 16 \\ 1 & 4 & 3 & 18 \\ 1 & 3 & 4 & 19 \end{bmatrix} \begin{bmatrix} 1 & 3 & 3 & 16 \\ 0 & 1 & 0 & 2 \\ 0 & 0 & 1 & 3 \end{bmatrix} \begin{matrix} R_2 \rightarrow R_2 + (-1)R_1 \\ R_3 \rightarrow R_3 + (-1)R_1 \end{matrix}$,

$$\sim \begin{bmatrix} 1 & 0 & 3 & 10 \\ 0 & 1 & 0 & 2 \\ 0 & 0 & 1 & 3 \end{bmatrix} R_1 \rightarrow R_1 + (-3)R_2.$$

We have, $\mathbf{AX} = \mathbf{B}$; ***i.e.*** $\begin{bmatrix} 1 & 0 & 3 \\ 0 & 1 & 0 \\ 0 & 0 & 1 \end{bmatrix} \begin{Bmatrix} x \\ y \\ z \end{Bmatrix} = \begin{Bmatrix} 10 \\ 2 \\ 3 \end{Bmatrix}$,

$$\text{i.e., } x + 3z = 10, \; y = 2, \; z = 3.$$

Hence, the solution to the system is $\boldsymbol{x} = 1$, $\boldsymbol{y} = 2$, $\boldsymbol{z} = 3$ .

2.11.24 Solve the following system of equations

$$27x + 6y - z = 85; \;\; x + y + 54z = 110; \;\; 6x + 15y + 2z = 72.$$

*Solution*:

Rewriting the given equation as

$27x + 6y - z = 85$; $6x + 15y + 2z = 72$; $x + y + 54z = 110$, $\therefore$ we have

$$x = \tfrac{1}{27}(85 - 6y + z) \tag{2.26}$$

$$y = \tfrac{1}{15}(72 - 6x - 2z) \tag{2.27}$$

$$z = \tfrac{1}{54}(110 - x - y) \tag{2.28}$$

Let $x^{(0)} = y^{(0)} = z^{(0)} = 0$ be the initial approximation.

(a) Gauss–Jacobian method

*Iteration-1*:

Substituting $x^{(0)} = y^{(0)} = z^{(0)} = 0$ in the R.H.S of Eqs. (2.26), (2.27), and (2.28), we get
$x^{(1)} = \frac{85}{27} = 3.148$; $y^{(1)} = \frac{72}{15} = 4.8$; and $z^{(1)} = \frac{110}{54} = 2.037$.

*Iteration-2*:

Substituting $x^{(1)} = 3.148$, $y^{(1)} = 4.8$, and $z^{(1)} = 2.037$ in the R.H.S of Eqs. (2.26), (2.27), and (2.28), we get $x^{(2)} = 2.157$, $y^{(2)} = 3.269$, and $z^{(2)} = 1.890$.
Continuing like this, we get:

$$x^{(3)} = 2.492,\ y^{(3)} = 3.685,\ z^{(3)} = 1.937.$$
$$x^{(4)} = 2.401,\ y^{(4)} = 3.545,\ z^{(4)} = 1.923.$$
$$x^{(5)} = 2.432,\ y^{(5)} = 3.583,\ z^{(5)} = 1.927.$$
$$x^{(6)} = 2.433,\ y^{(6)} = 3.570,\ z^{(6)} = 1.926.$$
$$x^{(7)} = 2.426,\ y^{(7)} = 3.574,\ z^{(7)} = 1.926.$$
$$x^{(8)} = 2.425,\ y^{(8)} = 3.573,\ z^{(8)} = 1.926.$$
$$x^{(9)} = 2.426,\ y^{(9)} = 3.573,\ z^{(9)} = 1.926.$$
$$x^{(10)} = 2.426,\ y^{(10)} = 3.573,\ z^{(10)} = 1.926.$$

As the two successive approximations converge, the solution is

$$x = 2.426,\ y = 3.573,\ and\ z = 1.926.$$

*Gauss–Seidel Method*

*Iteration-1*:

Substituting $y^{(0)} = 0$, $z^{(0)} = 0$ in the R.H.S of Eq. (2.26) we get, $x^{(1)} = 3.148$.
Substituting $x^{(1)} = 3.148$, $z^{(0)} = 0$ in the R.H.S of Eq. (2.27) we get, $y^{(1)} = 3.541$.
Substituting $x^{(1)} = 3.148$, $y^{(1)} = 3.541$ in the R.H.S of Eq. (2.28) we get, $z^{(1)} = 1.913$.
Hence, we have, $x^{(1)} = 3.148$; $y^{(1)} = 3.541$; $z^{(1)} = 1.913$.

*Iteration-2*:

Substituting $y^{(1)} = 3.541$, $z^{(1)} = 1.913$ in Eq. (2.26), we get $x^{(2)} = 2.432$.
Substituting $x^{(2)} = 2.432$, $z^{(1)} = 1.913$ in Eq. (2.27), we get $y^{(2)} = 3.572$.
Substituting $x^{(2)} = 2.432$, $y^{(2)} = 3.572$ in Eq. (2.27), we get $z^{(2)} = 1.926$.
Hence, we have $x^{(2)} = 2.432$, $y^{(2)} = 3.572$, *and* $z^{(2)} = 1.926$.
Continuing like this, we get

$$x^{(3)} = 2.426, \quad y^{(3)} = 3.573, \quad z^{(3)} = 1.926.$$
$$x^{(4)} = 2.426, \quad y^{(4)} = 3.573, \quad z^{(4)} = 1.926.$$

Hence, the solution is $x = 2.426, \quad y = 3.573, \quad \text{and } z = 1.926.$

## 2.12 Exercise Problems

2.12.1. For which value of x, will the matrix given below becomes singular?

$$\mathbf{A} = \begin{bmatrix} 8 & \boldsymbol{x} & 0 \\ 4 & 0 & 2 \\ 12 & 6 & 0 \end{bmatrix}.$$

2.12.2. Find the value of x, y, z, a given $\begin{bmatrix} \boldsymbol{x}+3 & 2\boldsymbol{y}+\boldsymbol{x} \\ z-1 & 4\boldsymbol{a}-6 \end{bmatrix} = \begin{bmatrix} 0 & -8 \\ 5 & 2\boldsymbol{a} \end{bmatrix}$.

2.12.3. If $\boldsymbol{A} = \begin{bmatrix} 1 & 2 \\ -2 & 3 \end{bmatrix}, \boldsymbol{B} = \begin{bmatrix} 2 & 1 \\ 2 & 3 \end{bmatrix}, \boldsymbol{C} = \begin{bmatrix} -3 & 1 \\ 2 & 0 \end{bmatrix}$.
Verify that (AB) C = A (BC) and A (B + C) = AB + AC.

2.12.4. Find the symmetric part of the matrix, $A = \begin{bmatrix} 1 & 2 & -1 \\ 3 & 5 & 4 \\ -2 & 3 & 1 \end{bmatrix}$.

2.12.5. Factorize, $A = \begin{bmatrix} 1 & 2 & 1 \\ 2 & 3 & 2 \\ 1 & 1 & 2 \end{bmatrix}$ into the form LU where L in unit lower triangular matrix and U in the upper triangular matrix?

2.12.6. Show that, $\boldsymbol{A} = \begin{bmatrix} -\frac{1}{3} & \frac{2}{3} & \frac{2}{3} \\ \frac{2}{3} & -\frac{1}{3} & \frac{2}{3} \\ \frac{2}{3} & \frac{2}{3} & -\frac{1}{3} \end{bmatrix}$ is orthogonal?

2.12.7. Show that the vectors (4, 1, 2, 0), (1, 2, –1, 0), (1, 3, 1, 2), and (6, 1, 0, 1) are linearly independent.

2.12.8. Show that the vectors X1 = (1, 2, −1, 4), X2 = (2, 4, 3, 5), and X3 = (−1, −2, 6, −7) form a linearly dependent system. Find the relation connecting them.

2.12.9. Given, $A = \begin{bmatrix} 1 & 1 & -1 \\ 2 & -3 & 4 \\ 3 & -2 & 4 \end{bmatrix}$ and $B = \begin{bmatrix} -1 & -2 & -1 \\ 6 & 12 & 6 \\ 5 & 10 & 5 \end{bmatrix}$ then the rank of (A + B) is ?

2.12.10. Find the rank of the matrix, $\mathbf{A} = \begin{bmatrix} 1 & 2 & 1 & 2 \\ 1 & 3 & 2 & 2 \\ 2 & 4 & 3 & 4 \\ 3 & 7 & 4 & 6 \end{bmatrix}$.

2.12.11. The rank of the matrix, $\mathbf{A} = \begin{bmatrix} 1 & 5 & x \\ 5 & 1 & -1 \\ 1 & 2 & 1 \end{bmatrix}$ is 2 then $x = ?$

2.12.12. Show that the vectors $X_1 = (1, 2, 2)$, $X_2 = (2, 1, -2)$, and $X_3 = (2, -2, 1)$ are linearly independent.

2.12.13. Examine if the system of equation $x_1 + x_2 + x_3 = 0$, $x_1 + 2x_2 - x_3 = 0$, $2x_1 + x_2 + 3x_3 = 0$ possesses a non-trivial solution.

2.12.14. If the system of equation $x - 2y + 2z = 0$, $3x + y + \lambda z = 0$, $x + y + z = 0$, possesses a non-trivial solution, then $\lambda = ?$

2.12.15. The equation $\begin{bmatrix} 2 & -2 \\ 1 & -1 \end{bmatrix}\begin{bmatrix} x_1 \\ x_2 \end{bmatrix} = \begin{bmatrix} 0 \\ 0 \end{bmatrix}$ has solution?

2.12.16. The system of equation $x + y + z = 6$, $x + 2y + 3z = 10$, $x + 2y + 3z = 5$ will have solution?

2.12.17. The value of C for which the following system of linear equation has an infinite number of solutions in $\begin{bmatrix} 1 & 2 \\ 1 & 2 \end{bmatrix}\begin{bmatrix} x \\ y \end{bmatrix} = \begin{bmatrix} C \\ 4 \end{bmatrix}$.

2.12.18. Show that the equation $x + y + z = a$, $3x + 4y + 5z = b$, $2x + 3y + 4z = c$ have no solution if a = b = c = 1; (ii) have many solution if $a = \frac{b}{2} = c = 1$.

2.12.19. Show that the Eq. $2x - y + 3z = 9$, $x + y + z = 6$, and $x - y + z =$ $2x - y + 3z = 9$, $x + y + z = 6$, and $x - y + z = 2$ are consistent and solve them.

2.12.20. Consider a $2 \times 2$ matrix, $A = [x]$, where $x$ is a constant. If the eigenvalue of the matrix A are $(\sigma + J\omega)$ and $(\sigma - J\omega)$ then $x$ is equal to?

2.12.21. Find the eigenvalue and eigenvector of the matrix, $\mathbf{A} = \begin{bmatrix} 2 & 2 & 1 \\ 1 & 3 & 1 \\ 1 & 2 & 2 \end{bmatrix}$.

2.12.22. Find the eigenvalues and number of linearly independent eigenvectors of $A = \begin{bmatrix} 2 & 1 & 0 \\ 0 & 2 & 0 \\ 0 & 0 & 3 \end{bmatrix}$.

2.12.23. If 3 and 15 are two eigenvalues of $A = \begin{bmatrix} 8 & -6 & 2 \\ -6 & 7 & -4 \\ 2 & -4 & 3 \end{bmatrix}$. Find the eigenvalues of **A, A-5I**, and $\mathbf{A}^2$

2.12.24. Find, the eigenvalues, eigenvectors, and modal matrix of the matrix, $A = \begin{bmatrix} 5 & 4 \\ 1 & 2 \end{bmatrix}$.

2.12.25. Find the index and signature of the quadratic form $x_1^2 + 2x_2^2 - 3x_3^2$.

2.12.26. Find the nature of the quadratic form $x^2 + 5y^2 + z^2 + 2xy + 2yz + 6zx$.

2.12.27. If $A = \begin{bmatrix} 1 & 2 & 3 \\ 0 & 3 & 1 \\ 0 & 0 & 1 \end{bmatrix}$ then obtain the matrix P such that $P^{-1}AP$ is a diagonal matrix.

2.12.28. Find the eigenvalues, eigenvectors, and the modal matrix for $A = \begin{bmatrix} 1 & 0 & 0 \\ 0 & 3 & -1 \\ 0 & -1 & 3 \end{bmatrix}$.

2.12.29. Find a matrix P such that $P^{-1}AP = D$, a diagonal matrix if $A = \begin{bmatrix} -1 & 3 \\ -2 & 4 \end{bmatrix}$.

2.12.30. Reduce matrix $A = \begin{bmatrix} 8 & -6 & 2 \\ -6 & 7 & -4 \\ 2 & -4 & 3 \end{bmatrix}$ to a diagonal form by orthogonal reduction.

2.12.31. Reduce the quadratic form $6x^2 + 3y^2 + 3z^2 - 4xy - 2yz + 4xz$ into canonical form by an orthogonal reduction. Find its nature, rank, index, and signature.

2.12.32. Prove that $6x_1^2 + 3x_2^2 + 14x_3^2 + 4x_1x_2 + 18x_1x_3 + 4x_2x_3$ is indefinite.

2.12.33. Diagonalize the matrix $A = \begin{bmatrix} 4 & 1 \\ 2 & 3 \end{bmatrix}$.

2.12.34. Find a matrix **B** which transfers the matrix $A = \begin{bmatrix} 1 & 0 & -1 \\ 1 & 2 & 1 \\ 2 & 2 & 3 \end{bmatrix}$ to the diagonal form. Hence, find $\mathbf{A}^4$.

2.12.35. Factorize and hence find the solution of the system of equation

$$5x - 2y + z = 4, 7x + y - 5z = 8, 3x + 7y + 4z = 10.$$

2.12.36. Solve the following system of equations: $5x_1 + x_2 + x_3 + x_4 = 4, x_1 + 7x_2 + x_3 + x_4 = 12,$
$x_1 + x_2 + 6x_3 + x_4 = -5, x_1 + x_2 + x_3 + 4x_4 = 6$ by Gaussian elimination method.

2.12.37. Solve the following system:
$10x + 2y + z = 9; 2x + 20y - 2z = -44; -2x + 3y + 10z = 22$ by (i) Jacobis' method (ii) Gauss–Seidel method.

2.12.38. Solve the following system
$2x + y + z = 10; 3x + 2y + 3z = 18$ ; and $x + 4y + 9z = 16$ by (i) Gauss method (ii) by Gauss–Jordan Method.

2.12.39. State the condition for the convergence of Jacobi and Gauss–Seidel methods.

2.12.40. If $A = \begin{bmatrix} 1 & 2 \\ 3 & 4 \end{bmatrix}$ Compute $A^{-1}$ by direct method and check the result by direct multiplication with A.

## Bibliography

Kreyszig (2011) Advanced engineering mathematics, 10th edn. John Willy & Sons Inc
Grewal BS (2014) Higher engineering mathematics, 43rd edn. Khanna Publication, New Delhi
Sastry SS (2005) Introductory method of numerical analysis, 4th edn. PHI, India
Vashista AR (2003) Matrices. Krishna PrakashanaMandir
Venkataraman MK (2005) Numerical method in science and engineering. National Publishing Co
Narayan S, Mithal PK (2013) A text book of matrices, 5th edn. Chandand Co Pvt, New Delhi
Mathur AB, Jaggi VP (2013) Advanced engineering mathematics. Khanna Publication, India
Ramana BV (2008) Higher engineering mathematics, 6th edn. Tata Mcgraw-Hill Publication Company Ltd

# Chapter 3
# Classical Approximate Solutions

**K. Jayakumar**

*Measure what is measurable, and make measurable what is not so—Galileo Galilei*

## 3.1 Introduction

The topic of variational methods in applied mathematics forms basis of determining numerical solutions to mathematical models. Before we try to understand what finite element method is, let us first appreciate how modern science works. We observe several phenomena in nature and try to comprehend them by relating principal quantities involved to the net effect. In other words, qualitative observations are translated to quantitative mathematical statements. These statements take the form of polynomial expressions, differential equations (ordinary or partial), integro-partial differential equations, etc. Such expressions are formulated based on some hypothesis, which concentrates on the cardinal influencing factors and what is being sought. As an example consider the Euler–Bernoulli[1] beam. This is based on a small deflection hypothesis, which essentially means that the differential equation represents linear behavior and this mathematical model should seldom be used to estimate large deflection of beams.

It is important to remember that the mathematical models are only an approximation to reality. Verifying mathematical models against real physical behavior, from either laboratory-based experiments or in-situ observations and updating the models based on experiments is an essential part of design process.

[1] $EI\frac{d^4w}{dx^4} = q(x)$; $EI$ is the flexural rigidity; $w$ is the transverse deflection; $q(x)$ is the loading function; $0 \leq x \leq L$ where $L$ is the length of the beam.

K. Jayakumar (✉)
PSLV/VSSC, Thiruvananthapuram 695022, India
e-mail: k_jayakumar@vssc.gov.in



S. U. Nair and S. Somanath (eds.), *Introduction to Finite Element Analysis*,
https://doi.org/10.1007/978-981-19-7989-7_3

Finding a solution to these mathematical models is the next logical step. Constraints are identified before resorting to finding solutions. Wherever or whenever it is possible, a closed-form solution is preferred option because it gives the interrelationship of various parameters through an expression, which can be easily understood and solved through a finite number of standard operations. Analytical solutions based on variational principles are resorted to when closed-form solutions are not possible, which is true in many situations. But variational methods have their own limitations as they can give solutions only over simple domains. In many real-life situations, mathematical models are available, but particular closed-form solutions to mathematical models can be obtained only for a limited class of problems. It, therefore, becomes necessary to get an analytical approximate solution to the given model to understand underlying mechanics of the problem. Mathematical principles rooted in calculus of variations aid achieving this objective, i.e. finding approximate analytical solutions from a function space. While solving differential equations using the variational method, the equations are not solved directly. An equivalent problem is formulated to obtain an approximate analytical solution. Analytical functions, albeit approximate, offer insights not only into the problem but also into nature of solution. At times it is the only way out.

Although many classical approximate solution techniques are available, most in use can generally be categorized into variational or weighted residual methods. These methods form different classes of solutions to the mathematical model. The emphasis of this chapter is to introduce to the reader classical approximate solution techniques and how these ideas form foundation of numerical methods such as finite element method. Numerical methods are used for finding solutions when closed-form solution is impossible or analytical solutions are inadequate and cumbersome. The finite element method is one of the numerical methods to solve partial differential equations. The principle of approximate solution finding is illustrated in Fig. 3.1. Here, perimeter of a circle is approximated by successively increasing the number of sides of inscribed (or circumscribed) polygon. The chosen polygon may be regular or of one or many sides of different known lengths. In this way, perimeter is estimated in terms of the known lengths of inscribed (or circumscribed of desired) polygon approximately. When solving a system of complicated equations (differential equations), the sought solution is expressed similarly in terms of combinations of known solutions or functions. While working out procedures for finding an approximate solution, the underlying focus is on minimizing error between approximate and exact solutions.

Many real-life applications involve finding extremum values of a mathematical statement expressed usually in an integral form. Extremum values of interest are maximum, minimum and stationary. The following equations and illustrations are from standard published literature, which is reproduced here to explain the subject contained in this chapter. Readers are encouraged to look into advanced books related to this domain as given in the bibliography.

Mathematical problem definitions of real-life phenomena generally end up as integral statements. Form of an integral statement in general involves not only the independent and dependent variables but also derivatives of dependent variables. For

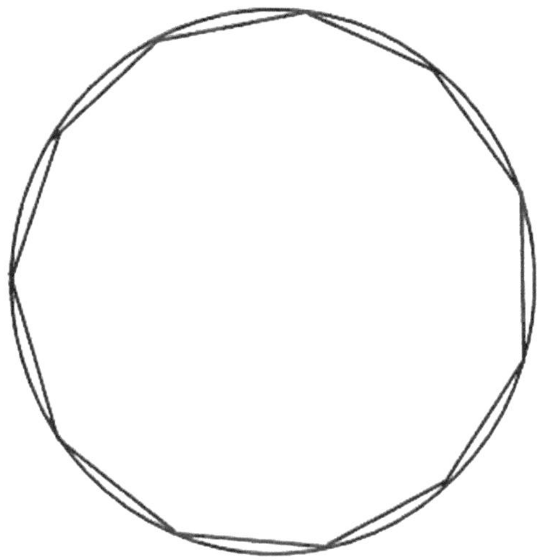

**Fig. 3.1** Approximation of perimeter of circle by progressively increasing number of sides of an inscribed regular polygon

one independent variable, say $x$, this can be expressed as

$$I = \int_{x_1}^{x_2} f\left(x, y, y'\right)dx. \tag{3.1}$$

Equation (3.1) clearly shows that the integral $I$ is a function of $y = y(x)$. The integral $I$ depends on families of functions and is therefore known as a functional. In other words, integral $I$ is a function of functions. Furthermore, for every choice of $y = y(x)$, the integral has a specific value. Prime over $y$ denotes the derivative with respect to $x$.

It is of interest to determine the extremum of $I$, which indeed depends on infinite choice of $y = y(x)$ from a function space.

Consider Fig. 3.2; points **P** and **Q** are connected by the required curve $y = y(x)$, which makes the integral extremum. Equation (3.1) is a varied or neighboring curve of $y = y(x)$ joining **P** and **Q**.

$$y = y(x) + \varepsilon\ \eta(x). \tag{3.2}$$

Clearly, the variation $\eta(x) = 0$ at **P** and **Q**.

Therefore, Eq. (3.1) corresponding to varied curve or function, given by Eq. (3.2), is

$$I = \int_{x_1}^{x_2} f\left\{x, y(x) + \varepsilon\ \eta(x), y'(x) + \varepsilon\ \eta'(x)\right\}dx. \tag{3.3}$$

As Eq. (3.3) corresponds to neighboring path, $I = I(\varepsilon)$, the integral is therefore a function of $\varepsilon$ that defines variation in path.

Hence, for integral $I$ to be maximum or minimum, the condition is

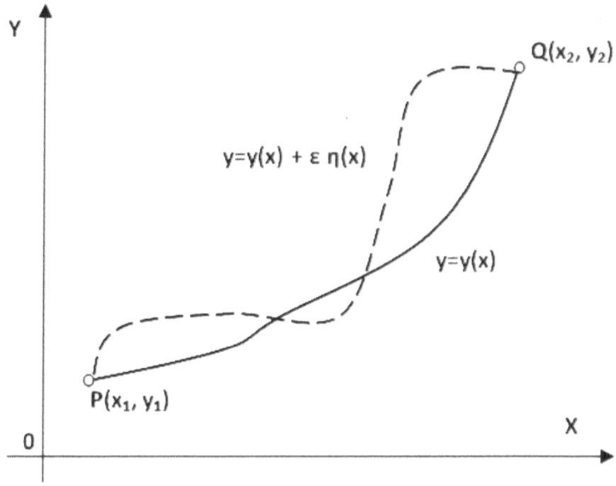


**Fig. 3.2** Function and its variation between terminal points

$$\frac{\mathrm{d}I}{\mathrm{d}\varepsilon} = 0 \text{ at } \varepsilon = 0. \tag{3.4}$$

Applying this condition to Eq. (3.3) gives the total derivative

$$\frac{\mathrm{d}I}{\mathrm{d}\varepsilon} = \frac{\partial f}{\partial x}\frac{\mathrm{d}x}{\mathrm{d}\varepsilon} + \frac{\partial f}{\partial y}\frac{\mathrm{d}y}{\mathrm{d}\varepsilon} + \frac{\partial f}{\partial y'}\frac{\mathrm{d}y'}{\mathrm{d}\varepsilon}. \tag{3.5}$$

It is clear that $\boldsymbol{\varepsilon}$ is a factor which distinguishes or identifies neighborhood functions of a function space, and therefore the variable $\boldsymbol{x}$ is independent of ε implying $\frac{\partial x}{\partial \varepsilon} = 0$. Hence,

$$\frac{\mathrm{d}I}{\mathrm{d}\varepsilon} = \int_{x_1}^{x_2} \left\{ \frac{\partial f}{\partial y}\eta(x) + \frac{\partial f}{\partial y'}\eta'(x) \right\} dx. \tag{3.6}$$

Integrating, the second term of Eq. (3.6) by parts,

$$\frac{\mathrm{d}I}{\mathrm{d}\varepsilon} = \int_{x_1}^{x_2} \frac{\partial f}{\partial \mathrm{y}}\eta(x)\mathrm{d}x + \left\{ \left| \frac{\partial f}{\partial \mathrm{y}'}\eta'(x) \right|_{x_1}^{x_2} - \int_{x_1}^{x_2} \frac{\mathrm{d}}{\mathrm{d}x}\left(\frac{\partial f}{\partial \mathrm{y}'}\right)\eta(x)\mathrm{d}x \right\}. \tag{3.7}$$

In Eq. (3.7), the boundary term $\left| \frac{\partial f}{\partial \mathrm{y}'}\eta'(x) \right|_{\mathrm{x}_1}^{\mathrm{x}_2}$ is zero at **P** and **Q**, i.e. at $x_1$ and $x_2$, because $\eta(x) = 0$. Rewriting Eq. (3.7),

$$\int_{x_1}^{x_2} \left\{ \frac{\partial f}{\partial \mathrm{y}} - \frac{\mathrm{d}}{\mathrm{d}x}\left(\frac{\partial f}{\partial \mathrm{y}'}\right) \right\} \eta(x)\mathrm{d}x = 0. \tag{3.8}$$

But, the varied path $\eta(x)$ cannot be zero, therefore

$$\frac{\partial f}{\partial y} - \frac{d}{dx}\left(\frac{\partial f}{\partial y'}\right) = 0. \tag{3.9}$$

Equation (3.9) is known as Euler's equation, which gives necessary condition for the functional to be extremum.

Similar expressions exist for many dependent variables, as a function of more than one independent variable.

Consider an integral form

$$I_1 = \iint f\left(x, y, u, v, u_{,x}, v_{,x}, u_{,y}, v_{,y}, u_{,xx}, v_{,yy}, u_{,xy}, v_{,xy}\right) dxdy. \tag{3.10}$$

The functional in Eq. (3.10) has two independent variables $x$, $y$ and dependent variables $u$, $v$, and their derivatives. Corresponding Euler equations that are required to determine the extremal of $I_1$ are given as

$$\frac{\partial f}{\partial x} - \frac{\partial}{\partial x}\frac{\partial f}{\partial u_{,x}} - \frac{\partial}{\partial y}\frac{\partial f}{\partial u_{,y}} + \frac{\partial^2}{\partial x^2}\frac{\partial f}{\partial u_{,xx}} + \frac{\partial^2}{\partial x \partial y}\frac{\partial f}{\partial u_{,xy}} + \frac{\partial^2}{\partial y^2}\frac{\partial f}{\partial u_{,yy}} = 0 \tag{3.11}$$

$$\frac{\partial f}{\partial y} - \frac{\partial}{\partial x}\frac{\partial f}{\partial v_{,x}} - \frac{\partial}{\partial y}\frac{\partial f}{\partial v_{,y}} + \frac{\partial^2}{\partial x^2}\frac{\partial f}{\partial v_{,xx}} + \frac{\partial^2}{\partial x \partial y}\frac{\partial f}{\partial v_{,xy}} + \frac{\partial^2}{\partial y^2}\frac{\partial f}{\partial v_{,yy}} = 0 \tag{3.12}$$

where $u_{,x} = \frac{\partial u(x,y)}{\partial x}$; $u_{,y} = \frac{\partial u(x,y)}{\partial y}$; $u_{,xx} = \frac{\partial^2 u(x,y)}{\partial x^2}$; $u_{,xy} = \frac{\partial^2 u(x,y)}{\partial x\, \partial y}$; $u_{,yy} = \frac{\partial^2 u(x,y)}{\partial y^2}$
$v_{,x} = \frac{\partial v(x,y)}{\partial x}$; $v_{,y} = \frac{\partial v(x,y)}{\partial y}$; $v_{,xx} = \frac{\partial^2 v(x,y)}{\partial x^2}$; $v_{,xy} = \frac{\partial^2 v(x,y)}{\partial x\, \partial y}$; $v_{,yy} = \frac{\partial^2 v(x,y)}{\partial y^2}$.
These ideas will be utilized to explain some underlying concepts concerning solution of approximate analytical solutions.

The following sub-sections briefly describe three classical problems that initiated foundations of variational calculus. Modern computational techniques utilized for approximate solutions of physical problems are one way or the other based on principles of variational methods. There are many standard textbooks on this subject that the reader can refer to.

### 3.1.1 *Brachistochrone*

Origins of calculus of variations can be traced to seventeenth century when Johann Bernoulli posed the *brachistochrone* problem. *Brachis* in Greek language means 'small' and *chrone* means something related to time. This problem involves determining locus of a particle between two points in space under some conditions. One such condition can be uniform gravitational field. This statement is a generalization of the shortest distance between two points in space when time is considered. It can be shown that the locus is a cycloid.

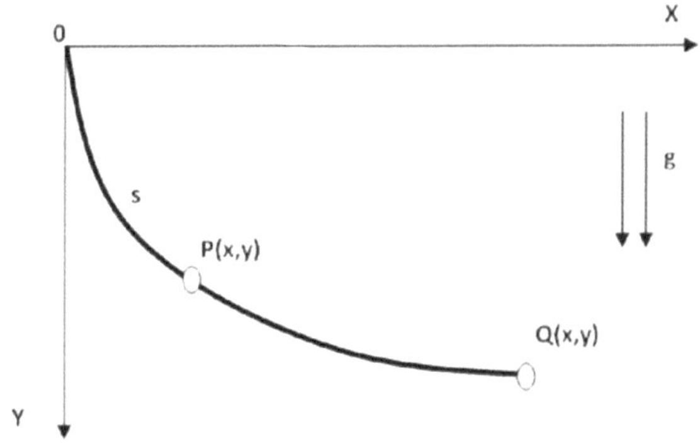


**Fig. 3.3** Locus of particle under uniform gravitational field

Consider Fig. 3.3; a particle with mass $m$, under uniform gravitational field is traversing from origin to **Q** without friction with zero initial velocity. If particle in time $t_1$ is at **P**, tracing an arc $s$, then by conservation of energy, the difference in kinetic energy of the particle at origin and **P** should be equal to work done in displacing it from origin to **P**

Quantitatively,

KE at P − KE at origin = Work done in arc $s$

$$\frac{1}{2}m\left(\frac{\mathrm{d}s}{\mathrm{d}t}\right)^2 - 0 = m\,g\,y \tag{3.13}$$

The total time required for traversing from origin to **P** is

$$\mathrm{T} = \int_0^{\mathrm{T}} \mathrm{d}t = \int_{\mathrm{P}}^{\mathrm{Q}} \frac{\mathrm{d}s}{\mathrm{V}} = \int_{\mathrm{P}}^{\mathrm{Q}} \frac{\sqrt[2]{\mathrm{d}x^2 + \mathrm{d}y^2}}{\mathrm{V}} = \frac{1}{\sqrt{2g}} \int_0^{x_1} \sqrt{\frac{1+\left(y'\right)^2}{y}}\mathrm{d}x. \tag{3.14}$$

The integral **T** in Eq. (3.14) is a functional as it is dependent on function $y = y(x)$. Applying Euler's equation and solving for $y'$

$$y' = \frac{dy}{dx} = \sqrt{\frac{k-y}{y}} \tag{3.15}$$

where $k$ is the constant of integration. Equation (3.15) can be represented by parameter $\theta$. Therefore, substituting $y = k\sin^2(\theta)$ and simplifying

$$x = k\int_0^{\theta} \{1 - \cos(2\theta)\}\,\mathrm{d}\theta = \frac{k}{2}\{2\theta - \sin(2\theta)\} \tag{3.16}$$

$$y = \frac{k}{2}\{1 - \cos(2\theta)\}. \tag{3.17}$$

Equations (3.16) and (3.17) are parametric representations of cycloid, the parameter being $\theta$. The constant $k$ can be determined if coordinates of a point on the curve are known.

### 3.1.2 Geodesic

On a general curved surface, between two points, it is of interest to determine the shortest distance. This locus is known as a geodesic. On Earth, which for the purpose of analysis can be approximated as a sphere, the minimum distance between two points is constrained by Earth's curvature. This problem is not confined to theoretical interest but is used on a daily basis by the airline and shipping industries. An airplane cannot travel between two points on Earth by using 'conventional' wisdom of following a straight line for the shortest distance when destinations are separated by great distances. It will result in wastage of expensive aircraft turbine fuel. Under these circumstances, the airplane follows a geodesic, which is nothing but a segment of a circular arc of a greater circle passing through these points. Note that the greater circular planes of a sphere share its center.

This concept can be understood by proving that a curve joining any two points on a plane is a straight line, which is a geodesic. Consider Fig. 3.4; points **P** and **Q** are on plane **R**.

Length of a curve from elementary calculus is

$$s = \int_{x_1}^{x_2} ds = \int_{x_1}^{x_2} \sqrt{1 + (y')^2} dx \tag{3.18}$$

Equation (3.18) is independent of any expression that is a function of the dependent variable $y = y(x)$. Therefore, $\frac{\partial f}{\partial y} = 0$. Applying the Euler-Lagrange Eq. (3.9) on (3.18) gives condition for an extremum of the functional as

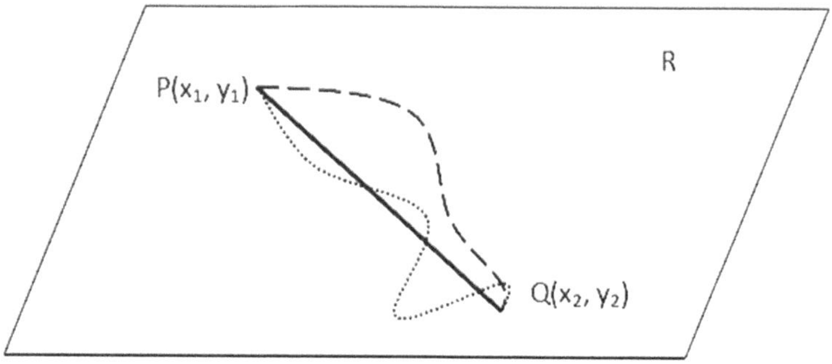


**Fig. 3.4** Curves on plane

$$\frac{\mathrm{d}}{\mathrm{d}x}\left\{\frac{\partial}{\partial y'}\left(\sqrt{1+(y')^2}\right)\right\} = 0. \tag{3.19}$$

On solving (3.19),

$$y' = k. \tag{3.20}$$

While integrating this, Eq. (3.21) is obtained, which represents a straight line

$$y = kx + c. \tag{3.21}$$

Constants appearing in Eq. (3.21) can be determined using coordinate values of two points lying on the line.

### 3.1.3 *Isoperimetric*

This problem involves determining shape of a closed curve enclosing maximum area. This idea can be extended to higher dimensions. If an endless string, i.e. a loop of string is placed on a plane, and its geometry is continuously altered without intersections, it can be proven that the area encompassed is maximum when it assumes a circular shape.

Consider Fig. 3.5; let C be constant length of the curve between points U, V.

This problem involves maintaining length of the curve as constant and area enclosed by it as a varying quantity. Length of the curve is given by Eq. (3.18) and area enclosed between curve and $x$-axis is

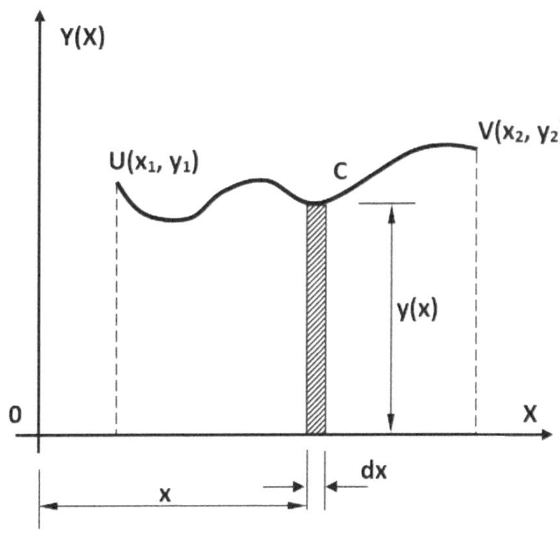


**Fig. 3.5** Curve of constant length U − V = $K$

$$A = \int_{x_1}^{x_2} y dx. \tag{3.22}$$

The infinitesimal length, Eq. (3.23) of a planar curve is given as

$$ds = \sqrt{1 + \left(\frac{dy}{dx}\right)^2} dx = \sqrt{1 + (y')^2} dx. \tag{3.23}$$

From multivariable calculus, Lagrange multiplier $\lambda$ is applied on constant parameter, which in this case is the perimeter of closed curve. Recall that when a stationary value (maxima, minima, or saddle point) of a function is sought under some constraints/conditions, the Lagrange multiplier technique is adopted. For example, if it is required to minimize the cost function $f(u, v)$ of a car with a given set of features represented by $g(u, v) = c$, then we seek a solution to the expression

$$G(u, v, \lambda) = f(u, v) + \lambda\{g(u, v) - c\}. \tag{3.24}$$

Solution of Eq. (3.24) is sought by obtaining particular values $u_0$, $v_0$, and $\lambda_0$ satisfying the given condition. Readers may refer to standard textbooks for more details on Lagrange multipliers.

Continuing with discussion on isoperimetric problem, the applicable functional can be expressed as

$$F = y + \lambda\sqrt{1 + (y')^2}. \tag{3.25}$$

The necessary condition for integral $I$ in this context is

$$I = \int_{x_1}^{x_1} F dx. \tag{3.26}$$

For Eq. (3.26) to be extremum, Euler-Lagrange condition is

$$\frac{\partial F}{\partial y} - \frac{d}{dx}\left(\frac{\partial F}{\partial y'}\right) = 0. \tag{3.27}$$

Using Eq. (3.25) in (3.27), we get

$$y' = \frac{x - k_1}{\sqrt{\lambda^2 - (x - k_1)^2}}. \tag{3.28}$$

Solving Eq. (3.28) for y,

$$y = \sqrt{\lambda^2 - (x - k_1)^2} + k_2. \tag{3.29}$$

Rearranging Eq. (3.29),

$$(x - k_1)^2 + (y - k_2)^2 = \lambda^2 \tag{3.30}$$

where $k_1$ and $k_2$ are constants.

Clearly, Eq. (3.30) is an expression representing a circle. Therefore, a closed planar curve encompasses maximum area, when it assumes a circular shape.

It should be noted that these three cases belong to a class of problems that has solutions under certain constraining conditions. This is true in most of the observed physical problems. The reader is encouraged to refer to several published literature and online resources on the above problems, which laid foundations for variational methods.

This chapter will attempt to introduce the background of determining approximate solutions. These ideas form basis for numerical methods.

Approximate analytical methods can be broadly classified into

1. Variational method.
2. Weighted residual method.

Popular among variational and weighted residual techniques are the Rayleigh–Ritz and Galerkin methods, respectively. Commercially available finite element software based on displacement field (in structural engineering and solid mechanics area) utilizes the Rayleigh–Ritz scheme along with polynomial basis functions. It has to be remembered that the finite element method is a numerical method for solving partial differential equations and basically is piecewise application of variational or weighted residual methods.

## 3.2 Mathematical Preliminaries

The following sections will briefly describe certain mathematical concepts which are related to most of the engineering application situations. Interested readers are requested to refer to specialized publications for more details.

## 3.3 Classification of Partial Differential Equations

Most of the physical problems of interest fortunately can be modeled using linear second-order partial differential equations. A general form of two-dimensional linear partial differential operator is described below

$$\mathrm{L}(\phi) = f \tag{3.31}$$

where

$$\mathrm{L}(\phi) = a\frac{\partial^2\phi}{\partial x^2} + b\frac{\partial^2\phi}{\partial x\partial y} + c\frac{\partial^2\phi}{\partial y^2} + \mathrm{F}\left(x, y, \phi, \frac{\partial\phi}{\partial x}, \frac{\partial\phi}{\partial y}\right).$$

For linear operator L, $a$, $b$, and $c$ are constants

The operator is classified as

Elliptic if

$$b^2 < 4ac \tag{3.32}$$

Hyperbolic if

$$b^2 > 4ac \tag{3.33}$$

and Parabolic if

$$b^2 = 4ac. \tag{3.34}$$

Classification of above operators expressed by the conditions (3.32)–(3.34) help in strategizing solution scheme or schemes for a problem posed. In many cases, this classification aids in identifying analogous solutions encountered within and outside the domain of concern. Specific solution differs only with respect to the physical constants accounted as coefficients in differential operators.

The following are real-life examples where above-mentioned partial differential equations are utilized for modeling physical phenomena involved.

Equation (3.35) refers to an electrostatic phenomenon

$$-\nabla^2\Phi = \frac{\rho}{\epsilon} \tag{3.35}$$

where $\Phi$ is induced electrostatic potential required to be determined in a medium having $\boldsymbol{\epsilon}$ as permittivity (property of the medium) when a distributed charge density $\boldsymbol{\rho}$ is present. For three-dimensional space and assuming rectangular Cartesian coordinates, the operator $\nabla \equiv \mathrm{i}\frac{\partial}{\partial x} + \mathrm{j}\frac{\partial}{\partial y} + \mathrm{k}\frac{\partial}{\partial z}$. Expression (3.35) is an elliptic partial differential equation. This class of equation models steady state phenomena, i.e. governing variables do not vary with time; in the present case, electric potential $\Phi$ does not vary with time. In structural engineering/solid mechanics analysis, governing equation of a plate is

$$\mathrm{D}\nabla^4 w(x, y) = q(x, y) \tag{3.36}$$

Equation (3.36) too is an elliptic differential equation which relates the time invariant transverse deflection $w(x, y)$ of a plate of flexural rigidity $\mathrm{D} = \frac{E\,t^3}{12(1-\nu^2)}$ to an applied distributed load $q(x, y)$. $E$, $t$, and $\nu$ are Young's modulus, thickness and Poisson's ratio of the plate, respectively.

Most of wave propagation problems encountered in many physical situations are modeled using hyperbolic partial differential equations. For three-dimensional cases in a rectangular Cartesian coordinate system, it is expressed as

$$\nabla^2 u = \frac{1}{c^2}\frac{\partial^2 u}{\partial\, t^2}. \tag{3.37}$$

The one-dimensional form of Eq. (3.37) is

$$\frac{\partial^2 u}{\partial x^2} = \frac{1}{c^2}\frac{\partial^2 u}{\partial t^2} \tag{3.38}$$

where $u = u(x, y, z, t)$ for Eq. (3.37) and $u = u(x, t)$ for Eq. (3.38) is transverse displacement of the medium with respect to wave propagation direction. The wave velocity is represented by $c$.

In solid mechanics, the dynamic response of a plate is given by

$$\mathrm{D}\nabla^4 w(x, y, t) + \rho\frac{\partial^2 w(x, y, \mathrm{t})}{\partial t^2} = q(x, y, t). \tag{3.39}$$

Equation (3.39) describes the temporal response of any point on a plate with flexural rigidity D and density per unit area $\rho$ when a time-varying load of $q(x, y, t)$ is applied. Equation (3.39) is of hyperbolic type.

Diffusion phenomena such as flow through porous media and conduction type heat transfer are some of the problems that can be modeled using a parabolic partial differential equation. The three-dimensional heat conduction equation is given as

$$\nabla^2 \mathrm{T} = \frac{1}{\mathrm{k}}\frac{\partial \mathrm{T}}{\partial t}. \tag{3.40}$$

In Eq. (3.40), $\mathrm{T} = \mathrm{T}(x, y, z, t)$ is temperature distribution on the medium or object as a function of time $t$ with uniform thermal conductivity, k.

The above cases are some of few real-life examples that can be modeled to quantitatively relate governing parameters involved. For engineering applications, this knowledge is paramount for gaining insights into the phenomena to realize workable and reliable systems.

The cardinal objective of classical approximate solution techniques is to find analytical solutions to a mathematical model when closed-form solution is not possible. Furthermore, they form basis of obtaining numerical solutions to complex domains represented in most situations by above-discussed types of partial differential equations.

## 3.4 Boundary and Initial Conditions

To determine dependent variables appearing in governing differential equation that generally assume one of the forms discussed above, boundary and initial values need to be specified, to determine unknown constants appearing in solution function. Initial conditions are specific values of dependent variable at an instance of time. This is applicable to parabolic and hyperbolic partial differential equations.

There are three types of boundary conditions, namely, Dirichlet, Neumann, and mixed boundary conditions. Consider Fig. 3.6, which represents a domain $\Omega(\mathbf{x}, \mathbf{y})$ in two-dimensions with a closed boundary curve $\boldsymbol{\Gamma}$.

In Dirichlet boundary conditions, values of dependent variables are specified at boundary of the problem domain. In Neumann boundary conditions, gradient or normal derivative of dependent variable is specified at the boundary, which usually is associated with specification of inputs or applied fields, such as loads in structural engineering problems. In mixed mode, as the name implies, both types will be specified as a linear combination. Mixed boundary condition is also known as Robin boundary condition and finds applications in heat transfer and electromagnetic problems.

Consider Eq. (3.31) which is reproduced below

$$\mathrm{L}(\phi) = f$$

where $\phi = \phi(x, y)$ and $f = f(x, y)$ are defined over $\mathbb{R}^2$, i.e. a two-dimensional real space bounded by a curve $\boldsymbol{\Gamma}$. Let the following homogenous boundary conditions be satisfied by $\phi$ on $\Gamma$:

$$\phi = 0 \tag{3.41}$$

$$\frac{\partial \phi}{\partial n} = 0 \tag{3.42}$$

$$\frac{\partial \phi}{\partial n} + \phi\, s(\mathrm{s}) = 0. \tag{3.43}$$

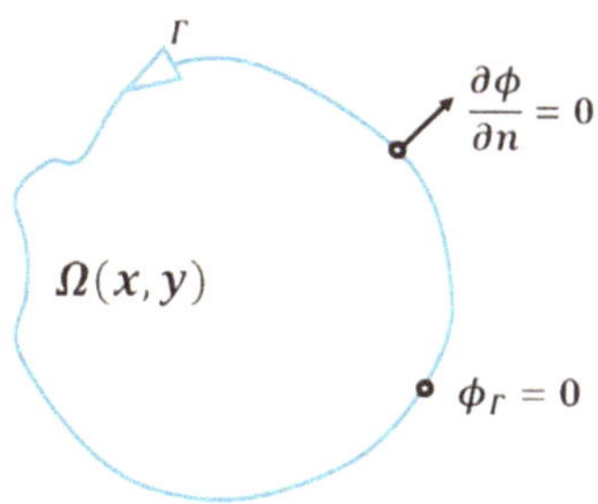


**Fig. 3.6** Dirichlet, Neumann, and mixed boundary conditions

Equations (3.41)–(3.43) correspond to Dirichlet, Neumann, and Mixed boundary conditions, respectively. Furthermore, $s \subset \boldsymbol{\Gamma}$ and $\frac{\partial}{\partial n}$ is normal derivative, where $\subset$ is the symbol for subset. In other words, $s$ corresponds to a certain chosen part of $\boldsymbol{\Gamma}$ where the value of normal derivative is specified.

The above definitions are required for determining functions that minimize functional corresponding to the partial differential equations. In solid mechanics, the functional is total energy of the system. This fortunately can be derived and can be related to physics of the problem. A similar analogy may not be possible in other domains. It is possible to determine the functional directly from governing differential equations. It is to be noted that **L** is a differential operator and that the functions on which it operates are continuous up to the highest order present in the operator.

## 3.5 Weak Formulation of Differential Equations and Rayleigh–Ritz Method

A differential equation is a strong statement defined over a field. It means solution function should satisfy this statement at each and every point in the domain. This strict condition on dependent variables imposes difficulty in finding a practical solution for real-life situations. For example, if a square plate with dimension 1 × 1 m has a hole of diameter 0.0001 m (10 μ) somewhere within its extent subjected to traverse loads, the deflection function will not be sensitive to such discontinuities. But attempting to satisfy this discontinuity can result in complicated expressions without any significant advantage in estimating the solution. A method which imposes less continuity requirements on solution function, such that it satisfies the differential equation in an average sense within the domain, forms basic idea behind approximate solution finding techniques such as weak formulation. In weak formulation, continuity requirement on the solution function is reduced. This aids in determining approximate solutions from practical considerations.

Consider Eq. (3.31) reproduced below

$$\mathrm{L}(\phi) = f.$$

When solution function is approximate $\phi_{\text{approx}} \cong \phi$, then right- and left-hand sides of the equality are not same. Hence, we define a measure, residue R

$$\mathrm{R} = \mathrm{L}\left(\phi_{\text{approx}}\right) - f. \tag{3.44}$$

For a two-dimensional case, the solution is a defined function of $x$ and $y$, i.e. $\phi = \phi(x, y)$. If the solution is exact, then $\mathrm{R} = 0$. When an approximate solution is sought, then $\mathrm{R} \neq 0$. Approximate solutions contain unknown coefficients, which on substituting directly into the governing equations will result in an inconsistent set of relations among unknown coefficients. Determination of these unknown coefficients is the very purpose of method being discussed. If $\boldsymbol{n}$ is the number of such unknowns, to determine them uniquely $\boldsymbol{n}$ conditions are required. Hence, the governing differential equations need to be recast appropriately to obtain necessary conditions for finding unknown coefficients. Mathematical statement of weak formulation in continuum sense is the integral of scalar product between residue and weighting function over the domain. This is to ensure that unknown coefficients in assumed solution will attain appropriate values to make the residue minimum.

Specifically, let us consider the following second-order differential equation in solid mechanics context. For higher order partial differential equations, readers are encouraged to refer to texts given in the reference:

$$-\frac{\mathrm{d}}{\mathrm{d}x}\left\{a(x)\frac{\mathrm{d}u}{\mathrm{d}x}\right\} = q(x)\ 0 < x < L. \tag{3.45}$$

In Eq. (3.45), $a(x)$ corresponds to system parameters such as material and geometry, and $q(x)$ is the applied distributed loads, using which the axial displacement $u(x)$ is determined. The boundary conditions for Eq. (3.45) are

$$u(0) = u_{0,}\ \left(a\frac{\mathrm{d}u}{\mathrm{d}x}\right)\bigg|_{x=L} = \mathrm{Q}_0 \tag{3.46}$$

where $u_0$ and $\mathbf{Q}_0$ in Eq. (3.46) are specified displacement and applied axial force at two ends of a bar-like configuration, respectively.

An approximate solution, Eq. (3.47), known as the trial function is assumed as

$$u(x) \approx \mathrm{U}_n(x) = \sum_{j=1}^{n} c_j\phi_j(x) + \phi_0(x). \tag{3.47}$$

Weighted residual statement of the differential equation is

$$0 = \int_0^L w(x)\left[-\frac{\mathrm{d}}{\mathrm{d}x}\left\{a(x)\frac{\mathrm{dU}_n(x)}{\mathrm{d}x}\right\} - q(x)\right]\mathrm{d}x \tag{3.48}$$

where $w(x)$ in Eq. (3.48) is the weight function or the test function. The terms in square bracket correspond to the non-zero residue, when approximate solution $\mathrm{U}_n(x)$ is substituted in Eq. (3.45). If $\phi$ contains $n$ unknowns, then $n$ linearly independent functions are chosen. From solid mechanics point of view, $w(x)$ corresponds to virtual displacement, and Eq. (3.48) is a statement of total virtual work. Integrating Eq. (3.48) by parts,

$$0 = \int_0^L \left\{\frac{\mathrm{d}w(x)}{\mathrm{d}x}a(x)\frac{\mathrm{dU}_n(x)}{\mathrm{d}x} - w(x)q(x)\right\}\mathrm{d}x - w(x)a(x)\frac{\mathrm{dU}_n(x)}{\mathrm{d}x}\bigg|_0^L. \tag{3.49}$$

We observe that the continuity condition requirement on $u(x) \cong \mathrm{U}(x)$ is reduced in Eq. (3.49). If the assumed approximate solution function contains terms differentiable up to order of differentiation available in governing equation. In the weak form this is reduced. For Eq. (3.45), the order of differentiation is two. On integrating weighted residual expression Eq. (3.48), by parts, differentiation order of solution function $u(x) \cong \mathrm{U}(x)$ is reduced, as notable in Eq. (3.49), implying reduced continuity requirements. This translates to fewer unknown coefficients in the assumed solution, which is a significant aspect when large-order numerical problems are considered. Hence, the term weak (weakening of continuity) formulation.

In boundary terms, $a(x)\frac{\mathrm{dU}_n(x)}{\mathrm{d}x}$ is the secondary variable and its specification is the natural or Neumann boundary condition. Expressing $u(x)$, at the boundary, in the same form as $w(x)$ is the primary variable and its specification is the essential or Dirichlet boundary condition. At the boundary, the weight function $w(x)$ vanishes. Specifying $u(x)$ at any point in the domain results in $w(x) = 0$. This means the variation about that point on the function $u(x)$ is zero. Therefore,

$$w(0) = 0 \text{ as } u(0) = u_0 \tag{3.50}$$

and

$$\left\{a(x)\frac{\mathrm{d}u(x)}{\mathrm{d}x}\right\}n_x\bigg|_{x=L} = \mathrm{Q}_0. \tag{3.51}$$

It is to be remembered that only one of the conjugate pairs of secondary variable and primary variable at the boundary can be specified example (Displacement, Force), (Temperature, Heat). Readers can refer to these aspects in any applied mathematics textbooks.

Weak equivalent form to the differential equation with natural boundary condition is

$$0 = \int_0^L \left\{\frac{\mathrm{d}w(x)}{\mathrm{d}x}a(x)\frac{\mathrm{dU}_n(x)}{\mathrm{d}x} - w(x)q(x)\right\}\mathrm{d}x - w(L)\mathrm{Q}_0. \tag{3.52}$$

Equation (3.52) is expressed in operator form for clarity as

$$0 = \mathrm{A}(w, u) - \mathrm{h}(w) \tag{3.53}$$

where

$$\mathrm{A}(w, u) = \int_0^L \left\{a(x)\frac{\mathrm{d}w(x)}{\mathrm{d}x}\frac{\mathrm{d}u(x)}{\mathrm{d}x}\right\}\mathrm{d}x \tag{3.54}$$

$$\mathrm{h}(w) = \int_0^L \{w(x)q(x)\}\mathrm{d}x + w(L)\mathrm{Q}_0. \tag{3.55}$$

It is required to minimize Eq. (3.53), a functional, by choosing the trial and test functions $u(x)$ and $w(x)$, respectively. Choice of $w(x)$ is the essence, where various approximate methods subtly differ. In this chapter, a brief introduction to popular Rayleigh–Ritz and Galerkin's method is given as the former is a variational method and the latter is a weighted residual method.

The quadratic functional associated with the differential Eq. (3.45) can be stated as

$$\mathrm{I}(u) = \frac{1}{2}\mathrm{A}(u, u) - \mathrm{h}(u). \tag{3.56}$$

Every differential equation admits a weighted-integral statement, Eq. (3.56); weak form exists provided the equation is of order two or more. For a functional to exist, the associated bilinear form $\mathrm{A}(w, u)$ should be symmetric in its arguments. Variational methods and finite element method do not require a functional; an integral statement or a weak form of the equation to be solved is sufficient.

For higher dimensions, the weak form can be obtained on similar lines by using appropriate differential operators.

Rayleigh–Ritz method uses an approximate field to estimate solutions over the entire domain of interest. For using this method, we need to have a functional. A functional is an integral expression that implicitly contains differential equation, which describes the system. These functionals will be used to formulate finite element problems. It is based on the minimization of potential energy and involves the construction of an assumed displacement field from a solid mechanics point of view.

The Eq. (3.53), which is a variational statement, is restated as

$$\mathrm{A}(w, u) = \mathrm{h}(w).$$

In Rayleigh–Ritz method, the approximate solution/trial function given by Eq. (3.47) is substituted in (3.53) and the weighing/test function is chosen from the same space of trial function, as given below

$$\mathrm{A}\left(\phi_i, \sum_{j=1}^{n} c_j\phi_j(x) + \phi_0(x)\right) = \mathrm{h}(\phi_i),\ i = 1, 2, \ldots, n. \tag{3.57}$$

Corresponding quadratic functional, i.e. Eq. (3.56) assumes the following form:

$$\mathrm{I}(u) = \mathrm{I}(c_1, c_2, \ldots, c_n). \tag{3.58}$$

These unknown constants can be determined by using the following conditions:

$$\frac{\partial \mathrm{I}}{\partial c_1} = \frac{\partial \mathrm{I}}{\partial c_2}, \ldots, = \frac{\partial \mathrm{I}}{\partial c_n} = 0. \tag{3.59}$$

This is the condition for minimization of total potential I($u$). These conditions result in simultaneous algebraic equations in unknowns $c_1, c_2, \ldots, c_n$. Upon solving for the unknowns, they are substituted back into Eq. (3.47) for obtaining the approximate solution.

The following examples illustrate Rayleigh–Ritz technique for determining approximate solutions for boundary value problems.

**Example 1** Consider the differential equation

$$\frac{d^2\phi}{dx^2} + x^2 = 0, \ 0 < x < 1 \tag{3.60}$$

with boundary conditions $\phi(0) = \phi(1) = 0$.

For the second-order differential operator, let us obtain a solution from polynomial space. Keeping continuity requirement of chosen operator, the polynomial order should be three. A complete third-order polynomial is

$$\phi = c_0 + c_1x + c_2x^2 + c_3x^3. \tag{3.61}$$

On applying given boundary conditions, assumed solution assumes the following form:

$$\phi = k_1x(1-x) + k_2x^2(1-x). \tag{3.62}$$

Using Eqs. (3.53) and (3.62),

$$A_{11} = \int_0^1 (1-2x)^2 dx = \frac{1}{3}$$

$$A_{12} = A_{21} = \int_0^1 (1-2x)\left(2x-3x^2\right) dx = \frac{1}{6}$$

$$A_{22} = \int_0^1 \left(2x-3x^2\right)^2 dx = \frac{2}{15}$$

$$h_1 = \int_0^1 \left(x-x^2\right) x^2 dx = \frac{1}{20}$$

$$h_2 = \int_0^1 \left(x^2-x^3\right) x^2 dx = \frac{1}{30}.$$

Unknowns $k_1$ and $k_2$ can be determined using Rayleigh–Ritz equations

$$[A]\{k\} = \{h\}.$$

On solving $k_1 = \frac{1}{15}$ and $k_2 = \frac{1}{6}$ therefore the approximate solution is

$$\phi = \frac{1}{15}x(1 - x) + \frac{1}{6}x^2(1 - x).$$

The exact solution is $\phi = \frac{x}{12}(1 - x^3)$.

Figure 3.7 shows the plots of exact and approximate solutions. The approximate solution's error can be minimized by increasing more terms or using functions other than polynomials, such as trigonometric functions.

**Example 2** This example illustrates use of the Rayleigh–Ritz method for solving deflection of a simply supported square plate, as shown in Fig. 3.8. Before solving the problem, recall the principle of minimum potential energy.

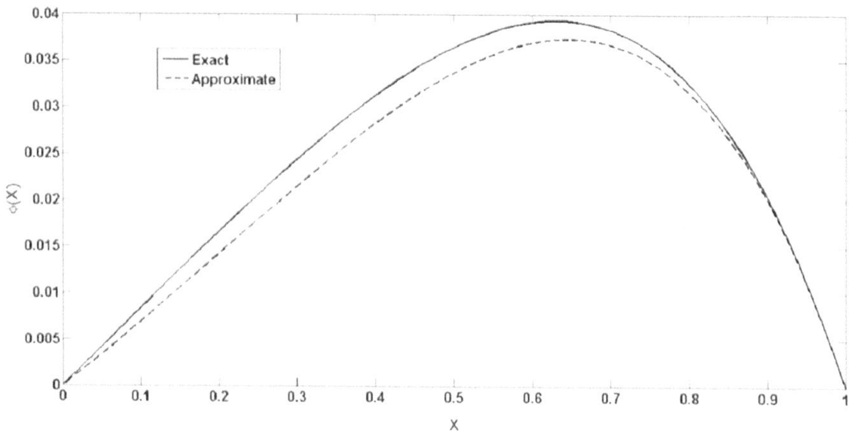


**Fig. 3.7** Comparison of exact and approximate solution; Rayleigh–Ritz method

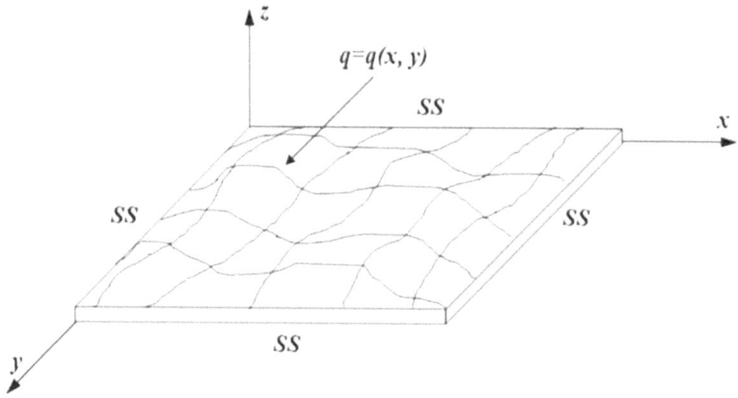


**Fig. 3.8** A simply supported square plate with load varying in $x$- and $y$-directions

The minimum potential energy states that, "*of all displacement configurations which satisfy compatibility and kinematic boundary conditions, the one corresponding to stable equilibrium has the least potential energy*". Compatibility implies preservation of continuum of the structure. Kinematic boundary conditions refer to Dirichlet/displacement/essential or geometric boundary condition.

The derivations of following principal expressions for strain energies, external work done and equilibrium equations are beyond the scope of this book. The objective is to determine solution of a given governing equation, using the Rayleigh–Ritz method. Remember that in this example instead of deriving functional from governing equation, it is directly derived from total potential energy of the plate. Classical plate theory (CPT) based on the Kirchhoff–Poisson hypothesis results in following governing differential equation, which can be referred to in many standard textbooks on plates and shells:

$$\mathrm{D}\nabla^4 w = q(x, y) \tag{3.63}$$

where $\mathrm{D} = \frac{E\,t^3}{12(1-\nu^2)}$ is the flexural rigidity,
$E$ is Young's modulus,
$\nu$ is Poisson's ratio,
$t$ is the thickness of the plate, and
$w(x, y)$ is the transverse deflection of the plate due to load $q(x, y)$.
$\nabla \equiv \mathrm{i}\frac{\partial}{\partial \mathrm{x}} + \mathrm{j}\frac{\partial}{\partial \mathrm{y}}$ in a rectangular Cartesian coordinate system, and **i** and **j** are the unit vectors along a respective coordinate axis.

The strain energy **U**, due to the deformation of the plate given by the integral over its area A, is

$$\mathrm{U} = \frac{\mathrm{D}}{2}\iint\limits_{\mathrm{A}} \left\{ \left(\nabla^2 w\right)^2 - 2(1-\nu)\left(\frac{\partial^2 w}{\partial x^2}\frac{\partial^2 w}{\partial y^2} - \frac{\partial^2 w}{\partial x\, y}\right) \right\} \mathrm{dA}. \tag{3.64}$$

Equation (3.65) gives the net potential energy of a distributed load $q(x, y)$ over the plate as

$$\mathrm{V} = -\iint\limits_{\mathrm{A}} q\, w\, \mathrm{d}x\, \mathrm{d}y. \tag{3.65}$$

Total potential of the system is

$$\Pi = \mathrm{U} + \mathrm{V}. \tag{3.66}$$

Load function is assumed as the expression given by Eq. (3.67) for a square plate with side '$a$' as

$$q(x, y) = q_0 \sin\left(\frac{m\,\pi x}{a}\right) \sin\left(\frac{n\,\pi y}{a}\right) \tag{3.67}$$

where $(m, n) \in Z$ are wave numbers. Sinusoidal load variation is assumed as many real-life load cases can additively decomposed into sine functions of certain frequencies and amplitudes.

Simply supported conditions are

$$w = 0 \text{ at } x = 0, a \text{ and } y = 0, a. \tag{3.68}$$

The moments at the boundaries are given by

$$\mathrm{M}_x = 0 \text{ at } x = 0, a$$

$$\mathrm{M}_y = 0 \text{ at } y = 0, a.$$

An admissible function, which satisfies the essential boundary condition (3.68), is

$$w(x, y) = \mathrm{W}x(x - a)y(y - a) \tag{3.69}$$

where W is the unknown constant to be determined.

The expressions in Eqs. (3.64) and (3.65) are evaluated using Eqs. (3.67) and (3.69) as given below

$$\mathrm{U} = \frac{11}{45}\mathrm{D}a^6\mathrm{W}^2 \tag{3.70}$$

$$\mathrm{V} = \frac{-16}{\pi^6}q_0 a^6\mathrm{W}. \tag{3.71}$$

Substituting Eqs. (3.70) and (3.71) in (3.66), the total potential energy is

$$\Pi = \frac{11}{45}\mathrm{D}a^6\mathrm{W}^2 + \frac{-16}{\pi^6}q_0 a^6\mathrm{W}. \tag{3.72}$$

The condition for total potential energy to be minimum is given by

$$\frac{\mathrm{d}\Pi}{\mathrm{dW}} = 0 \text{ and } \frac{\mathrm{d}^2\Pi}{\mathrm{dW}^2} > 0. \tag{3.73}$$

The first condition of Eq. (3.73) yields

$$\mathrm{W} = \frac{360\mathrm{q}_0}{11\pi^6\mathrm{D}}. \tag{3.74}$$

Therefore, Rayleigh–Ritz solution for the transverse deflection function is obtained by substituting Eq. (3.74) into (3.69)

$$w_{RR}(x, y) = \frac{360\, q_0}{11\pi^6 D} x(x-a)y(y-a). \tag{3.75}$$

The exact solution is

$$w(x, y) = \frac{q_0\, a^4}{4\pi^4 D} \sin\left(\frac{\pi x}{a}\right) \sin\left(\frac{\pi y}{a}\right). \tag{3.76}$$

The reader is advised to plot the Rayleigh–Ritz and exact solutions for comparison of results.

## 3.6 Weighted Residual Technique and Galerkin's Method

Galerkin's method is used to convert a continuous operator problem such as a differential equation to a discrete problem for ease of solving.

It has to be noted that the approximate analytical solution/s $\overline{\phi}$ result is residual as given by Eq. (3.44), reproduced here as

$$R = L(\phi_{\text{approx}}) - f.$$

In Galerkin's method, the weighted residual integral is minimized or rather equated to zero to obtain an approximate solution instead of a weak form representation obtained by integration by parts. To determine unknown coefficients in assumed approximate solution, the following integration is carried out:

$$\int_\Omega \left\{ L\left(\overline{\overline{\phi}}\right) - f \right\} \overline{\phi}\, d\Omega = 0 \tag{3.77}$$

where $\overline{\overline{\phi}} = \sum_{j=1}^{n} c_j \overline{\phi}_j + \overline{\phi}_0$ is the assumed approximate solution described over the domain $\Omega$ under consideration.

The weighted integral statement can be interpreted not only as virtual work but also as orthogonalization process, by which the approximate solution is determined by minimizing projection of error over the solution by finding appropriate coefficients of $\overline{\phi}_j$. Galerkin's method is more general than the Rayleigh–Ritz method as the operator need not be even ordered or linear. Furthermore, the assumed approximate solution should satisfy both essential and natural boundary condition, *a priori*. It is to be noted that in the Rayleigh–Ritz method, the assumed solution function needs to satisfy only essential boundary conditions. In the weak formulation phase of the Rayleigh–Ritz method, the natural boundary conditions emerge during the process of integration by parts and it is specified/satisfied at this stage.

**Worked Example 1** Consider the differential equation

$$\frac{d^2\phi}{dx^2} - \phi + x = 0, \quad 0 < x < 1, \ \phi(0) = \phi(1) = 0. \tag{3.78}$$

Let the approximate solution be assumed as

$$\overline{\overline{\phi}} = cx(1 - x). \tag{3.79}$$

Note that $\overline{\phi} = x(1 - x)$.
Residual equation is

$$\mathrm{R} = \frac{d^2\overline{\overline{\phi}}}{dx^2} - \overline{\overline{\phi}} + x. \tag{3.80}$$

The weighted integral statement is
$\int_{\Omega}\left\{\mathrm{L}\left(\overline{\overline{\phi}}\right) - f\right\}\overline{\phi}\mathrm{d}\Omega = 0$, which for the present case assumes the form

$$\int_0^1\left(\frac{d^2\overline{\overline{\phi}}}{dx^2} - \overline{\overline{\phi}} + x\right)x(1 - x)dx = 0 \qquad \text{or} \tag{3.81}$$

$$\int_0^1\{-2c - c\left(x - x^2\right) + x\}x(1 - x)dx = 0.$$

On solving,

$$c = \frac{5}{22}.$$

Therefore, the approximate solution is

$$\overline{\overline{\phi}}(x) = \frac{5}{22}x(1 - x). \tag{3.82}$$

The exact solution is

$$\phi(x) = x - \frac{e^x - e^{-x}}{e - e^{-1}}. \tag{3.83}$$

Figure 3.9 shows the comparison between the exact solution and that obtained using the Galerkin method. As discussed earlier, the errors in the approximate function can be minimized by considering more number of higher order terms or choosing trigonometric functions, etc.

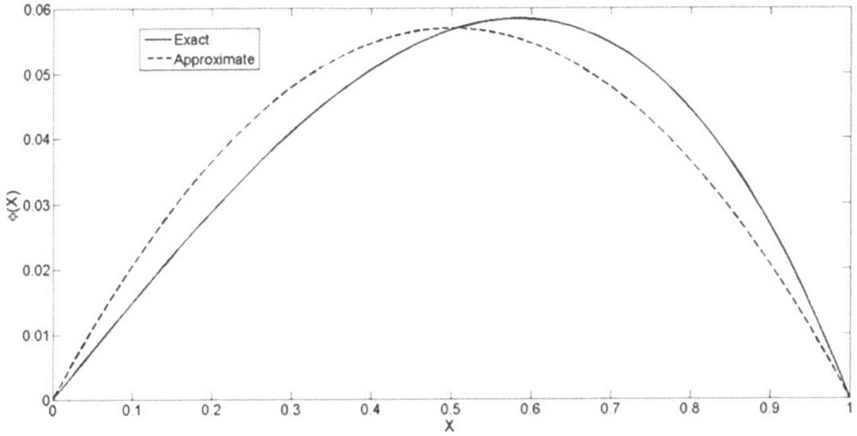


**Fig. 3.9** Comparison of exact and approximate solution; Galerkin method

## 3.7 Link Between Classical Approximate Solution and Finite Element Method

Ideas conveyed in this chapter are very fundamental without loss of generality albeit expressed in a simple manner. The author has used only essential mathematics required while explaining concepts. If the reader wants to solve real-life problems, aid of higher mathematical techniques is definitely needed. This section will furnish a big-picture view of the association between classical approximate solutions such as Rayleigh–Ritz and Galerkin methods and finite element/numerical methods.

In the previous sections, the solutions for the differential equations were obtained for simple domains and these solutions were an analytic function for the whole domain, although approximate. That is, continuous solutions over simple geometries were determined for every point within it. In real-life situations, the solution is sought for complex domains. Machine elements, complex structural geometries such as automobiles and aircraft, bridges and wind turbine blades are some of the examples. Solutions with simple function representations for such constructs are a far cry, as it is nearly impossible to represent a mathematical model for such systems. In other words, it is not possible to have a differential equation/s for the whole of, say, an aircraft's displacement field when subjected to externally applied loads. Consider the complex geometry of a typical aircraft structure as shown in Fig. 3.10. Determining a differential equation for the entire geometry to assess the displacement field for given constraints and tractions (external loads) is nearly impossible.

The strategy adopted will be to solve for dependent variable of interest on simple geometries using any of the approximate methods on a triangle or a quadrilateral (any convenient geometrical shape) using a suitable mathematical model. Furthermore, solution sought will be on certain locations on the chosen simple geometries rather than on the whole of a triangle or quadrilateral. Such points or locations where

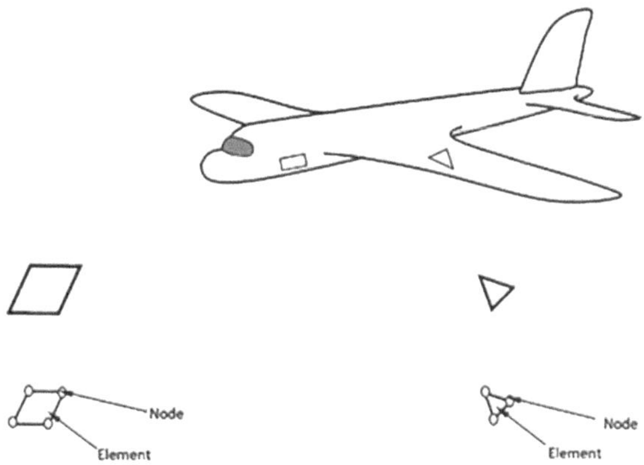


**Fig. 3.10** A structure of arbitrary shape subjected to external load and constraints; element and nodes

solution is required are known as nodes and the simple geometries on which differential equation is solved are called elements. In Fig. 3.10, triangular and quadrilateral elements with nodes are shown. Usually, the nodes are located at element boundaries. Expression/s for the solution is determined by applying Neumann boundary condition. Such elements are tessellated over the entire domain for approximating extent of arbitrary geometry and solution. While using these elements, continuity is preserved at these node locations between elements. This is the reason at element level solution only Neumann condition is applied, not Dirichlet condition. The process by which the arbitrary geometry is approximated or represented by collection of elements is known as meshing. On arranging corresponding element equations, this mathematically translates into a set of $n$ equations in $n$ unknowns. The solution vector gives values of unknowns at the nodes. To find the values of unknowns within an element, convenient interpolation functions are used.

Let us consider a simple one-dimensional problem as shown in Fig. 3.11. The solution is sought at the two ends (points) of the straight line.

Therefore, the assumed solution has the form

$$u(x) = a_0 + a_1 x. \tag{3.84}$$

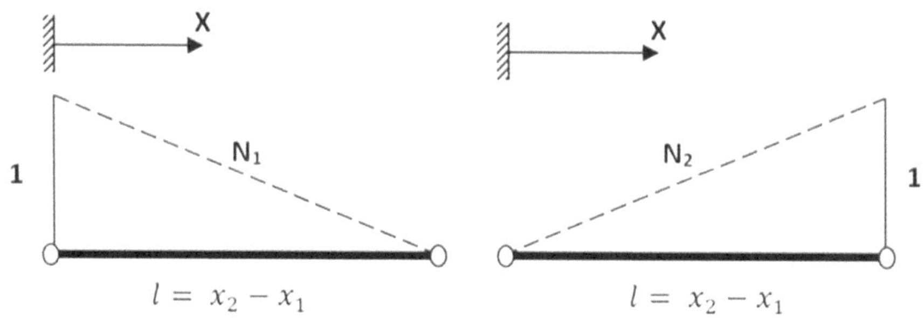


**Fig. 3.11** Variation of interpolation function

At the end of solution process, as explained in previous sections, the unknowns $a_0$ and $a_1$ in Eq. (3.84) are determined. But these constants have no physical association with the element shown in Fig. 3.11, even though using (3.84) solution on any point between the nodes can be determined by substituting various values of $x$. In finite element analysis, these 'node-less' quantities $a_0$ and $a_1$ are related to nodal unknowns $u_1$ and $u_2$ by expressing solution function as

$$u(x) = \mathrm{N}_1 u_1 + \mathrm{N}_2 u_2. \tag{3.85}$$

If $l$ is the length of the domain, i.e. $l = x_2 - x_1$, then comparing (3.84) and (3.85), $\mathrm{N}_1 = \frac{l-x}{l}$ and $\mathrm{N}_2 = \frac{x}{l}$.

$\mathrm{N}_i$'s are known as interpolation functions and convey variation of $u(x)$ with respect to $x$ within an element, which is shown in Fig. 3.11.

This is the essence of link between variational method and finite element method. Other secondary details can be found in published literature as well as finite element chapters of this textbook.

## 3.8 Exercise

Determine the solution of following problems (1–5) using the Rayleigh–Ritz and/or Galerkin methods. Compare the solutions with exact solutions by plotting them wherever possible.

3.8.1 A beam simply supported at both ends, with flexural rigidity EI and length **L**, is subjected to uniform moment, $\mathrm{M}(x) = \mathrm{M}_0$. The governing equation and the boundary conditions are

$$\mathrm{EI}\frac{\mathrm{d}^2 w}{\mathrm{d}x^2} - \mathrm{M}(x) = 0;\ \ 0 \le x \le \mathrm{L}$$

$w(0) = 0$ and $w(\mathrm{L}) = 0$.

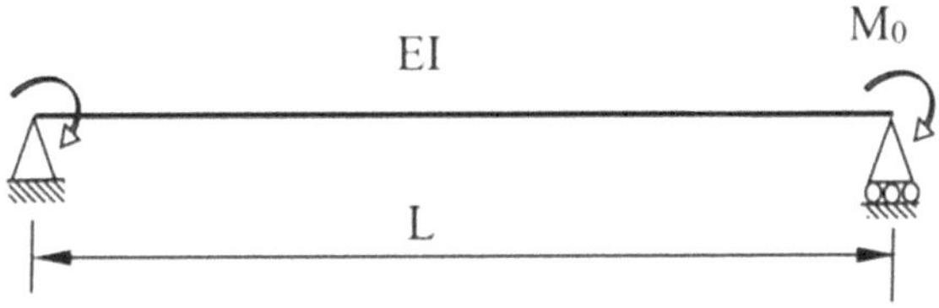


where the solution $w = w(x)$ corresponds to transverse displacement along the length.

3.8.2 Determine the solution for the following second-order linear differential equation:

$$\frac{d^2w}{dx^2} + w + x = 0;\ 0 \leq x \leq 1.$$

Given, $w(0) = 0$ and $w(1) = 0$.

3.8.3 Solve the nonlinear differential equation using Galerkin's method:

$$\frac{d^2w}{dx^2} + \frac{dw}{dx} + 2x(1 - x) = 0;\ 0 \leq x \leq 1.$$

Given, $w(0) = 0$ and $w(1) = 0$.

3.8.4 Find exact and approximate solutions for the differential equation and compare solutions by plotting it

$$\frac{d^2w}{dx^2} - \frac{dw}{dx} + x = 0;\ 0 \leq x \leq 1.$$

Given, $w(0) = 0$ and $w(1) = 0$.

3.8.5 A rod of length **L**, cross-sectional area **A,** and composed of a material whose Young's modulus is **E**, as shown in the figure, is subjected to an axial load **P(x)** **= 5x²**. The governing differential equation is given by $\left(\mathrm{AE}\frac{\mathrm{du}}{\mathrm{dx}}\right) - \mathrm{P(x)} = 0$, where u(x) is axial displacement function in variable x, along the length of beam. The boundary conditions are $u(0) = u(\mathrm{L}) = 0$.

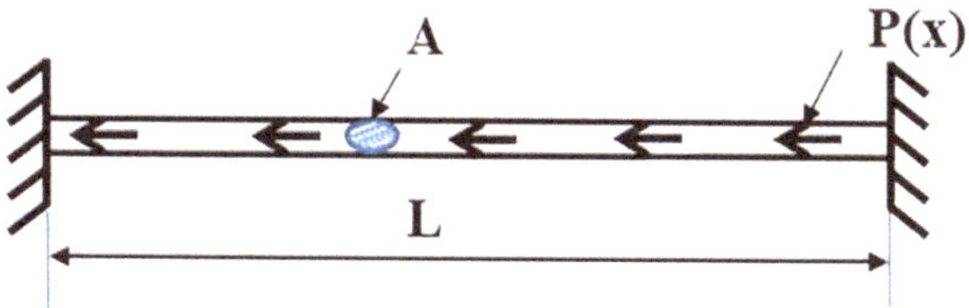


# Bibliography

Cook RD, Malkus DS, Plesha ME (1989) Concepts and applications of finite element analysis, 3rd edn. Wiley, NY

Davies AJ (1980) The finite element method: a first approach. Oxford University Press

Rao JS (1992) Advanced theory of vibrations. Wiley Eastern Limited, New Delhi

Reddy JN (1993) An introduction to the finite element method. McGraw-Hill Inc., NY

Rektorys K (1977) Variational methods in mathematics, science and engineering. Reidel, Boston

Varadan TJ, Bhaskar K (1999) Analysis of plates: theory and problem. Narosa, N. Delhi

Wagg D, Neild S (2010) Nonlinear vibration with control, for flexible and adaptive structures. Springer, Dordrecht

Wunderlich W, Pilkey WD (2003) Mechanics of structures, variational and computational methods, 2nd edn. CRC Press, NY

# Chapter 4
# Elementary Concepts in Elasticity

**A. K. Asraff**

*Design is not how it looks like and feels like. Design is how it works—Steve Jobs*

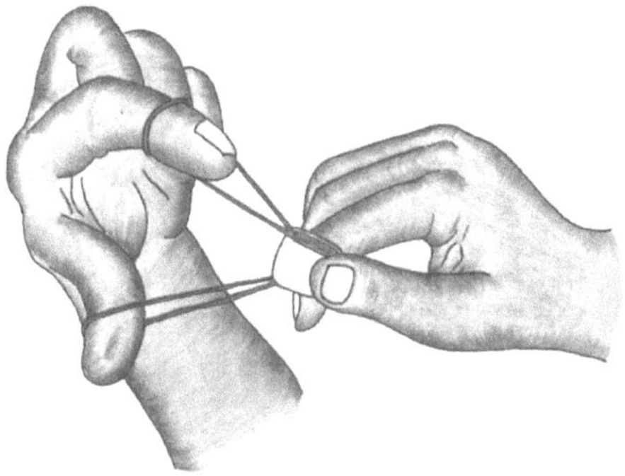

Bodies that bear loads in some form or other are called structures. Elasticity is defined as the ability of a structure to resist loads and return to its original shape when the load is removed. Robert Hooke in 1660 discovered one of the most important laws in elasticity, known as Hooke's law, which states that the elongation of a body is directly proportional to the load applied on it. This was later extended to three-dimensional elastic bodies. The contributions of Augustin Louis Cauchy and Claude Louis Marie Henri Navier finally resulted in laying the basic foundations of the theory of elasticity. The Navier–Cauchy equations consist of 15 coupled partial differential equations: three equations of equilibrium, six strain–displacement relationships, and six compatibility conditions. However, these are quite complicated equations, and there are no closed form solutions available for them.

Four important concepts in elasticity are (i) equations of equilibrium, (ii) strain–displacement relationships, (iii) compatibility conditions, and (iv) constitutive relationships. The equations of equilibrium indicate that the body is in equilibrium under

A. K. Asraff (✉)
DD, MDA/LPSC, Valiamala, Thiruvananthapuram 695547, India
e-mail: ak_asraff@lpscv.dos.gov.in



S. U. Nair and S. Somanath (eds.), *Introduction to Finite Element Analysis*,
https://doi.org/10.1007/978-981-19-7989-7_4

the action of different forces and moments, and there are no net forces or moments in the body. The relationship that connects the displacement field with the strains developed in the structure is called the strain–displacement relationship. Compatibility conditions are those mathematical equations which have to be satisfied to ensure that gaps/overlaps do not develop when a continuous body is deformed. Constitutive relationships indicate how the stresses in the structure are related to the strains induced.

The following sections elaborate on some of the above concepts in detail.

## 4.1 Equilibrium Equations

The state of stress in an elemental volume of a loaded structure is shown in Fig. 4.1. For a true representation of the state of stress in such a three-dimensional body, our old concept of stress being unit force acting over an elemental area is insufficient. Therefore, we adopt here a better mathematical expression to define the stress called Tensor. A tensor is a quantity which has magnitude, direction, and a plane in which it acts. Both stress and strain are three-dimensional tensors. A zero-order tensor is called a Scalar and a first-order tensor is known as a Vector. Higher order tensors are denoted by their respective order.

For example, stress can be conveniently defined as a second-order tensor called the stress tensor:

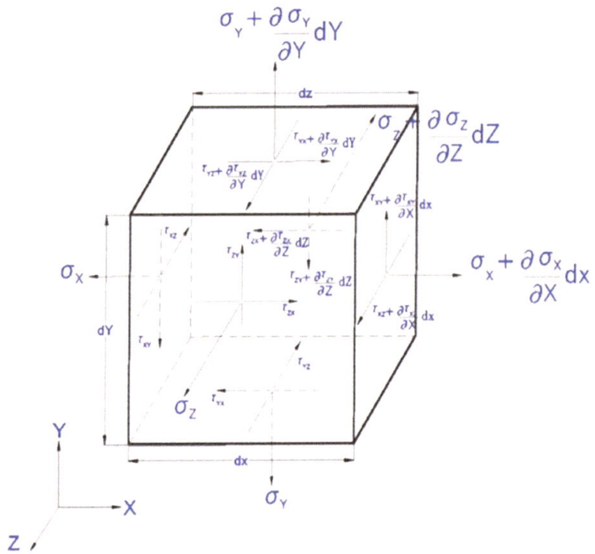


**Fig. 4.1** State of stress in an elemental volume of a body

$$[\sigma] = \begin{bmatrix} \sigma_{xx} & \tau_{xy} & \tau_{xz} \\ \tau_{yx} & \sigma_{yy} & \tau_{yz} \\ \tau_{zx} & \tau_{zy} & \sigma_{zz} \end{bmatrix} \tag{4.1}$$

where $\sigma_{xx}$, $\sigma_{yy}$, $\sigma_{zz}$ are the normal stress components and $\tau_{xy}$, $\tau_{yz}$, $\tau_{zx}$ are the shear stress components. Stresses acting on the positive face of the elemental volume in a positive coordinate direction are treated as positive. Those acting on the negative face in the negative direction are also considered as positive. All other components are taken as negative. A positive face is one on which a normal vector directed outwards from the element points in a positive direction.

The equations of equilibrium which must be satisfied at every point in the body under consideration are obtained by writing the force equilibrium in all three dimensions and then simplifying. The final equations in the three directions are given below

$$\frac{\partial \sigma_{xx}}{\partial x} + \frac{\partial \tau_{xy}}{\partial y} + \frac{\partial \tau_{xz}}{\partial z} + \overline{X} = 0 \tag{4.2}$$

$$\frac{\partial \sigma_{yy}}{\partial y} + \frac{\partial \tau_{yz}}{\partial z} + \frac{\partial \tau_{yx}}{\partial x} + \overline{Y} = 0 \tag{4.3}$$

$$\frac{\partial \sigma_{zz}}{\partial z} + \frac{\partial \tau_{zx}}{\partial x} + \frac{\partial \tau_{zy}}{\partial y} + \overline{Z} = 0 \tag{4.4}$$

where $\overline{X}$, $\overline{Y}$, and $\overline{Z}$ are the body forces (body forces are forces which act on the bulk material of a body such as inertial forces and thermal loads as opposed to surface forces which act only on the surface of a body viz. stress, pressure, frictional forces, surface traction, etc.).

Considering the symmetry of the stress tensor (due to conservation of angular momentum), it can also be expressed in a convenient vector form (not to be confused with a vector being defined as a quantity having both magnitude and direction), called the stress vector, which is well suited for finite element analysis as

$$\{\sigma\} = \begin{Bmatrix} \sigma_{xx} \\ \sigma_{yy} \\ \sigma_{zz} \\ \tau_{xy} \\ \tau_{yz} \\ \tau_{zx} \end{Bmatrix} = \begin{Bmatrix} \sigma_{x} \\ \sigma_{y} \\ \sigma_{z} \\ \tau_{xy} \\ \tau_{yz} \\ \tau_{zx} \end{Bmatrix} \tag{4.5}$$

for the sake of simplicity.

If the coordinate axes are taken as the principal axes (principal planes do not carry any shear stresses), the stress tensor and stress vector become

$$[\sigma] = \begin{bmatrix} \sigma_1 & 0 & 0 \\ 0 & \sigma_2 & 0 \\ 0 & 0 & \sigma_3 \end{bmatrix} \text{ and } \{\sigma\} = \begin{Bmatrix} \sigma_1 \\ \sigma_2 \\ \sigma_3 \\ 0 \\ 0 \\ 0 \end{Bmatrix} \quad (4.6)$$

An invariant is a quantity which does not change with the rotation of axes or which remains unaffected under transformation from one set of axes to another. The stress tensor has three invariants denoted as $J_1$, $J_2$, and $J_3$ as given below

$$J_1 = \sigma_x + \sigma_y + \sigma_z \quad (4.7)$$

$$J_1 = \sigma_1 + \sigma_2 + \sigma_3 \quad (4.8)$$

(where $J_1$ is expressed in terms of principal stresses)

$$J_2 = \sigma_x\sigma_y + \sigma_y\sigma_z + \sigma_z\sigma_x - \tau_{xy}^2 - \tau_{yz}^2 - \tau_{zx}^2 \quad (4.9)$$

$$J_2 = \sigma_1\sigma_2 + \sigma_2\sigma_3 + \sigma_3\sigma_1 \quad (4.10)$$

(where $J_2$ is expressed in terms of principal stresses)

$$J_3 = \sigma_x\sigma_y\sigma_z + 2\tau_{xy}\tau_{yz}\tau_{zx} - \sigma_x\tau_{yz}^2 - \sigma_y\tau_{zx}^2 - \sigma_z\tau_{xy}^2 \quad (4.11)$$

$$J_3 = \sigma_1\sigma_2\sigma_3 \quad (4.12)$$

(where $J_3$ is expressed in terms of principal stresses).

The stress tensor can also be represented as the sum of the deviatoric and hydrostatic stress components as

$$[\sigma] = [\sigma]_D + [\sigma]_H \quad (4.13)$$

where $[\sigma]_D$ is called the deviatoric stress tensor given by the expression:

$$[\sigma]_D = \begin{bmatrix} \sigma_1 - \sigma_m & 0 & 0 \\ 0 & \sigma_2 - \sigma_m & 0 \\ 0 & 0 & \sigma_3 - \sigma_m \end{bmatrix} \quad (4.14)$$

and $[\sigma]_H$ is called the hydrostatic or volumetric or spherical stress tensor given by the expression:

$$[\sigma]_H = \begin{bmatrix} \sigma_m & 0 & 0 \\ 0 & \sigma_m & 0 \\ 0 & 0 & \sigma_m \end{bmatrix} \tag{4.15}$$

where $\sigma_m$ is known as the mean stress given by the expression:

$$\sigma_m = \frac{\sigma_1 + \sigma_2 + \sigma_3}{3} \tag{4.16}$$

The hydrostatic component of the stress tensor produces only volume changes without any change of shape, while the deviatoric component produces only a distortion or change of shape. These components are important since they are used extensively in theories of failure of a material such as the Octahedral shear stress theory.

Another important concept in solid mechanics is the octahedral plane and the stresses acting on that plane. An octahedron is a solid having eight identical faces as shown in Fig. 4.2. All these faces are equilateral triangles and equally inclined to the X, Y, and Z axes. These faces are termed as octahedral planes. The stresses acting on this plane can be resolved as a normal stress and a shear stress, which are called the Octahedral normal stress and Octahedral shear stress respectively as shown in Fig. 4.3. They are given by the following equations:

$$\sigma_{oct} = \frac{2}{3}\sqrt{(\sigma_1 - \sigma_2)^2 + (\sigma_2 - \sigma_3)^2 + (\sigma_3 - \sigma_1)^2} \tag{4.17}$$

$$\tau_{oct} = \frac{1}{3}\sqrt{(\sigma_1 - \sigma_2)^2 + (\sigma_2 - \sigma_3)^2 + (\sigma_3 - \sigma_1)^2}. \tag{4.18}$$

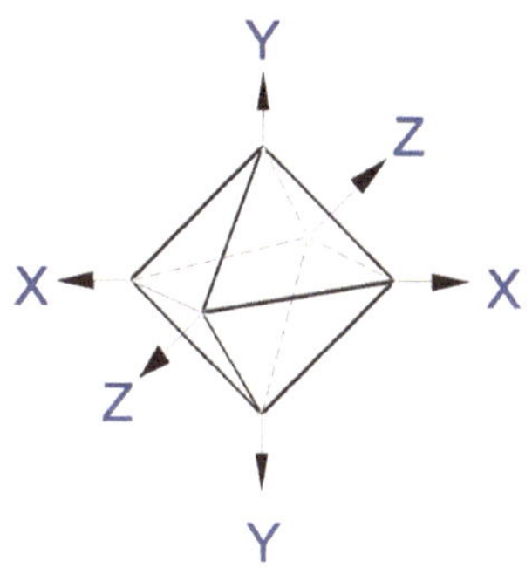


**Fig. 4.2** Octahedron

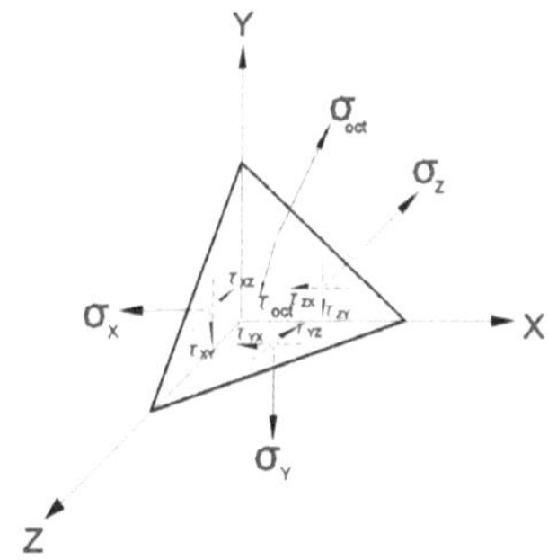


**Fig. 4.3** Stresses on an octahedral plane

## 4.2 Kinematic Conditions

Kinematics deals with the study of the motion of a structure. In this section, we would focus on the strains developed in the body under the action of stresses. Similar to the stress tensor cited in Eq. (4.1), the state of strain can be expressed as a second-order tensor called the strain tensor and strain vector as

$$[\varepsilon] = \begin{bmatrix} \varepsilon_{xx} & \gamma_{xy} & \gamma_{xz} \\ \gamma_{yx} & \varepsilon_{yy} & \gamma_{yz} \\ \gamma_{zx} & \gamma_{zy} & \varepsilon_{zz} \end{bmatrix} \text{ and } \{\varepsilon\} = \begin{Bmatrix} \varepsilon_{xx} \\ \varepsilon_{yy} \\ \varepsilon_{zz} \\ \gamma_{xy} \\ \gamma_{yz} \\ \gamma_{zx} \end{Bmatrix} = \begin{Bmatrix} \varepsilon_{x} \\ \varepsilon_{y} \\ \varepsilon_{z} \\ \gamma_{xy} \\ \gamma_{yz} \\ \gamma_{zx} \end{Bmatrix} \{\varepsilon\} = \begin{Bmatrix} \varepsilon_{1} \\ \varepsilon_{2} \\ \varepsilon_{3} \\ 0 \\ 0 \\ 0 \end{Bmatrix} \tag{4.19}$$

Similar to the three stress invariants, the three invariants of the strain tensor are denoted as $I_1$, $I_2$, and $I_3$, respectively. They are as given below

$$I_1 = \varepsilon_x + \varepsilon_y + \varepsilon_z \tag{4.20}$$

$$I_1 = \varepsilon_1 + \varepsilon_2 + \varepsilon_3 \tag{4.21}$$

(where $I_1$ is expressed in terms of principal strains)

$$I_2 = \varepsilon_x\varepsilon_y + \varepsilon_y\varepsilon_z + \varepsilon_z\varepsilon_x - \frac{1}{4}\left(\gamma_{xy}^2 + \gamma_{yz}^2 + \gamma_{zx}^2\right) \tag{4.22}$$

$$I_2 = \varepsilon_1\varepsilon_2 + \varepsilon_2\varepsilon_3 + \varepsilon_3\varepsilon_1 \tag{4.23}$$

(where $I_2$ is expressed in terms of principal strains)

$$I_3 = \varepsilon_x \varepsilon_y \varepsilon_z + \frac{1}{4}\left(\gamma_{xy}\gamma_{yz}\gamma_{zx} - \varepsilon_x \gamma_{yz}^2 - \varepsilon_y \gamma_{zx}^2 - \varepsilon_z \gamma_{xy}^2\right) \tag{4.24}$$

$$I_3 = \varepsilon_1 \varepsilon_2 \varepsilon_3 \tag{4.25}$$

(where $I_3$ is expressed in terms of principal strains).

Here, the factor of 1/4 arises from the difference in the definitions of engineering strain and tensorial strain, for example, $\gamma_{xy} = 2\varepsilon_{xy}$ (where $\gamma_{xy}$ is the tensorial strain and $\varepsilon_{xy}$ is the engineering strain).

The strain at a point can also be expressed as a sum of its deviatoric and hydrostatic components

$$[\varepsilon] = [\varepsilon]_D + [\varepsilon]_H \tag{4.26}$$

where $[\varepsilon]_D$ is the deviatoric strain tensor given by the expression

$$[\varepsilon]_D = \begin{bmatrix} \varepsilon_1 - \varepsilon_m & 0 & 0 \\ 0 & \varepsilon_2 - \varepsilon_m & 0 \\ 0 & 0 & \varepsilon_3 - \varepsilon_m \end{bmatrix} \tag{4.27}$$

and $[\varepsilon]_H$ is called the hydrostatic or volumetric or spherical strain tensor given by the expression

$$[\varepsilon]_H = \begin{bmatrix} \varepsilon_m & 0 & 0 \\ 0 & \varepsilon_m & 0 \\ 0 & 0 & \varepsilon_m \end{bmatrix} \tag{4.28}$$

where $\varepsilon_m$ is the mean strain

$$\varepsilon_m = \frac{\varepsilon_1 + \varepsilon_2 + \varepsilon_3}{3}. \tag{4.29}$$

Octahedral normal strain and Octahedral shear strain are given by the following equations:

$$\varepsilon_{oct} = \frac{1}{3}(\varepsilon_1 + \varepsilon_2 + \varepsilon_3) \tag{4.30}$$

$$\gamma_{oct} = \frac{2}{3}\sqrt{(\varepsilon_1 - \varepsilon_2)^2 + (\varepsilon_2 - \varepsilon_3)^2 + (\varepsilon_3 - \varepsilon_1)^2}. \tag{4.31}$$

## 4.3 Strain–Displacement Equations

Strains and displacements in a body are not one and the same. The mathematical relationships connecting the strain developed to the deformations of the body are termed strain–displacement equations. The relation between the components of strain and that of the displacement *u*, *v*, and *w* at a point along the *x*, *y*, and *z* axes are given by the following 6 equations:

$$\varepsilon_x = \frac{\partial u}{\partial x} + \frac{1}{2}\left[\left(\frac{\partial u}{\partial x}\right)^2 + \left(\frac{\partial v}{\partial x}\right)^2 + \left(\frac{\partial w}{\partial x}\right)^2\right] \tag{4.32a}$$

$$\varepsilon_y = \frac{\partial v}{\partial y} + \frac{1}{2}\left[\left(\frac{\partial u}{\partial y}\right)^2 + \left(\frac{\partial v}{\partial y}\right)^2 + \left(\frac{\partial w}{\partial y}\right)^2\right] \tag{4.32b}$$

$$\varepsilon_z = \frac{\partial w}{\partial z} + \frac{1}{2}\left[\left(\frac{\partial u}{\partial z}\right)^2 + \left(\frac{\partial v}{\partial z}\right)^2 + \left(\frac{\partial w}{\partial z}\right)^2\right] \tag{4.32c}$$

$$\gamma_{xy} = \frac{\partial v}{\partial x} + \frac{\partial u}{\partial y} + \frac{\partial u}{\partial x}\frac{\partial u}{\partial y} + \frac{\partial v}{\partial x}\frac{\partial v}{\partial y} + \frac{\partial w}{\partial x}\frac{\partial w}{\partial y} \tag{4.32d}$$

$$\gamma_{yz} = \frac{\partial w}{\partial y} + \frac{\partial v}{\partial z} + \frac{\partial u}{\partial y}\frac{\partial u}{\partial z} + \frac{\partial v}{\partial y}\frac{\partial v}{\partial z} + \frac{\partial w}{\partial y}\frac{\partial w}{\partial z} \tag{4.32e}$$

$$\gamma_{zx} = \frac{\partial u}{\partial z} + \frac{\partial w}{\partial x} + \frac{\partial u}{\partial z}\frac{\partial u}{\partial x} + \frac{\partial v}{\partial z}\frac{\partial v}{\partial x} + \frac{\partial w}{\partial z}\frac{\partial w}{\partial x}. \tag{4.32f}$$

The expressions for strain components can be simplified by considering only the first-order terms as

$$\varepsilon_x = \frac{\partial u}{\partial x},\ \varepsilon_y = \frac{\partial v}{\partial y},\ \varepsilon_z = \frac{\partial w}{\partial z},\ \gamma_{xy} = \frac{\partial v}{\partial x} + \frac{\partial u}{\partial y},\ \gamma_{yz} = \frac{\partial w}{\partial y} + \frac{\partial v}{\partial z},\ \gamma_{zx} = \frac{\partial u}{\partial z} + \frac{\partial w}{\partial x}. \tag{4.33}$$

The relations in Eq. (4.33) are considered valid if the body experiences only small deformations, that is, each derivative in Eqs. (4.32a)–(4.32f) is much smaller than unity. If the body experiences large or finite deformations or strains, higher order terms must be retained, as in Eqs. (4.32a)–(4.32f). These terms represent significant changes in the geometry of the body and thus are called geometric nonlinearities.

## 4.4 Constitutive Relations

Constitutive relations connect the stress and corresponding strain in a body. As indicated earlier, the simplest constitutive relation is the well-known Hooke's law.

Hooke's law for uniaxial loading states that the strain is proportional to the stress. In the more general case of three-dimensional loading, six components of stress and strain will be present as discussed in Sects. 4.1 and 4.2. As a natural extension of Hooke's law, each of the six stress components may be expressed as a linear function of the six components of strain and vice versa. This is called the generalized Hooke's law:

$$\begin{Bmatrix} \sigma_{xx} \\ \sigma_{yy} \\ \sigma_{zz} \\ \tau_{xy} \\ \tau_{yz} \\ \tau_{zx} \end{Bmatrix} = \begin{bmatrix} C_{11} & C_{12} & C_{13} & C_{14} & C_{15} & C_{16} \\ C_{21} & C_{22} & C_{23} & C_{24} & C_{25} & C_{26} \\ C_{31} & C_{32} & C_{33} & C_{34} & C_{35} & C_{36} \\ C_{41} & C_{42} & C_{43} & C_{44} & C_{45} & C_{46} \\ C_{51} & C_{52} & C_{53} & C_{54} & C_{55} & C_{56} \\ C_{61} & C_{62} & C_{63} & C_{64} & C_{65} & C_{66} \end{bmatrix} \begin{Bmatrix} \varepsilon_{xx} \\ \varepsilon_{yy} \\ \varepsilon_{zz} \\ \gamma_{xy} \\ \gamma_{yz} \\ \gamma_{zx} \end{Bmatrix} \tag{4.34}$$

$$\{\sigma\} = [C]\{\varepsilon\} \tag{4.35}$$

and

$$\begin{Bmatrix} \varepsilon_{xx} \\ \varepsilon_{yy} \\ \varepsilon_{zz} \\ \gamma_{xy} \\ \gamma_{yz} \\ \gamma_{zx} \end{Bmatrix} = \begin{bmatrix} D_{11} & D_{12} & D_{13} & D_{14} & D_{15} & D_{16} \\ D_{21} & D_{22} & D_{23} & D_{24} & D_{25} & D_{26} \\ D_{31} & D_{32} & D_{33} & D_{34} & D_{35} & D_{36} \\ D_{41} & D_{42} & D_{43} & D_{44} & D_{45} & D_{46} \\ D_{51} & D_{52} & D_{53} & D_{54} & D_{55} & D_{56} \\ D_{61} & D_{62} & D_{63} & D_{64} & D_{65} & D_{66} \end{bmatrix} \begin{Bmatrix} \sigma_{xx} \\ \sigma_{yy} \\ \sigma_{zz} \\ \tau_{xy} \\ \tau_{yz} \\ \tau_{zx} \end{Bmatrix} \tag{4.36}$$

$$\{\varepsilon\} = [D]\{\sigma\} \tag{4.37}$$

where $\boldsymbol{C}$ is called the constitutive matrix and $\boldsymbol{D}$ the compliance matrix.

Equations (4.35) and (4.36) represent the constitutive law for a linear elastic anisotropic and homogeneous material. Though there are 36 constants each in the matrices $[C]$ and $[D]$, because of their symmetric nature, a complete constitutive description of a general anisotropic solid requires experimental evaluation of only 21 elastic constants. Certain materials exhibit further symmetry with respect to directions within the body, so the number of material constants can be reduced from the 21 required in the anisotropic case. For example, Eq. (4.35) for an orthotropic material (material having different elastic properties along the three coordinate axes) is expressed in terms of nine constants as

$$\begin{Bmatrix} \sigma_{xx} \\ \sigma_{yy} \\ \sigma_{zz} \\ \tau_{xy} \\ \tau_{yz} \\ \tau_{zx} \end{Bmatrix} = \begin{bmatrix} C_{11} & C_{12} & C_{13} & 0 & 0 & 0 \\ C_{21} & C_{22} & C_{23} & 0 & 0 & 0 \\ C_{31} & C_{32} & C_{33} & 0 & 0 & 0 \\ 0 & 0 & 0 & C_{44} & 0 & 0 \\ 0 & 0 & 0 & 0 & C_{55} & 0 \\ 0 & 0 & 0 & 0 & 0 & C_{66} \end{bmatrix} \begin{Bmatrix} \varepsilon_{xx} \\ \varepsilon_{yy} \\ \varepsilon_{zz} \\ \gamma_{xy} \\ \gamma_{yz} \\ \gamma_{zx} \end{Bmatrix}. \tag{4.38}$$

The constitutive relations for the orthotropic material can be written in terms of Young's moduli and Poisson's ratios in three directions as

$$\varepsilon_x = \frac{\sigma_x}{E_x} - \frac{\upsilon_{yx}}{E_y}\sigma_y - \frac{\upsilon_{zx}}{E_z}\sigma_z \tag{4.39}$$

$$\varepsilon_y = \frac{\sigma_y}{E_y} - \frac{\upsilon_{xy}}{E_x}\sigma_x - \frac{\upsilon_{zx}}{E_z}\sigma_z \tag{4.40}$$

$$\varepsilon_z = \frac{\sigma_z}{E_z} - \frac{\upsilon_{xz}}{E_x}\sigma_x - \frac{\upsilon_{yz}}{E_y}\sigma_y \tag{4.41}$$

$$\gamma_{xy} = \frac{\tau_{xy}}{G_{xy}}, \gamma_{yz} = \frac{\tau_{yz}}{G_{yz}}, \gamma_{zx} = \frac{\tau_{zx}}{G_{zx}}. \tag{4.42}$$

Some of the commonly used equations in the strength of materials are reproduced below

$$\text{Bulk modulus} = \mathrm{K} = \frac{\mathrm{E}}{3(1-2\nu)} \tag{4.43}$$

$$\text{Shear modulus, G} = \frac{\mathrm{E}}{2(1+\nu)} \tag{4.44}$$

$$\text{Lame's constant, } \lambda = \frac{\nu \mathrm{E}}{(1+\nu)(1-2\nu)}. \tag{4.45}$$

## 4.5 Exercise Problems

4.5.1 Find the relationship between shear modulus, bulk modulus, and Poisson's ratio.

4.5.2 What is Poisson's ratio for a perfectly incompressible linear elastic material?

4.5.3 What is the range of values for material parameters like Young's modulus, Poisson's ratio, and Lame's constant $\lambda$?

4.5.4 What is the physical meaning of coefficients like Young's modulus, Poisson's ratio, shear modulus, and bulk modulus?

4.5.5 What is the relationship between Dilatational stress and Volumetric strain?

4.5.6 Which is more compressible—rubber or steel?

4.5.7 If Poisson's ratio of an elastic material is 0.33, what would be the ratio of the modulus of rigidity to Young's modulus?

4.5.8 Why stress tensor is called a second-order tensor?

4.5.9 How many independent elastic constants are required to describe the behavior of a homogenous, isotropic material?

4.5.10 Under the condition that Young's modulus, E, is positive, determine Lame's constant, shear modulus, and bulk modulus for Poisson's ratios of 0, 0.25, and 0.5.

4.5.11 What pressure should be applied to a lead block to reduce its volume by 15%, with bulk modulus for lead being 7 GPa?

4.5.12 Which law leads to symmetry of the stress tensor at a point in the body under equilibrium?

## Bibliography

Chandrakant S Desai, John F Abel (1987) Introduction to finite element method: a numerical method for engineering analysis. CBS Publishers & Distributors

Drucker DC (1967) Introduction to mechanics of deformable solids. McGraw-Hill Book Co, New York

Fung YC (1965) Foundations of solid mechanics. Prentice-Hall, Englewood Cliffs

Love AEH (1944) A treatise on the mathematical theory of elasticity. Dover Publications, New York

Popov EP (1968) Introduction to mechanics of solids. PrenticeHall, Englewood Cliffs

Rao JS (2011) History of rotating machinery dynamics. Springer

Sokolnikoff IS (1956) Mathematical theory of elasticity. McGraw-Hill Book Co, New York

Timoshenko S, Goodier JN (1951) Theory of elasticity. McGrawHill Book Co, New York

Wang CT (1953) Applied elasticity. McGraw-Hill Book Co., New York

# Chapter 5
# Finite Element Formulations

**P. Raveendranath**

*Nothing in life is to be feared, it is only to be understood. Now is the time to understand more, so that we may fear less*
*—Marie Curie*

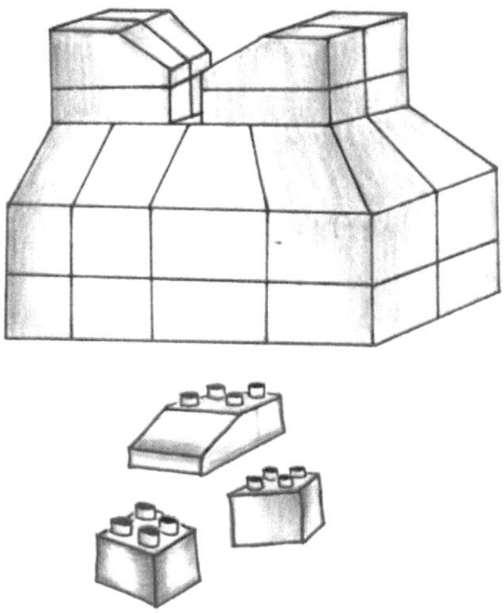

## 5.1 Introduction

Shape, size, and material completely define a structural component for engineering applications. The aim of structural design is to finalize the size, shape, and material

P. Raveendranath (✉)
Department of Aerospace Engineering, Indian Institute of Space Science and Technology, Valiamala, Thiruvananthapuram, Kerala 695547, India
e-mail: raveendranath@iist.ac.in; raveendranathiist@gmail.com



S. U. Nair and S. Somanath (eds.), *Introduction to Finite Element Analysis*,
https://doi.org/10.1007/978-981-19-7989-7_5

of a component, which performs a set of desired functions, when subjected to a set of loads and constraints on the boundary, under service conditions. Structural analysis and structural testing are employed to assess the suitability and reliability of a particular design to meet specific requirements.

*Analysis versus Testing*: In the iterative design cycle of an engineering product, analysis and testing have a crucial role in validating the design and suggesting modifications. However, testing may not be practicable in certain cases. Also, testing provides only a limited amount of information, and the process of testing is expensive and slow, causing a delay in the design cycle. Analysis can provide a quick and wide range of information pertaining to the validity of the design in each cycle of modifications.

Theoretical methods for structural analyses are based on classical solutions and are applicable only to simple geometries and simple loadings. Numerical methods have been employed to solve engineering problems involving arbitrary structural shapes and loads. Among the numerical methods in the field of structural analysis, finite element methods (FEM) have emerged as the most popular and powerful tool.

## 5.2 Functional Overview of a Finite Element Solver

A typical application of FEM to an engineering problem is to determine the distribution of a primary field variable (such as displacement in solid mechanics, temperature in heat transfer, voltage in electrical conduction, and field potential in electrostatics problems) when the continuum is subjected to some prescribed 'loads' (like force, heat-flux, current, etc.). A general procedure for FE analysis to solve a continuum problem consists of the following steps.

***Discretization of the continuum***: The basis of the finite element method is the representation of an arbitrarily shaped continuum as an assemblage of a large number of regular shaped pieces. The process of dividing the continuum into small units is called discretization. Figure 5.1 shows an assemblage of triangular and quadrilateral segments to form a continuum. These small segments are called the finite elements which are interconnected at the discrete corner points called nodes. Thus, each element can be defined by a connectivity sequence of numbered nodes, as shown in Fig. 5.1.

***Selection of the field interpolation functions***: By virtue of their regular shapes, formulation of governing equations is simpler for these individual elements. Simple functions (usually polynomials) are chosen to approximate the distribution of the primary field variable over each element.

***Computation of element matrices***: Applying appropriate variational methods or weighted residual methods pertaining to the field of application, the equilibrium equations can be formed for an individual element, in the form of matrices, in terms of unknown values of the field variable at the nodes of that element.

***Assembly of element matrices***: The equations (contributions) of individual elements can be assembled to form a global set of equations (which represent the

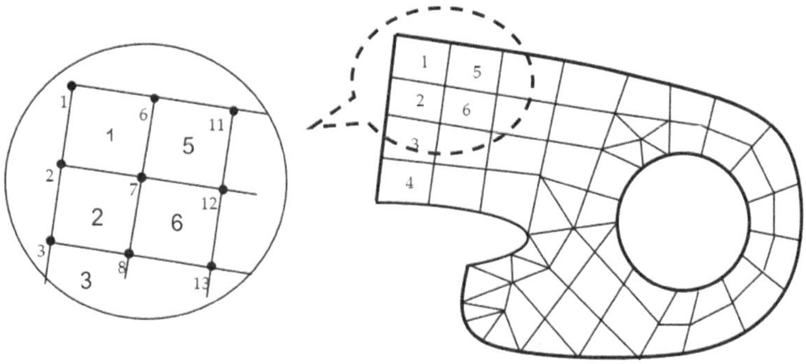


**Fig. 5.1** Discretization of a continuum into a number of elements defined by discrete node points: Element 1: (1, 2, 7, 6); Element 2: (2, 3, 8, 7); Element 5: (6, 7, 12, 11)

whole structure) on which any prescribed boundary conditions (i.e. known values of the field variable on some boundary nodes) can be imposed. By assembling the individual element stiffness matrices, the global stiffness matrix of the entire structure is obtained.

***Solve global equations***: The resulting set of equations is now solved for the unknown values of the field variable at nodes over the whole structure.

***Computation of derived quantities***: Having solved the values of the primary field variable at nodes for the problem, other derived quantities (strains, stresses, resultant forces or temperature-gradient, etc.) at any point within a finite element in the continuum can be computed from the values of the primary variable at nodes for that particular element, using appropriate equations applicable for the domain.

In the sections that follow, a detailed description of basic equations and finite element formulations for structural stress analysis are presented.

## 5.3 The Domain of Linear Structural Analysis

***Requirements and design criteria***: For any structural design to meet the functional requirements safely and reliably, it is a common practice to design structural components to satisfy some criteria. Typical examples of design criteria are as follows.

The deflection of the structure should be within a prescribed limit and the maximum stress developed at any point of the structure should not exceed a particular limit (usually yield stress of the material). Also known as static analysis, stress analysis is carried out to determine the distribution of displacements, strains, and stresses in the structure for the given loads and boundary conditions. Displacements are the primary output of the analysis. Strains are computed from the spatial derivative of displacements. Stresses are computed using strains and material properties of the element. The basic equation solved here is $[K]\{u\} = \{F\}$, where [K] is the global stiffness matrix, {F} is the global load vector, and {u} is the global vector of displacements at nodes.

The frequency of free vibration of the structure should be considerably far from the frequencies of external sources of excitation or disturbance, encountered by the structure under operating conditions. Free vibration analysis computes the natural frequencies of vibration and corresponding mode shapes. Design constraints on the frequencies are imposed to keep the natural frequencies of the structure well separated from the frequencies of external excitations or disturbances. The FE procedures compute the stiffness matrix [K] and the mass matrix [M] for the structure and solve the eigenvalue problem ([K] - λ[M]) {u} = 0. The first non-zero eigenvalue λ represents $4\pi^2 f^2$, where f is the fundamental frequency of free vibration. The eigenvector {u} indicates the shape of the vibration mode.

The structure under the applied loads should not become unstable and catastrophically collapse due to stresses developed in the structure. Buckling is a mode of structural failure characterized by a sudden change in the intended shape/configuration of slender structures when subjected to excessive compressive loads. This can happen even when the stresses are below the yield stress of the material. When buckling occurs, the normal deformed shape (the equilibrium state) under the action of applied loads spontaneously changes and the structure acquires a new equilibrium state. Under this condition, the structure is considered as functionally failed. Buckling analysis makes use of the stress distribution obtained from previous stress analysis to compute Stress stiffness matrix $[K_\sigma]$, which is a measure of the weakening of the structure due to induced stresses under the applied load. Then, an eigenvalue problem is solved as $([K] + \lambda [K_\sigma]) \{u\} = 0$, where the eigenvalue λ is the critical load factor. The product of the applied loads with λ gives the critical load, beyond which the structure will fail due to buckling. If $\lambda > 1$, the structure is safe under the applied load level. $0 < \lambda < 1$ indicates that the applied load is already higher than the critical load and the structure cannot withstand it. $\lambda < 0$ indicates that the structure will fail if the applied load is reversed in direction. The eigenvector {u} represents displacements at nodes (usually normalized) which representatively indicate the buckled shape of the structure.

## 5.4 Displacement-Based FE Formulations

The basis of the method is the approximate representation of any arbitrary-shaped body by an assemblage of regular-shaped elements, which are called finite elements. The vertex points of these finite elements are called nodes. For each regular shaped element, a relationship between a vector of forces at nodes and a vector of displacements at nodes can be related through a stiffness matrix computed for the element. The element stiffness matrix and the element load vector for each element are separately computed and assembled into a system of global linear algebraic equations which represents the behavior of the complete structure.

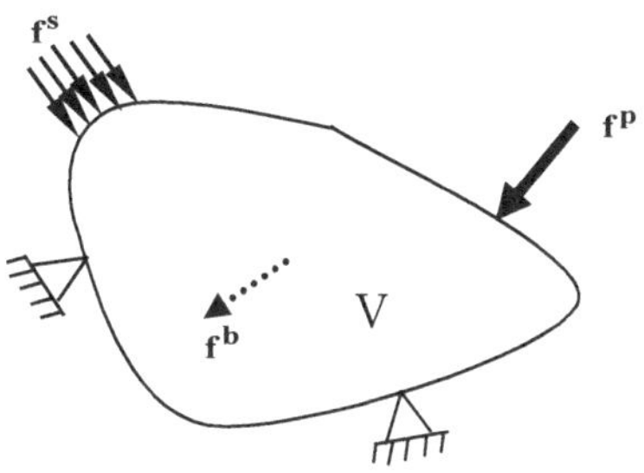


**Fig. 5.2** A body of arbitrary shape subjected to externally applied loads

The formulation of the element stiffness matrix is based on the theory of elasticity. The accuracy of the discrete values of solved displacement at nodes depends on the total number of spatial density of finite elements used for discretizing the structure. Very coarse mesh or very distorted shapes of elements can produce inaccurate results. In the event of very fine meshes of fairly regular shaped elements, finite element results converge to the exact solutions. The use of the natural coordinate system and the concept of isoparametric elements (elements with the same shape functions or interpolation functions used for defining both deformed shape and displacement field in the element) facilitate modeling of curved boundaries of the structure by a smaller number of elements. Here, the geometry as well as displacements of the element are interpolated in terms of spatial functions of the same order. Numerical integration procedures are used for the computation of element stiffness matrices and consistent element load vectors.

Consider a three-dimensional body supported at some points on the boundary as shown in Fig. 5.2. The body is subjected to externally applied loads, as shown. Now, the meaning of a few terms used in this chapter are given below.

***Loads***: Loads are what 'force' a structure to deform. Depending on the physical nature, loads can be classified as follows:

***Point load*** ($\mathbf{f}^{\mathbf{p}}$): A load concentrated on a very small area compared to the surface area of the structure. For all practical purposes, it is assumed to act at a point on the structure. Locally applied forces and reaction forces from small area supports are typically considered as point loads on a structure. An example is a person standing at the free end of a diving board. Point load is expressed in the units of force (Newton, in SI units).

***Surface load*** (t): Often known as 'traction', it is a distributed load acting over a considerably large area of the surface of the structure. Wind loads on tall structures and aerodynamic loads on aircraft and launch vehicles are typical examples of surface loads. It is expressed as force per unit area (Newton per square-metre, in SI units).

***Body Load*** (b): It is a load distributed throughout the volume of the body. A common example of this kind is gravitational force. Other examples include magnetic force and inertial force. Every particle of the structure is acted upon by these forces. It is expressed as force per unit volume (Newton per cubic-metre, in SI units).

***Displacements***: Structures deform under the action of forces. The deformation can be described by displacement fields expressed as a function of the spatial coordinates X–Y–Z. The components of the displacement field along the three coordinate

**Fig. 5.3** Change of shape during extensional and shear deformations

directions X, Y, and Z are *u*(x, y, z), *v*(x, y, z), and *w*(x, y, z), respectively. The displacement fields within the structure can be represented by the vector:

$$\{u\}^T = \{u, v, w\}. \tag{5.1}$$

***Strains***: Strains are measures of the rate of change of displacement fields w.r.t. spatial coordinates. There are two types of strains: (i) Extensional strain and (ii) Shear strain. The corresponding deformations can be shown as given in Fig. 5.3, in the simple case of a two-dimensional deformation.

For small deformations, the three-dimensional state of strain at a point can be represented by a vector containing the six strain components as

$$\{\varepsilon\}^T = \left\{\varepsilon_x, \varepsilon_y, \varepsilon_z, \gamma_{xy}, \gamma_{yz}, \gamma_{zx}\right\} \tag{5.2}$$

where $\varepsilon_x, \varepsilon_y, \varepsilon_z$ are the extensional strains given by

$$\varepsilon_x = \frac{\partial u}{\partial x}; \quad \varepsilon_y = \frac{\partial v}{\partial y}; \quad \varepsilon_z = \frac{\partial w}{\partial z} \tag{5.3}$$

and $\gamma_{xy}, \gamma_{yz}, \gamma_{zx}$ are the shear strains given by

$$\gamma_{xy} = \frac{\partial u}{\partial y} + \frac{\partial v}{\partial x}; \quad \gamma_{yz} = \frac{\partial v}{\partial z} + \frac{\partial w}{\partial y}; \quad \gamma_{zx} = \frac{\partial w}{\partial x} + \frac{\partial u}{\partial z}. \tag{5.4}$$

***Stresses***: Corresponding to the above strain components, the three-dimensional state of stress at a point can be represented by a vector containing the six stress components as

$$\{\sigma\}^T = \left\{\sigma_x, \sigma_y, \sigma_z, \tau_{xy}, \tau_{yz}, \tau_{zx}\right\}. \tag{5.5}$$

***Material constitutive matrix***: Material constitutive matrix [E] relates the state of stress to the state of strain at a point in the body. They are also known as stress–strain relations, as

$$\{\sigma\} = [\mathrm{E}]\{\varepsilon\}. \tag{5.6}$$

For a linear elastic isotropic material, [E] is a function of the two independent material properties viz. Young's modulus $E$ and Poisson's ratio ($\upsilon$) and is given by

$$[\mathrm{E}] = \frac{E}{(1+\upsilon)(1-2\upsilon)}\begin{bmatrix} 1-\upsilon & \upsilon & \upsilon & 0 & 0 & 0 \\ & 1-\upsilon & \upsilon & 0 & 0 & 0 \\ & & 1-\upsilon & 0 & 0 & 0 \\ & & & \frac{1-2\upsilon}{2} & 0 & 0 \\ & sym. & & & \frac{1-2\upsilon}{2} & 0 \\ & & & & & \frac{1-2\upsilon}{2} \end{bmatrix}. \tag{5.7}$$

***Strain Energy***: The elastic strain energy (U) stored in a body is given by

$$\mathrm{U} = \frac{1}{2}\int_{\upsilon}\{\varepsilon\}^{T}\{\sigma\}\mathrm{dv}. \tag{5.8}$$

***Potential Energy***: The potential energy $\Pi_p$ of a strained body is the algebraic sum of the strain energy stored in the body and the potential lost by the loads due to deformation. Hence, it can be written as

$$\Pi_p = \frac{1}{2}\int_{\upsilon}\{\varepsilon\}^{T}\{\sigma\}\mathrm{dv} - \int_{\upsilon}\{u\}^{T}\{\mathrm{b}\}\mathrm{dv} - \int_{s}\{u\}^{T}\{\mathrm{t}\}\mathrm{ds} - \{u_i\}^{T}\left\{\mathrm{f}^{P}\right\} \tag{5.9}$$

where {b} represents the vector of body-load component fields acting over the volume $v$, {t} represents the vector of surface-load (traction) component fields acting over the surface $s$, and $\mathrm{f^p}$ represents the vector of point-load components acting at discrete points on the structure.

## 5.5 Classification of Structural Finite Elements

Based on the geometry of the structure and application, the most commonly used finite elements can be classified as follows:

### *5.5.1 Three-Dimensional Elements*

These are the most general finite elements, which are employed to model bulk structures, where simplification of the geometric modeling by means of one- and

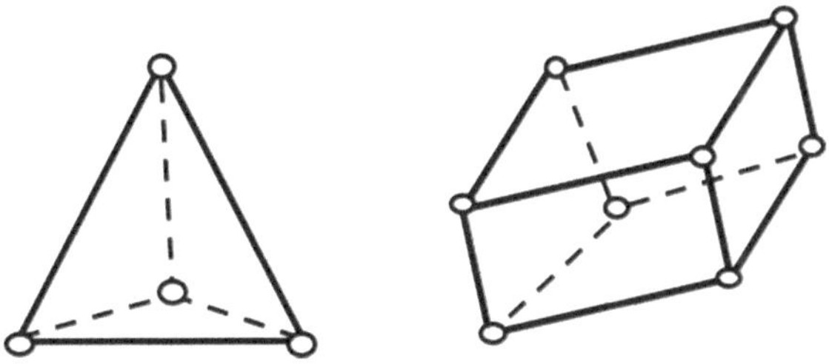

**Fig. 5.4** Four-node tetrahedron and eight-node hexahedron elements

two-dimensional elements is not meaningful. The whole volume of the structure is modeled using 3D-shaped elements. The elements can be oriented in three-dimensional space (X–Y–Z).

The tetrahedron and hexahedron elements shown in Fig. 5.4 are the most common elements of this kind.

### 5.5.2 Two-Dimensional Elements

The following specializations exist for two-dimensional elements:

***Plane stress elements***: These elements are used to model thin structures characterized by very small dimensions in the Z-direction, in comparison with the structural dimensions in X- and Y-directions. A thin plate stressed by loads in its plane is a typical example of this kind. The state of stress does not vary across the thickness. The complete structure is modeled by an assemblage of plane stress elements in the X–Y plane.

***Plane strain elements***: These elements are used to model long bodies whose geometry of cross section and loading do not vary significantly in the longitudinal direction (Z). A long cylindrical pipe carrying pressurized fluid is a typical example of this kind. Only a representative cross section of such structures, in the transverse X–Y plane, needs to be discretized and analyzed to understand the deformation and stress distribution in the structure.

***Axisymmetric solid elements***: These elements are used to model structures whose geometry can be generated by revolving a planar cross section, say in an X–Z plane, around an axis (called the axis of revolution), usually taken as the vertical Z-axis. The horizontal X-axis is also known as the radial axis. A vertical conical nozzle and

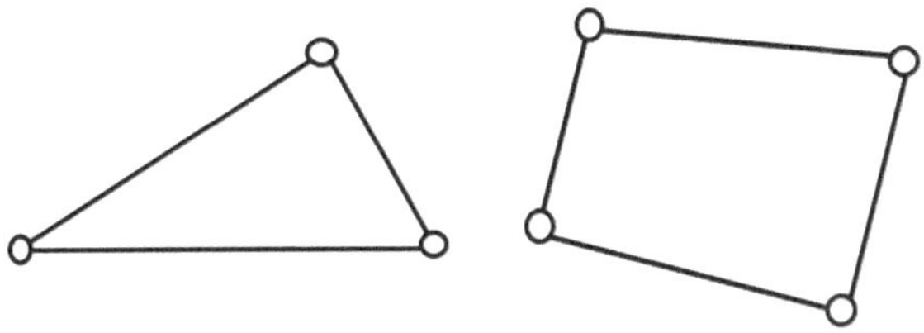

**Fig. 5.5** 3-node triangular and 4-node quadrilateral two-dimensional elements

a vertical cylindrical tube are typical examples of this kind. These structures can be generated by revolving their wall cross section about their longitudinal (vertical) axis, in the circumferential direction (θ). Such a structure is said to be axisymmetric about Z-axis. X–θ–Z forms a right-handed coordinate system, θ being tangential to the circular surface formed by the revolution of cross section about the Z-axis. If the loads acting on the structure are also axisymmetric about Z-axis, then the deformations, strains, and stresses in the structure will also be axisymmetric. A deformation at a point on this cross section represents (being equal) the deformation of all the points on the same circumferentially revolved line at that point of the structure. In other words, the radial and longitudinal deformations do not vary at any point, in θ-direction. Consequently, only the planar cross section in the X–Z plane needs to be modeled to analyze the structure.

***Plate bending elements***: These elements are used to model thin flat/curved structures oriented in three-dimensional space subjected to in-plane and out-of-plane loads. Structures like pressure vessels, aircraft wings, etc. which are made out of sheet metals, are typical examples of this kind.

### 5.5.3 *One-Dimensional Elements*

These elements are used to model slender structural members, which have a predominant dimension (called axial length, in the local x-direction). A cut (in the local y–z plane) normal to this longitudinal axis at any point reveals its cross section which has a much smaller dimension compared to its length. These are the simplest structural elements denoted by a line, representing the centroid axis of the slender structural component. They can be oriented in three-dimensional space. Examples of this type of elements are as follows.

***Truss elements***: Also referred to as link or bar elements, their ends can be connected to ground or adjacent members through pin or hinge or ball joints. Hence, in a framework of truss elements where the loads are applied at the joints, the truss members take only tensile or compressive forces along their centroid axis and always remain straight, even after deformation. They cannot sustain moments or shear loads. Each truss element shall be designated by node coordinates of the two end-nodes and the area of cross section. When used for analyzing a structure oriented in an X–Y–Z global coordinate system, it shall have only the translational degrees of freedom (U, V, W) at each node.

**Fig. 5.6** A two-node one-dimensional element

***Beam elements***: Beam elements can take tensile, compressive, bending, shear, and torsional loads. It is convenient to visualize a beam as a structure made by extruding a small planar area (cross section of the beam) along a line (axis of the beam) normal to the area and passing through its centroid. This longitudinal axis forms the x-axis for the local element coordinate system defined at any cross section of the beam. The beam element is described by its end (node) coordinates and cross-sectional properties, typically (i) Area, (ii) Torsional constant of the area about the x-axis, and (iii) Second moments of area of cross section about chosen local transverse axis (y, z) directions on the cross-sectional plane. It should be noted that all the cross-sectional properties are described in the local element coordinate system x–y–z. The most popular beam element is the 2-node (linear) beam element. The 3-node beam element shall have its third node at the mid-point of its axis and can represent curved beams more accurately.

## 5.6 Assumed Displacement Fields and Shape (Interpolation) Functions

The concept of shape functions is evolved from the basic requirement of expressing the distribution of a field variable within a finite element, in terms of node values of the same field variable. For example, in the case of a 4-node plane stress finite element, the displacement field $u(x, y)$ can be expressed as

$$u(x, y) = N_1(x, y)u_1 + N_2(x, y)u_2 + N_3(x, y)u_3 + N_4(x, y)u_4. \tag{5.10}$$

The expressions $N_i(x, y)$ are called the shape functions and $u_1, u_2, u_3, u_4$ are the values of $u(x, y)$ at the four nodes. Generally, these shape functions are derived for each type of finite element, based on an appropriately assumed displacement field for $u(x, y)$. The same shape functions can be used for expressing $v(x, y)$ in a similar manner, in terms of $v_1, v_2, v_3,$ and $v_4$.

The assumed displacement fields are chosen based on the dimension of the analysis domain (1- or 2- or 3D) and the number of nodes available in the element. Generally, polynomial functions of the appropriate degree are most widely used for this purpose. This representation may not be exact for the domain (size) of a finite element over which it is applied. However, as the number of finite elements used for idealizing the structure is increased, the size of each finite element decreases and the assumed field naturally approaches the accuracy of the exact distribution of the displacement field. Hence, in the finite element method, which is a numerical technique, the solution progressively converges to the exact solution as the size of the elements is decreased by increasing the number of elements used for meshing the structure.

**Fig. 5.7** A 2-node line element

For example, for a one-dimensional 2-node line element of length $L$ as shown in Fig. 5.7, a linear polynomial displacement field is chosen as

$$u(x) = a_0 + a_1 x. \tag{5.11}$$

If we assign $u_1$ and $u_2$ to denote the node displacements (values of the displacement field $u(x)$ at nodes 1 and 2, respectively), the arbitrary coefficients $a_0$ and $a_1$ can be computed using the coordinate value ($x_1 = 0$ and $x_2 = L$) of the nodes. Thus, we can write

$$u_1 = a_0 \qquad u_2 = a_0 + a_1 L. \tag{5.12}$$

Solving for $a_0$ and $a_1$, we get $a_0 = u_1$ and $a_1 = (u_2 - u_1)/L$. Substituting these values in Eq. (5.11) and rearranging the terms, we get

$$u(x) = N_1(x)u_1 + N_2(x)u_2 \tag{5.13}$$

where $N_1(x) = (1 - x)/L$ and $N_2(x) = x/L$.

The expressions $N_1$ and $N_2$ are called the shape functions, which relate a displacement field to its own node values and $L = x_2 - x_1$ is the length of the element.

Similarly, for a 4-node quadrilateral planar (2D) element, the appropriate assumed polynomial field for generating the shape functions shall be $u(x) = a_0 + a_1 x + a_2 y + a_3 xy$. The equations to be solved for calculating the unknown coefficients are generated by writing the equations for node displacements $u_1, u_2, u_3$, and $u_4$ using the coordinates $(x_1, y_1)$, $(x_2, y_2)$, $(x_3, y_3)$, and $(x_4, y_4)$. It may be noted that the number of terms in the assumed polynomial shall be equal to the number of nodes for the element so that the number of equations and the number of unknowns are equal and hence the unknown coefficients can be solved uniquely.

For convergence of the finite element method, the assumed polynomial field must satisfy the following requirements:

(i) **Continuity requirement**: While sub-dividing the entire structure into elements, it is important to ensure that the displacement field is continuous not only inside the element but also across the element boundaries. Based on the application of the element formulation (rod/link, beam, plate/shell, or solid element formulations), one order less than the highest order derivative appearing in the integral formulation of the element equations must be continuous across the element boundaries.

Based on this, for second-order problems of elasticity (rod, Timoshenko beam, and Mindlin Plate theories), where strain is defined as the first derivative of displacement fields, only the displacement fields need to be continuous across the nodes. The node values of the field variables are node degrees of freedom which are common to all the elements adjacent to the node. They are called $C^0$ continuous elements. Similarly, for fourth-order problems of elasticity (e.g. Euler–Bernoulli beam formulation), where strain depends on derivatives of displacement fields, both the transverse displacement field and its derivative need to be continuous across the element boundaries. They are called $C^1$ continuous elements.

(ii) **Completeness requirement**: The completeness requirement suggests that the assumed polynomials used for approximating the displacement field should be such that the resulting element can reproduce a state of rigid body motion (zero strain) and a state of constant strain, within the finite element. Satisfaction of this requirement ensures the smooth convergence of the finite element solution to the exact solution, as we increase the number of elements used for discretizing the structure. In a structure, there may be regions that undergo only rigid body displacements and hence remain unstrained. Also, as we increase the number of elements tending to infinity, the size of elements tends to zero, in which case the strain distribution tends to be constant within the element. These factors suggest the mandatory inclusion of a constant term and a linear term in the assumed displacement polynomial viz. 1 and $x$ in a one-dimensional $C^0$ problem and 1, $x$, $y$ in a two-dimensional $C^0$ problem.

In the procedure described above, the shape functions are derived using the actual coordinate values ($x$, $y$, $z$) of the nodes of the elements in the Cartesian coordinate system. This demands that the shape functions applicable to each finite element be computed using the node coordinates of the specific element. A better approach usually employed is the use of a local 'natural coordinate system' to express the shape functions in terms of normalized coordinates ($\xi$, $\eta$, $\zeta$), where the dimensions of the elements are normalized.

## 5.7 Natural Coordinate System

A natural coordinate system is a curvilinear coordinate system with its center at the geometric centroid of the element. Natural coordinates are local coordinates that can vary in a range between zero and unity. Natural coordinates are used to simplify the numerical evaluation of analytical integrals in the element equations.

For line, quadrilateral, and hexahedral-based elements, in each coordinate direction, its value always extends from $-1$ to 1, irrespective of the actual size (length) of the element in that direction. It is equivalent to mapping a length (1D) or an area (2D) or a volume (3D) in the Cartesian x–y–z coordinate system to a corresponding natural

**Fig. 5.8** One-dimensional mapping of a line to natural coordinate system

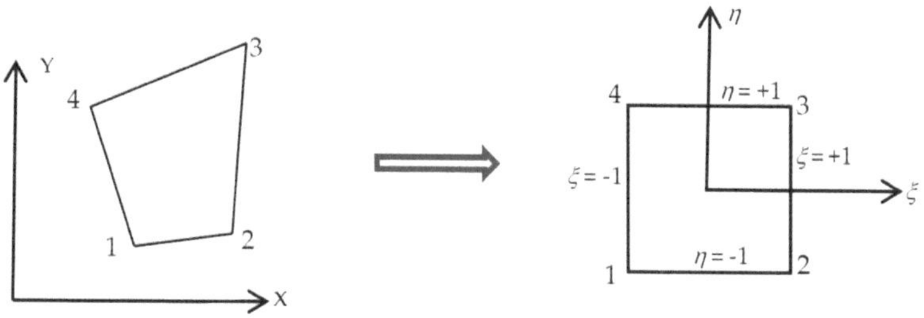


**Fig. 5.9** Two-dimensional mapping of a quadrilateral to natural coordinate system

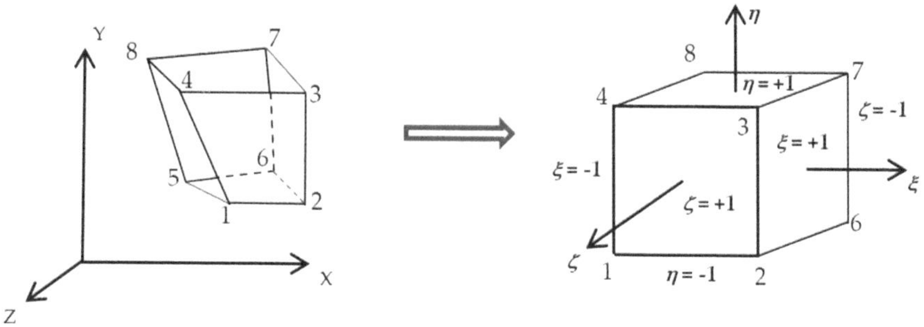


**Fig. 5.10** Three-dimensional mapping of hexahedron to natural coordinate system

coordinate system $\xi$-$\eta$-$\zeta$ as shown in Figs. 5.8, 5.9, and 5.10. It may be noticed that a general triangle is mapped into a right-angled triangle, a general quadrilateral is mapped into a square, and a general hexahedron is mapped into a cube in the natural coordinate system.

**Shape functions in natural coordinate system**:

Using the assumed displacement field in the natural coordinate system, and following similar procedures described in Sect. 5.6, a set of standard shape functions can be derived for field interpolation for a class of one-, two-, and three-dimensional finite elements, which are listed below (Table 5.1).

**Table 5.1** Assumed displacement field and shape functions in natural coordinate system

| Dimension | Assumed field | Shape functions | Element Type |
|---|---|---|---|
| 1D | $u = a_0 + a_1\xi$ | $N_1(\xi) = \frac{1}{2}(1-\xi)$; $N_2(\xi) = \frac{1}{2}(1+\xi)$ | Two-node Rod, Beam, and Axisymmetric shell |
| 2D 4-node | $u = a_0 + a_1\xi + a_2\eta + a_3\xi\eta$ | $N_1 = (1-\xi)(1-\eta)/4$<br>$N_2 = (1+\xi)(1-\eta)/4$<br>$N_3 = (1+\xi)(1+\eta)/4$<br>$N_4 = (1-\xi)(1+\eta)/4$ | Quadrilateral 4-node Plane stress, Plane strain, and Plate/Shell |
| 3D | $u = a_0 + a_1\xi + a_2\eta + a_3\zeta + a_4\xi\eta\zeta + a_5\xi^2 + a_6\eta^2 + a_7\zeta^2$ | $N_1 = (1-\xi)(1-\eta)(1-\zeta)/8$<br>$N_2 = (1+\xi)(1-\eta)(1-\zeta)/8$<br>$N_3 = (1+\xi)(1+\eta)(1-\zeta)/8$<br>$N_4 = (1-\xi)(1+\eta)(1-\zeta)/8$<br>$N_5 = (1-\xi)(1-\eta)(1+\zeta)/8$<br>$N_6 = (1+\xi)(1-\eta)(1+\zeta)/8$<br>$N_7 = (1+\xi)(1+\eta)(1+\zeta)/8$<br>$N_8 = (1-\xi)(1+\eta)(1+\zeta)/8$ | 8-node hexahedron (Brick) |

### Jacobian matrix of transformation

In finite element methods, a Cartesian coordinate system (global or local) is used for the definition of strains (e.g. $\partial u(x, y)/\partial x$ and $\partial v(x, y)/\partial y$). However, a natural coordinate system is used to generate shape functions (e.g. $N_1(\xi, \eta)$, $N_2(\xi, \eta)$) which interpolate a displacement field (e.g. $u(x, y)$,$v(x, y)$) in the Cartesian coordinate system in terms of its own node values (e.g. $u_1$, $u_2$, $v_1$, and $v_2$). This leads to the formation of associated matrices with terms in natural coordinates, which have to be integrated to get the stiffness matrix. The limits of these natural coordinates being $-1$ to $+1$, naturally facilitate the use of *numerical integration* procedures (described later) to compute the stiffness matrix for the element. However, these matrices to be integrated now involve derivative of shape functions in natural coordinates with respect to the Cartesian coordinates. This requires a relation between derivatives w.r.t. Cartesian coordinates and derivatives w.r.t. natural coordinates. This relation is essentially provided by the Jacobian matrix [J], which contains the derivative of Cartesian coordinates w.r.t. natural coordinates, as given below.

**For one-dimensional analysis**

$$\left\{\frac{\partial Ni}{\partial \xi}\right\} = [\mathrm{J}]\left\{\frac{\partial N_i}{\partial x}\right\}; \quad [\mathrm{J}] = \left[\frac{\partial x}{\partial \xi}\right]; \quad dx = |[\mathrm{J}]|d\xi = J\ d\xi. \tag{5.14}$$

**For two-dimensional analysis**

$$\begin{Bmatrix} \frac{\partial Ni}{\partial \xi} \\ \frac{\partial Ni}{\partial \eta} \end{Bmatrix} = [\mathrm{J}] \begin{Bmatrix} \frac{\partial N_i}{\partial x} \\ \frac{\partial N_i}{\partial y} \end{Bmatrix}; \quad [\mathrm{J}] = \begin{bmatrix} \frac{\partial x}{\partial \xi} & \frac{\partial y}{\partial \xi} \\ \frac{\partial x}{\partial \eta} & \frac{\partial y}{\partial \eta} \end{bmatrix}; \quad dxdy = |[\mathrm{J}]| d\xi d\eta = J\ d\xi d\eta. \tag{5.15}$$

**For three-dimensional analysis**

$$\begin{Bmatrix} \frac{\partial Ni}{\partial \xi} \\ \frac{\partial Ni}{\partial \eta} \\ \frac{\partial Ni}{\partial \eta} \end{Bmatrix} = [\mathrm{J}] \begin{Bmatrix} \frac{\partial N_i}{\partial x} \\ \frac{\partial N_i}{\partial y} \\ \frac{\partial N_i}{\partial y} \end{Bmatrix}; [\mathrm{J}] = \begin{bmatrix} \frac{\partial x}{\partial \xi} & \frac{\partial y}{\partial \xi} & \frac{\partial z}{\partial \xi} \\ \frac{\partial x}{\partial \eta} & \frac{\partial y}{\partial \eta} & \frac{\partial z}{\partial \eta} \\ \frac{\partial x}{\partial \zeta} & \frac{\partial y}{\partial \zeta} & \frac{\partial z}{\partial \zeta} \end{bmatrix}; dxdydz = |[\mathrm{J}]| d\xi d\eta d\zeta = J\ d\xi d\eta d\zeta. \tag{5.16}$$

## 5.8 Use of Principle of Minimum Potential Energy

The variational principle employed here to derive the finite element matrices is the *principle of minimum potential energy*, which states that "Of all possible displacement configurations that satisfy the constraints and kinematic boundary conditions, the equilibrium configuration assumed by a body is the one which makes the potential energy minimum". The potential energy for an elastically strained body subjected to body forces, surface forces, and point forces is computed as

$$\begin{Bmatrix} \frac{\partial Ni}{\partial \xi} \\ \frac{\partial Ni}{\partial \eta} \\ \frac{\partial Ni}{\partial \zeta} \end{Bmatrix} = [\mathrm{J}] \begin{Bmatrix} \frac{\partial N_i}{\partial x} \\ \frac{\partial N_i}{\partial y} \\ \frac{\partial N_i}{\partial z} \end{Bmatrix}; [\mathrm{J}] = \begin{bmatrix} \frac{\partial x}{\partial \xi} & \frac{\partial y}{\partial \xi} & \frac{\partial z}{\partial \xi} \\ \frac{\partial x}{\partial \eta} & \frac{\partial y}{\partial \eta} & \frac{\partial z}{\partial \eta} \\ \frac{\partial x}{\partial \zeta} & \frac{\partial y}{\partial \zeta} & \frac{\partial z}{\partial \zeta} \end{bmatrix}; dxdydz = |\mathrm{J}| d\xi d\eta d\zeta. \tag{5.17}$$

Considering a three-dimensional finite element, the assumed displacement fields *u(x,y,z), v(x,y,z), and w(x,y,z)* can be interpolated using standard *shape functions* ($N_i$, $i = 1$ to $n$, the number of nodes) in terms of their respective node values as

$$\{u\} = [\mathrm{N}]\{\mathrm{d}\} \tag{5.18}$$

In the above equations, $u^T = \{u, v, w\}$, the vector of displacement fields, and $\{\mathrm{d}\} = \{\mathrm{u}_1, \mathrm{v}_1, \mathrm{w}_1, \ldots\ldots\ldots\mathrm{u}_\mathrm{n}, \mathrm{v}_\mathrm{n}, \mathrm{w}_\mathrm{n}\}$, the vector of node displacements. [N] is called the shape function matrix.

The stress fields and the strain fields for the element can now be written as

$$\{\sigma\} = [\mathrm{E}]\{\varepsilon\} \text{ and } \{\varepsilon\} = [\mathrm{B}]\{\mathrm{d}\}. \tag{5.19}$$

In the above equation, [E] is the material constitutive matrix, which relates the vector of stress fields to vector strain fields. [B] is the strain–displacement matrix that relates the vector of strain fields to the node displacement vector. Naturally, the matrix [B] shall comprise the terms which are spatial derivatives of the shape functions.

Substituting Eqs. (5.18) and (5.19) in the expression for potential energy (5.17), we get

$$\pi_p = \frac{1}{2}\left\{\int_v \{d\}^T[B]^T[E][B]\{d\}dv - \right.$$
$$\left.\int_v \{d\}^T[N]^T\{b\}dv - \int_s \{d\}^T[N]^T\{t\}ds - \{d\}^T\{f^p\}\right\}. \quad (5.20)$$

By virtue of the principle of minimum potential energy, the first variation of total potential energy should be zero for every virtual displacement $\delta u_i$:

$$d\Pi_p = \delta d^T\left(\int_V [B]^T[E][B]\{u\}dv - \int_V [N]^T\{b\}dv - \int_S [N]^T\{t\}ds - \{f^p\}\right) = 0. \quad (5.21)$$

This condition demands that the terms within the bracket should vanish, which implies that

$$[k]\{u\} = \{f\} \quad (5.22)$$

where the element stiffness matrix $[k] = \int_v [B]^T[E][B]dv$, and the element load vector is $\{f\} = \int_v [N]^T\{b\}dv + \int_s [N]^T\{t\}ds + \{f^p\}$.

where {b} is the vector of body force component fields and {t} is the vector of traction component fields in the respective directions of the coordinate system. Evaluation of the integral expressions of [k] and {f} is carried out through the Gaussian numerical integration procedure. The *element stiffness matrix* and *element load vector* are computed for each finite element and assembled together to form the *global stiffness matrix* and the *global load vector*, respectively. The global matrix equations are solved for the unknown *global displacement vector at nodes.* Thus, the distribution of displacement fields in the whole structure is now available to us as their values at discrete points (nodes). As the number of elements employed for the discretization of the structure is increased, the finite element solution tends to converge to the exact solution. Strains and stresses at any point on the structure can now be computed using Eq. (5.19), using the strain–displacement matrix, [B] and the material constitutive matrix, [E].

## 5.9 Numerical Integration for Stiffness Matrix Computation

It is always advantageous to use a natural coordinate system to formulate finite element equations because the shape functions become independent of the geometric parameters/dimensions. This leads to the requirement of a technique to evaluate integral quantities (e.g. Stiffness matrix) for each element whose domain spans within

**Table 5.2** Gauss points and weights for Gaussian quadrature

| No. of points, $n$ *(order of integration)* | Location, $\xi_i$ | Weights, $W_i$ |
|---|---|---|
| $n = 1$ | $\xi_1 = 0.0$ | $W_1 = 2.0$ |
| $n = 2$ | $\xi_1 = +\sqrt{1/3}$ | $W_1 = 1.0$ |
| | $\xi_2 = -\sqrt{1/3}$ | $W_2 = 1.0$ |
| $n = 3$ | $\xi_1 = -\sqrt{3/5}$ | $W_1 = 5/9$ |
| | $\xi_2 = 0.0$ | $W_2 = 8/9$ |
| | $\xi_3 = +\sqrt{3/5}$ | $W_3 = 5/9$ |

the boundaries from $-1$ to $+1$, in each spatial dimension. In the case of computation of stiffness matrix, the integrands are made up of products of polynomials in terms of natural coordinates.

Consider the evaluation of one-dimensional integral of the following form:

$$I = \int_{-1}^{1} f(\xi) d\xi. \tag{5.23}$$

The numerical integration scheme aims to convert the analytical integration into a weighted summation of the value of the integrand evaluated at some optimally selected points. The number of selected points can be 1 or 2 or 3 or more. The optimum number of selected points depends on the degree of polynomial which has to be integrated. Gauss Quadrature scheme is a widely used numerical integration method, for which the standard Gauss point coordinates and the corresponding weights can be derived using mathematical procedures of minimization of errors. The coordinates and weights of these points are given in Table 5.2. An $n$-point Gaussian Quadrature can integrate a polynomial of degree $(2n - 1)$ exactly.

For example, for 2-point integration, the scheme becomes

$$I_2 = \int_{-1}^{1} f(\xi) d\xi = \sum_{i=1}^{2} f(\xi_i) \mathrm{W_i} \tag{5.24}$$

where $\xi_1, \xi_2$ are the Gauss point coordinates, and $W_1$, $W_2$ are the weights.

## 5.10 Transformation of Element Stiffness Matrices

The element stiffness matrix relates the element force vector to the element node displacement vector. The stiffness matrix is naturally formed in the element coordinate system in which the node displacement vector and the element force vector are defined. For three-dimensional (Tetrahedron and hexahedron elements) and two-dimensional (plane stress, plane strain, and axisymmetric solid elements) continuum finite elements, the element matrices can be directly formulated in the global coordinate system. For non-continuum elements (reduced dimension elements), the element coordinate system chosen for each element is different, based on their orientation in the global coordinate system. This is mainly applicable to rod, beam, and plate elements. For such elements, special displacement kinematics are formed to facilitate the dimensional reduction for the associated theories. In such cases, appropriate element coordinate system directions are chosen to facilitate the application of assumptions associated with the rod/beam/plate theory.

The element stiffness matrix is first computed in the element coordinate system and then transformed to the global coordinate system, before assembling into the global stiffness matrix. The element force vector is also transformed to the global coordinate system before assembly. The final solution vector obtained by solving the global equations shall be in the global coordinate system.

Suppose the global coordinate system is X–Y–Z and the element (local) coordinate system is x–y–z. The direction cosines of the local x-, y-, and z-axis with reference to the X–Y–Z system are $(l_1, m_1, n_1)$, $(l_2, m_2, n_2)$, and $(l_3, m_3, n_3)$, respectively. Now, the displacements of node $i$ in the local coordinate system $(u_i, v_i, w_i)$ can be transformed to $(U_i, V_i, W_i)$ in the global coordinate system, using the transformation matrix T, as shown below

$$\begin{Bmatrix} u_i \\ v_i \\ w_i \end{Bmatrix} = [\mathrm{T}_i] \begin{Bmatrix} U_i \\ V_i \\ W_i \end{Bmatrix}; \qquad [\mathrm{T}_i] = \begin{bmatrix} l_1 \; m_1 \; n_1 \\ l_2 \; m_2 \; n_2 \\ l_3 \; m_3 \; n_3 \end{bmatrix}. \tag{5.25}$$

Accordingly, the transformation matrix [T] for the whole element can be constructed for the node displacement vector, by arranging $[\mathrm{T}_i]$ matrices as diagonal blocks in [T]. Now the element stiffness matrix and the element force vector in the global coordinate system can be computed as

$$[\mathrm{K}] = [\mathrm{T}]^T[\mathrm{k}][\mathrm{T}]; \qquad \{\mathrm{F}\} = [\mathrm{T}]^T\{\mathrm{f}\}. \tag{5.26}$$

## 5.11 *Isoparametric* Finite Element Formulations

We need to form a coordinate system for each element to establish the associated mathematical relations and compute the element stiffness matrix. The shape functions $N_i$ in terms of natural coordinates are employed for interpolating the displacement fields. In what is called isoparametric formulations, the same shape functions are also used to interpolate the geometry of the element in terms of coordinates. It means that, to find out the global Cartesian coordinate of a point $(\xi, \eta, \zeta)$ in the element natural coordinate system, the following relations are used:

$$X(\xi,\eta,\zeta)=\sum_{1}^{n}N_i(\xi,\eta,\zeta)X_i;\quad Y(\xi,\eta,\zeta)=\sum_{1}^{n}N_i(\xi,\eta,\zeta)Y_i;$$
$$Z(\xi,\eta,\zeta)=\sum_{1}^{n}N_i(\xi,\eta,\zeta)Z_i \tag{5.27}$$

where $n$ is the number of nodes for the element.

This interpolation of coordinates helps in the computation of the terms in the Jacobian matrix where the derivative of the coordinate field (e.g. $\partial X/\partial\xi$, $\partial X/\partial\eta$) w.r.t. natural coordinates is required, as shown in Eqs. (5.14–5.16).

In the actual formulation and implementation of finite element analysis software, the following typical steps lead to the computation of the element stiffness matrix:

(i) Based on the type of the element, identify the local element coordinate system (x–y–z) in which the stiffness matrix is to be evaluated. For two-dimensional (planar and axisymmetric solid) elements and three-dimensional solid elements, the global coordinate system (X–Y–Z) itself can be used as the local element coordinates system. However, for rod, beam, plate, and axisymmetric shell elements, an appropriate element coordinate system (x–y–z) is formed based on the orientation of the element axis (for rod, beam and axisymmetric shell elements) or element mid-plane (for plate element). For these 'reduced dimension elements', the relevant strain components are identified using appropriate displacement kinematics conveniently defined in the local coordinate system.

(ii) Based on the type of element and the applicable loads, identify the displacement fields, strain fields, and stress fields in the element (local) coordinate system.

(iii) Identify the natural coordinate system and the shape functions appropriate for the element to interpolate the displacement fields in terms of its node values. This vector of node values forms the element displacement vector {d}.

(iv) Identify the order of the numerical integration scheme to be employed and the Gauss point coordinates. The numerical integration scheme demands that at each Gauss point, the strain–displacement matrix [B], the material constitutive matrix [E], and the Jacobian have to be calculated, for the computation of the element stiffness matrix.

(v) At each Gauss point, in the local coordinate system:

   a. Form the [B] matrix which relates the strain field vector to the node displacement vector {d}. The [B] matrix generally consists of shape functions and/or derivatives of the shape functions.
   b. Form the material constitutive matrix [E] which relates the stress vector to the strain vector, relevant to the type of the element.
   c. Compute the Jacobian matrix [J] and its determinant $J$. The Jacobian matrix generally consists of the derivative of local Cartesian coordinates (x,y,z) with respect to the natural coordinates $(\xi,\eta,\zeta)$.
   d. Calculate $[B]^T[E][B]$, multiply with the corresponding Gauss point weight (Wi), and the determinant of Jacobian matrix $(J)$.

e. Transform this stiffness matrix in the local coordinate system computed at the Gauss point to a global coordinate system using a transformation matrix ([T]) computed at that Gauss point. This step is required only if the local coordinate system x–y-z used for defining the strain field for the element is different from the global coordinate system.

(vi) Sum up all the transformed stiffness matrices computed at each of the Gauss points, which yields the element stiffness matrix in the global coordinate system.

The resulting element stiffness matrix is assembled in the global stiffness matrix, following standard assembly procedures, as illustrated in Chap. 6.

In the sections that follow, we identify the local coordinate system, the degrees of freedom, deformation kinematics (relevant only for rod, beam, and plate elements), and formulate the essential three matrices, namely [B], [E], and [J] for a few most commonly used finite elements. The material properties of the elements are assumed to be isotropic, with $E$ as Young's modulus, $G$ as Shear modulus, and $\nu$ as Poisson's ratio.

## 5.12 Stiffness Matrix for Rod/Truss/Link Finite Element

The rod/truss/link element is a 2-node uniaxial, tension–compression, one-dimensional element (line element) which can be used to model members of truss structures. The element may be oriented anywhere in three-dimensional global X–Y–Z coordinate system. The element has three global translational degrees of freedom per node ($U$, $V$, $W$).

When two or more of these elements are connected at their nodes to make a truss structure, they form a spherical ball joint at the node in a three-dimensional framework or a pin joint in a two-dimensional framework, such that only axial tension or compression are developed in these elements, irrespective of the global direction of the forces applied at the nodes. Since the pin/ball joints cannot resist moments, this element cannot sustain bending or transverse shear deformation. Hence, the element will always be straight, even after deformation. This indicates that in the local element coordinate system, only one strain component exists, namely the normal strain $\varepsilon_x$. Hence, it is convenient to first formulate the element stiffness matrix in the local coordinate system, in terms of the local node displacement components $u_1$ and $u_2$ and later transform it to the global coordinate system, in terms of global displacement vector $\{U_1, V_1, W_1, U_2, V_2, W_2\}^T$.

Figure 5.11 shows the geometry and coordinate systems for the rod element. The stiffness matrix of this element can be formed as follows:

**Global coordinate system**: X–Y–Z.

**Local coordinate system**: x (along the axis of the element), origin at the node 1.

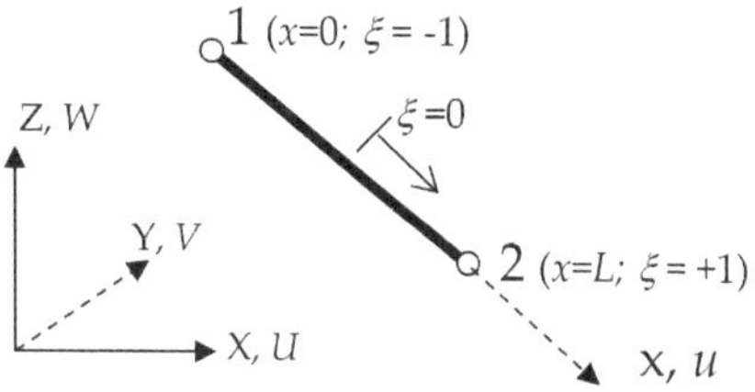


**Fig. 5.11** A 2-node rod/truss/link element

**Natural coordinate system**: $\xi$, along the axis of the element, origin at the mid-span centroid (x $= L/2$). At $x = 0$, $\xi = -1$ and at $x = L$, $\xi = +1$.

**Relevant displacement fields in local coordinate system**: $u$(x), the axial displacement.

**Local degrees of freedom at nodes**: $\{\mathrm{d}\} = \{u_1, u_2\}^T$.

**Global degrees of freedom at nodes**: $\{\mathrm{D}\} = \{U_1, V_1, W_1, U_2, V_2, W_2\}^T$.

**Shape functions**: $N_1(\xi) = \frac{1}{2}(1-\xi)$; $N_2(\xi) = \frac{1}{2}(1+\xi)$.

**Displacement Field interpolation**: $u(x) = N_1(\xi)u_1 + N_2(\xi)u_2$.

**Strain and Stress components in local coordinate system**: Since only axial force is developed in the element, only the normal strain ($\varepsilon_{\mathrm{x}}$) and the normal stress ($\sigma_{\mathrm{x}}$) need to be considered.

**Elastic strain energy**: The elastic strain energy of the element, in the local coordinate system, can be expressed as $\frac{1}{2}\int_L \varepsilon_x \sigma_x A dx$, where $L$ is the length of the element and $A$ is the area of cross section. In the case of a tapered rod, the area may be specified as a function of location ($x$) on the axis and a higher order Gauss integration rule should be used for computation of stiffness matrix.

**Constitutive matrix** [E]: The constitutive matrix (underlined in the equation below) for this element, which relates strain vector to the stress vector, is given by

$$\{\sigma_x\} = \underline{[E]}\{\varepsilon_x\}, \quad \text{where } E \text{ is Young's modulus of the material.}$$

**Strain–displacement matrix** [B]: The matrix [B] (underlined in the equation below) which relates the vector of strain field $\{\varepsilon_x\}$ to the node displacement vector {d} is given by

$$\{\varepsilon_x(x)\} = \underline{\begin{bmatrix} \frac{\partial N_1}{\partial x} & \frac{\partial N_2}{\partial x} \end{bmatrix}} \begin{Bmatrix} u_1 \\ u_2 \end{Bmatrix} ; \quad \frac{\partial N_1}{\partial x} = \frac{\partial N_1}{\partial \xi} . \frac{\partial \xi}{\partial x} = \frac{-1}{2} . \frac{2}{L} = \frac{-1}{L} \text{ and } \frac{\partial N_2}{\partial x} = \frac{1}{L}.$$

$$\text{Hence,} \quad [\mathrm{B}] = \begin{bmatrix} \frac{-1}{L} & \frac{1}{L} \end{bmatrix}. \tag{5.28}$$

**Jacobian matrix** [J]: The Jacobian matrix for this element is given by $[\mathrm{J}] = \left[\frac{\partial x}{\partial \xi}\right] = \left[\frac{L}{2}\right]$ where $L$ is the length of the element.

**Gauss integration rule**: 1-point rule ($\xi_1 = 0, \mathrm{W}_1 = 2.0$) for uniform cross section.

2-point rule $\left(\xi_1 = -\sqrt{1/3};\quad \xi_2 = +\sqrt{1/3};\quad W_1 = W_2 = 1.0\right)$ for tapered cross section.

**Element Stiffness Matrix in local coordinate system**: The element stiffness matrix can now be computed using Gauss numerical integration rule as

$$[k] = \int_{-1}^{1} [B]^T[E][B]J d\xi = \sum_{i=1}^{1} [B(\xi_i)]^T[E][B(\xi_i)]J(\xi_i)W_i. \tag{5.29}$$

The computed element stiffness matrix shall be $[k] = \frac{EA}{L}\begin{bmatrix} 1 & -1 \\ -1 & 1 \end{bmatrix}$.

## 5.13 Stiffness Matrix for Three-Dimensional Solid Finite Element

The three-dimensional finite elements are generally used for modeling bulk solid elements. The most popular elements of this family are the tetrahedron and hexahedron elements. The formulation for an 8-node hexahedron element is given below.

Figure 5.12 shows the geometry and coordinate systems for an 8-node hexahedral solid element.

The stiffness matrix of this element can be formed as follows:

**Global Coordinate system**: X–Y–Z.

**Local coordinate system**: The three-dimensional solid element does not employ a local coordinate system for defining the displacement fields. It does not employ any dimensional reduction or simplified kinematics for its formulation. The displacement fields defined in the global coordinate system are directly interpolated in terms of natural coordinates.

**Natural coordinate system**: A three-dimensional natural coordinate system $\xi$-$\eta$-$\zeta$, with origin at the element centroid, is employed here. For a given node numbering

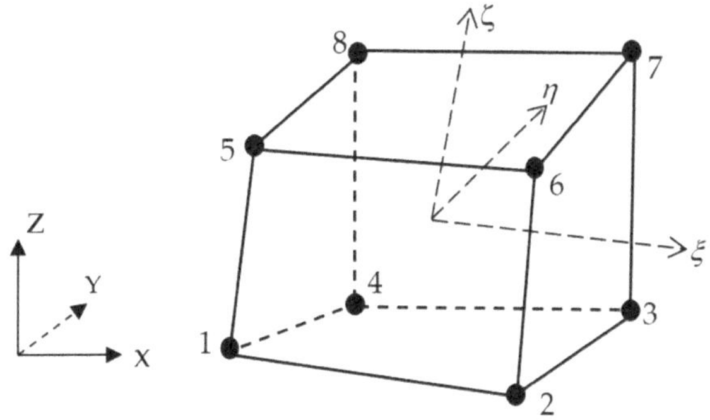


**Fig. 5.12** A 8-node hexahedron element

arrangement as shown in Fig. 5.12, the left and right faces are $\xi = -1$ plane and $\xi = +1$ plane, respectively. The front and back faces are $\eta = -1$ plane and $\eta = +1$ plane, respectively. The bottom and top faces are $\zeta = -1$ plane and $\zeta = +1$ plane, respectively.

**Relevant displacement fields in the global coordinate system**: $U$(X,Y,Z), $V$(X,Y,Z), and $W$(X,Y,Z).

**Global degrees of freedom**: $\{d\} = \{U_1, V_1, W_1, U_2, V_2, W_2 \ldots U_8, V_8, W_8\}^T$.

$$\textbf{Shape functions}: \; N_i = \frac{1}{8}(1 + \xi\xi_i)(1 + \eta\eta_i)(1 + \zeta\zeta_i). \tag{5.30}$$

$$\textbf{Displacement Field interpolation}: U(X, Y, Z) = \sum_1^8 N_i(\xi, \eta, \zeta)U_i$$

$$V(X, Y, Z) = \sum_1^8 N_i(\xi, \eta, \zeta)V_i$$

$$W(X, Y, Z) = \sum_1^8 N_i(\xi, \eta, \zeta)W_i. \tag{5.31}$$

**Strain and Stress components in global coordinate system**: Since forces in any direction can be applied to this element, all the three-dimensional strain and stress components shall be developed in the element:

$$\{\varepsilon\} = \left\{\varepsilon_x \; \varepsilon_y \; \varepsilon_z \; \gamma_{xy} \; \gamma_{yz} \; \gamma_{xz}\right\}^T; \qquad \{\sigma_x\} = \left\{\sigma_x \; \sigma_y \; \sigma_z \; \sigma_{xy} \; \sigma_{yz} \; \sigma_{xz}\right\}^T. \tag{5.32}$$

**Elastic strain energy**: The elastic strain energy of the element can be expressed as

$$\mathrm{U} = \frac{1}{2}\int_x\int_y\int_z (\varepsilon_x\sigma_x + \varepsilon_y\sigma_y + \varepsilon_z\sigma_z + \gamma_{xy}\tau_{xy} + \gamma_{xz}\tau_{xz} + \gamma_{yz}\tau_{yz})dxdydz. \tag{5.33}$$

**Constitutive matrix** [E]: The constitutive matrix for this element which relates strain vector to the stress vector is given by $\{\sigma\} = [\mathrm{E}]\{\varepsilon\}$, where, for an isotropic material,

$$[\mathrm{E}] = \frac{E}{(1+\upsilon)(1-2\upsilon)}\begin{bmatrix} 1-\upsilon & \upsilon & \upsilon & 0 & 0 & 0 \\ \upsilon & 1-\upsilon & \upsilon & 0 & 0 & 0 \\ \upsilon & \upsilon & 1-\upsilon & 0 & 0 & 0 \\ 0 & 0 & 0 & \frac{1-2\upsilon}{2} & 0 & 0 \\ 0 & 0 & 0 & 0 & \frac{1-2\upsilon}{2} & 0 \\ 0 & 0 & 0 & 0 & 0 & \frac{1-2\upsilon}{2} \end{bmatrix}. \tag{5.34}$$

**Strain–displacement matrix** [B]: The matrix [B] (underlined in the equation below) relates the vector of strain field $\{\varepsilon\}$ to the global node displacement vector **{D}**.

$$\begin{Bmatrix} \varepsilon_x \\ \varepsilon_y \\ \varepsilon_z \\ \gamma_{xy} \\ \gamma_{yz} \\ \gamma_{xz} \end{Bmatrix} = \begin{Bmatrix} \frac{\partial U}{\partial X} \\ \frac{\partial V}{\partial Y} \\ \frac{\partial W}{\partial Z} \\ \frac{\partial U}{\partial Y} + \frac{\partial V}{\partial X} \\ \frac{\partial V}{\partial Z} + \frac{\partial W}{\partial Y} \\ \frac{\partial U}{\partial Z} + \frac{\partial W}{\partial X} \end{Bmatrix} = \underline{\begin{bmatrix} \frac{\partial N_1}{\partial X} & 0 & 0 & \frac{\partial N_2}{\partial X} & 0 & 0 & \cdots\cdots & \frac{\partial N_8}{\partial X} & 0 & 0 \\ 0 & \frac{\partial N_1}{\partial Y} & 0 & 0 & \frac{\partial N_2}{\partial Y} & 0 & \cdots\cdots & 0 & \frac{\partial N_8}{\partial Y} & 0 \\ 0 & 0 & \frac{\partial N_1}{\partial Z} & 0 & 0 & \frac{\partial N_2}{\partial Z} & \cdots\cdots & 0 & 0 & \frac{\partial N_8}{\partial Z} \\ \frac{\partial N_1}{\partial Y} & \frac{\partial N_1}{\partial X} & 0 & \frac{\partial N_2}{\partial Y} & \frac{\partial N_2}{\partial X} & 0 & \cdots\cdots & \frac{\partial N_8}{\partial Y} & \frac{\partial N_8}{\partial X} & 0 \\ 0 & \frac{\partial N_1}{\partial Z} & \frac{\partial N_1}{\partial Y} & 0 & \frac{\partial N_2}{\partial Z} & \frac{\partial N_2}{\partial Y} & \cdots\cdots & 0 & \frac{\partial N_8}{\partial Z} & \frac{\partial N_8}{\partial Y} \\ \frac{\partial N_1}{\partial Z} & 0 & \frac{\partial N_1}{\partial X} & \frac{\partial N_2}{\partial Z} & 0 & \frac{\partial N_2}{\partial X} & \cdots\cdots & \frac{\partial N_8}{\partial Z} & 0 & \frac{\partial N_8}{\partial X} \end{bmatrix}} \begin{Bmatrix} U_1 \\ V_1 \\ W_1 \\ U_2 \\ V_2 \\ W_2 \\ \cdot \\ \cdot \\ U_8 \\ V_8 \\ W_8 \end{Bmatrix}. \tag{5.35}$$

The derivative of the shape functions with respect to global coordinates can be calculated using the inverse of Jacobian matrix given below.

**Jacobian matrix** [J]**:** The Jacobian for this element (underlined in the equation below) relates the derivatives in the natural coordinate system to the derivatives in the global coordinate system:

$$\begin{Bmatrix} \frac{\partial N_i}{\partial \xi} \\ \frac{\partial N_i}{\partial \eta} \\ \frac{\partial N_i}{\partial \zeta} \end{Bmatrix} = \underline{\begin{bmatrix} \frac{\partial X}{\partial \xi} & \frac{\partial Y}{\partial \xi} & \frac{\partial Z}{\partial \xi} \\ \frac{\partial X}{\partial \eta} & \frac{\partial Y}{\partial \eta} & \frac{\partial Z}{\partial \eta} \\ \frac{\partial X}{\partial \zeta} & \frac{\partial Y}{\partial \zeta} & \frac{\partial Z}{\partial \zeta} \end{bmatrix}} \begin{Bmatrix} \frac{\partial N_i}{\partial X} \\ \frac{\partial N_i}{\partial Y} \\ \frac{\partial N_i}{\partial Z} \end{Bmatrix}. \tag{5.36}$$

The derivative of global coordinates with respect to natural coordinates, at any point $(\xi, \eta, \zeta)$, is computed using shape function derivatives at that point and the node coordinates, e.g.:

$$\frac{\partial X}{\partial \xi} = \sum_{i=1}^{8} \frac{\partial N_i}{\partial \xi} X_i.$$

**Gauss integration rule**: 2 × 2 × 2 integration rule.

**Element Stiffness Matrix in global coordinate system**: The element stiffness matrix can now be computed using Gauss numerical integration rule as

$$\begin{aligned} [\mathrm{k}] &= \int_{-1}^{+1}\int_{-1}^{+1}\int_{-1}^{+1} [\mathrm{B}(\xi, \eta, \zeta)]^{\mathrm{T}}[\mathrm{E}][\mathrm{B}(\xi, \eta, \zeta)] J \; d\xi d\eta d\zeta \\ &= \sum_{i=1}^{2}\sum_{j=1}^{2}\sum_{k=1}^{2} \left[\mathrm{B}(\xi_{\mathrm{i}}, \eta_{\mathrm{j}}, \zeta_{\mathrm{k}})\right]^{T} [\mathrm{E}]\left[\mathrm{B}(\xi_{\mathrm{i}}, \eta_{\mathrm{j}}, \zeta_{\mathrm{k}})\right] J(\xi_i, \eta_j, \zeta_k) W_i W_j W_k. \end{aligned} \tag{5.37}$$

## 5.14 Stiffness Matrix for Planar Finite elements

There are two types of planar finite elements, plane *stress* and *plane strain* elements. Plane stress elements are used to model flat thin structures of arbitrary shape, lying in the global X–Y plane and subjected to loads only acting in the X–Y plane. A thin plate subjected to in-plane compressive, tensile, or shear loads is a typical example of a plane stress problem. Since the deformation in the Z-direction is zero, the mid-plane of the thin structure alone need to be modeled. The displacements, strains, and stresses in this two-dimensional element would be a subset of the general three-dimensional state of these quantities, explained in the previous section.

Plane strain elements are used to model structures whose geometry and loading do not vary significantly in the longitudinal direction (Z) and whose ends are constrained. A long cylinder carrying a pressurized fluid is a typical example of this kind. Only a representative cross section of the structure in the X–Y plane needs to be discretized.

It may be noted that the relevant displacement fields and the strain fields (which are active in the X–Y plane) for both plane stress and plane strain elements are the same. However, the constitutive matrix employed in plane stress and plane strain analyses is slightly different. For plane strain analysis, $\varepsilon_z = 0$, as the ends of the structure in the longitudinal direction are constrained. Hence, the constraining of Poisson's strain in plane strain analysis generates longitudinal stress, $\sigma_z \neq 0$, which can be computed from the knowledge of $\varepsilon_x$ and $\varepsilon_y$, after obtaining the finite element solution for the displacements. On the contrary, for plane stress analysis, the strain in thickness direction due to Poisson's effect, $\varepsilon_z \neq 0$. Since this strain is not constrained in any manner, $\sigma_z = 0$.

Figure 5.13 shows the geometry and coordinate systems for a 4-node planar element. The stiffness matrix of this element can be formed as follows:

**Global Coordinate system**: X–Y.

**Local coordinate system**: Two-dimensional solid elements do not employ a local coordinate system for defining the displacement fields. The displacement fields defined in the global coordinate system are interpolated using the natural coordinates within the element.

**Natural coordinate system**: A two-dimensional natural coordinate system $\xi$-$\eta$, with origin at the element centroid, is employed here. For a given node numbering scheme as shown in Fig. 5.13, the left and right edges are at $\xi = -1$ and $\xi = +1$, respectively. The bottom and top edges are $\eta = -1$ and $\eta = +1$, respectively.

**Fig. 5.13** A 4-node planar element

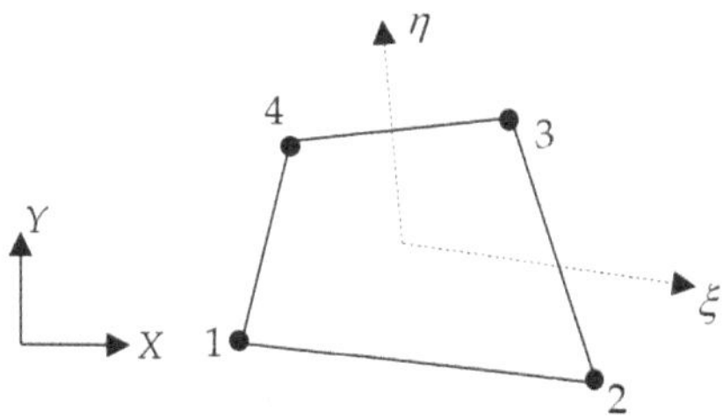

**Relevant displacement fields in the global coordinate system**: $U$(X,Y) and $V$(X,Y).

**Global degrees of freedom**: $\{\mathbf{D}\} = \{U_1, V_1, U_2, V_2, U_3, V_3, U_4, V_4\}^{\mathrm{T}}$.

**Shape functions**: $N_i = \frac{1}{4}(1 + \xi\xi_i)(1 + \eta\eta_i)$. Substituting the natural coordinate of each of the nodes, we get

$$N_1 = \frac{1}{4}(1-\xi)(1-\eta) \quad ; N_2 = \frac{1}{4}(1+\xi)(1-\eta)$$
$$N_3 = \frac{1}{4}(1+\xi)(1+\eta) \quad ; N_4 = \frac{1}{4}(1-\xi)(1+\eta). \tag{5.38}$$

$$\textbf{Displacement Field interpolation}: U(X,Y) = \sum_{i=1}^{4} N_i(\xi,\eta)U_i$$
$$V(X,Y) = \sum_{i=1}^{4} N_i(\xi,\eta)V_i. \tag{5.39}$$

**Strain and Stress components in global coordinate system**: Since only forces in the X–Y plane can be applied to the element, the strain and stress components developed in the element shall be

$$\{\varepsilon\} = \left\{\varepsilon_x\ \varepsilon_y\ \gamma_{xy}\right\}^T; \quad \{\sigma\} = \left\{\sigma_x\ \sigma_y\ \tau_{xy}\right\}^T. \tag{5.40}$$

**Elastic strain energy**: The elastic strain energy of the element can be expressed as.

$\Pi = \frac{h}{2}\int_x\int_y \left(\varepsilon_x\sigma_x + \varepsilon_y\sigma_y + \gamma_{xy}\tau_{xy}\right)dxdy$, where $h$ is the thickness of the structure, for the plane stress case. For the plane strain case, $h$ is taken as unity.

**Constitutive matrix [E]**: The constitutive matrix for this element which relates the strain vector to the stress vector is given by.

$\{\sigma\} = [\mathrm{E}]\{\varepsilon\}$, where, for an isotropic material,

For plane stress problems,

$$[\mathrm{E}] = \frac{E}{1-\upsilon^2}\begin{bmatrix} 1 & \upsilon & 0 \\ \upsilon & 1 & 0 \\ 0 & 0 & (1-\upsilon)/2 \end{bmatrix}. \tag{5.41}$$

For plane strain problems,

$$[\mathrm{E}] = \frac{E}{(1+\upsilon)(1-2\upsilon)}\begin{bmatrix} 1-\upsilon & \upsilon & 0 \\ \upsilon & 1-\upsilon & 0 \\ 0 & 0 & (1-2\upsilon)/2 \end{bmatrix}. \tag{5.42}$$

**Strain–displacement matrix [B]**: The matrix [B] (underlined in the equation below) relates the vector of strain field $\{\varepsilon\}$ to the global node displacement vector

**{D}**:

$$\begin{Bmatrix} \varepsilon_x \\ \varepsilon_y \\ \gamma_{xy} \end{Bmatrix} = \begin{Bmatrix} \frac{\partial U}{\partial X} \\ \frac{\partial V}{\partial Y} \\ \frac{\partial U}{\partial Y} + \frac{\partial V}{\partial X} \end{Bmatrix} = \underline{\begin{bmatrix} \frac{\partial N_1}{\partial X} & 0 & \frac{\partial N_2}{\partial X} & 0 & \frac{\partial N_3}{\partial X} & 0 & \frac{\partial N_4}{\partial X} & 0 \\ 0 & \frac{\partial N_1}{\partial Y} & 0 & \frac{\partial N_2}{\partial Y} & 0 & \frac{\partial N_3}{\partial Y} & 0 & \frac{\partial N_4}{\partial Y} \\ \frac{\partial N_1}{\partial Y} & \frac{\partial N_1}{\partial X} & \frac{\partial N_2}{\partial Y} & \frac{\partial N_2}{\partial X} & \frac{\partial N_3}{\partial Y} & \frac{\partial N_3}{\partial X} & \frac{\partial N_4}{\partial Y} & \frac{\partial N_4}{\partial X} \end{bmatrix}} \begin{Bmatrix} U_1 \\ V_1 \\ U_2 \\ V_2 \\ U_3 \\ V_3 \\ U_4 \\ V_4 \end{Bmatrix}. \tag{5.43}$$

The derivatives of the shape functions with respect to global coordinates are computed using the inverse of the Jacobian matrix given below.

**Jacobian matrix** [J]: The Jacobian matrix for this element (underlined in the equation below) relates the derivatives in the natural coordinate system to the derivatives in the global coordinate system:

$$\begin{Bmatrix} \frac{\partial N_i}{\partial \xi} \\ \frac{\partial N_i}{\partial \eta} \end{Bmatrix} = \underline{\begin{bmatrix} \frac{\partial X}{\partial \xi} & \frac{\partial Y}{\partial \xi} \\ \frac{\partial X}{\partial \eta} & \frac{\partial Y}{\partial \eta} \end{bmatrix}} \begin{Bmatrix} \frac{\partial N_i}{\partial X} \\ \frac{\partial N_i}{\partial Y} \end{Bmatrix}. \tag{5.44}$$

The derivative of global coordinates with respect to natural coordinates, at any point $(\xi,\eta)$, is computed using shape function derivatives at that point and the node coordinates, e.g.:

$$\frac{\partial X}{\partial \xi} = \sum_{i=1}^{4} \frac{\partial N_i}{\partial \xi} X_i.$$

**Gauss integration rule**: 2 × 2 integration rule.

**Element Stiffness Matrix in global coordinate system**: The element stiffness matrix can now be computed using Gauss numerical integration rule as

$$\begin{aligned} [\mathrm{k}] &= h \int_{-1}^{+1} \int_{-1}^{+1} [\mathrm{B}(\xi,\eta)]^T [\mathrm{E}][\mathrm{B}(\xi,\eta)]\, J \mathrm{d}\xi \mathrm{d}\eta \\ &= \sum_{i=1}^{2} \sum_{j=1}^{2} h \left[\mathrm{B}(\xi_i,\eta_j)\right]^T [\mathrm{E}] \left[\mathrm{B}(\xi_i,\eta_j)\right] J(\xi_i,\eta_j) \mathrm{W_i W_j}. \end{aligned} \tag{5.45}$$

For the plane stress element, $h$ shall be the actual thickness of the element. For the plane strain element, where all the loads are input per unit longitudinal length, the value of $h = 1$.

## 5.15 Stiffness Matrix for Axisymmetric Solid Element

An axisymmetric solid structure is formed by revolving a cross section in the R–Z plane (as shown in Fig. 5.14) about the Z-axis, by 360 degrees. The Z-axis is called the axis of revolution. Any point 'A' shown on this cross section at a distance $r$ from this axis of revolution represents a full circle of material, whose radius is $r$. A tangent to this circle at the point A, which is directed normal and into the page, is called the *circumferential* direction or *hoop* direction or θ-direction. The R–θ–Z forms a right-handed *Global coordinate system.* Each material point on the R–Z plane represents a 'ring of material' parallel to the R-θ plane. If such a structure is subjected to loads (a line load or a pressure load) which are also axisymmetric (distributed in the circumferential direction, without any change in magnitude), the deformations of a material point on the cross-sectional plane shall be confined on the R–Z plane. In other words, there will be no circumferential displacement for any point in the structure. However, a 'hoop strain', which is a normal strain, shall be developed in the circumferential direction due to radial expansion of the 'ring of material'. The displacements field shall be only in the 'radial' and 'axial' directions in the R-Z plane. Hence, only a cross section of the structure needs to be discretized using axisymmetric solid elements and analyzed for finding out the deformations, strains, and stress in the three-dimensional axisymmetric structures subjected to axisymmetric loads. In effect, each two-dimensional axisymmetric solid element used for meshing the cross section represents a three-dimensional ring structure evolved by revolving the element about the Z-axis.

Figure 5.14 shows the geometry and coordinate systems for a 4-node axisymmetric solid element. The stiffness matrix of this element can be formed as follows:

**Global Coordinate system**: R–Z.

**Local coordinate system**: Axisymmetric solid elements do not employ a local coordinate system for defining the displacement fields. The displacement fields are

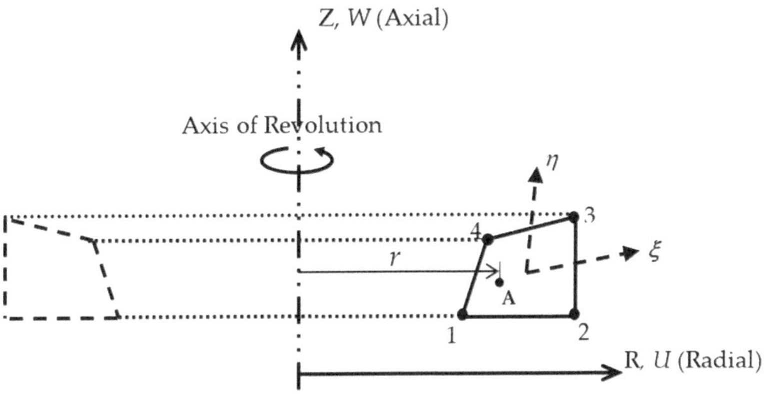


**Fig. 5.14** A 4-node Axisymmetric Solid element

defined in the global coordinate system and interpolated using the natural coordinates within the element.

**Natural coordinate system**: A two-dimensional natural coordinate system $\xi$-$\eta$, with origin at the element centroid, is employed here. For a given node numbering arrangement as shown in Fig. 5.14, the left and right edges are at $\xi = -1$ and $\xi = +1$, respectively. The bottom and top edges are $\eta = -1$ and $\eta = +1$ respectively.

**Relevant displacement fields in the global coordinate system**: $U$(R, Z) and $W$(R, Z).

**Global degrees of freedom**: $\{\mathrm{D}\} = \{U_1, W_1, U_2, W_2, \ldots\ldots\ldots U_4, W_4\}^{\mathrm{T}}$.

**Shape functions**: $N_i = \frac{1}{4}(1 + \xi\xi_i)(1 + \eta\eta_i)$.

Substituting the natural coordinate of each of the nodes, we get

$$N_1 = \frac{1}{4}(1 - \xi)(1 - \eta); \quad N_2 = \frac{1}{4}(1 + \xi)(1 - \eta)$$
$$N_3 = \frac{1}{4}(1 + \xi)(1 + \eta); \quad N_4 = \frac{1}{4}(1 - \xi)(1 + \eta). \tag{5.46}$$

**Displacement Field interpolation**:

$$U(R, Z) = \sum_{i=1}^{4} N_i(\xi, \eta)U_i$$
$$W(R, Z) = \sum_{i=1}^{4} N_i(\xi, \eta)W_i. \tag{5.47}$$

**Strain and Stress components in global coordinate system**: Apart from the basic planar strain components in the R-Z plane, the radial deformation of the structure introduces circumferential strain ($\varepsilon_\theta$) and stress ($\sigma_\theta$) in the axisymmetric solid elements. It may be noted that this circumferential strain at a point at a distance $r$ from the axis due to a radial displacement $U$ shall be equal to $U/r$. Hence, the strain and stress components for an axisymmetric solid element shall be

$$\{\varepsilon\} = \{\varepsilon_r \ \ \varepsilon_\theta \ \ \varepsilon_z \ \ \gamma_{rz}\}^T; \qquad \{\sigma\} = \{\sigma_r \ \ \sigma_\theta \ \ \sigma_z \ \ \tau_{rz}\}^T. \tag{5.48}$$

**Elastic strain energy**: The elastic strain energy of the element can be expressed as.

$\mathrm{U} = \frac{1}{2}\int_z \int_r (\varepsilon_r\sigma_r + \varepsilon_\theta\sigma_\theta + \varepsilon_z\sigma_z + \gamma_{rz}\tau_{rz})2\pi r\, dr\, dz$ where $r$ is the distance from the axis of revolution.

**Constitutive matrix [E]**: The stresses and strains of the $R$-$\theta$-$Z$ coordinate system perfectly form a three-dimensional state of stress/strain, and hence the relevant constitutive matrix which relates the strain vector to the stress vector of the axisymmetric solid element can be picked up from the three-dimensional constitutive matrix for solid elements (Eq. 5.34):

$\{\sigma\} = [\mathrm{E}]\{\varepsilon\}$, where, for an isotropic material,

$$[\mathrm{E}] = \frac{E}{(1+\upsilon)(1-2\upsilon)}\begin{bmatrix} 1-\upsilon & \upsilon & \upsilon & 0 \\ \upsilon & 1-\upsilon & \upsilon & 0 \\ \upsilon & \upsilon & 1-\upsilon & 0 \\ 0 & 0 & 0 & (1-2\upsilon)/2 \end{bmatrix}. \tag{5.49}$$

**Strain–displacement matrix [*B*]:** The matrix [B] (underlined in the equation below) relates the vector of strain field $\{\varepsilon\}$ to the global node displacement vector **{D}:**

$$\begin{Bmatrix} \varepsilon_x \\ \varepsilon_\theta \\ \varepsilon_z \\ \gamma_{rz} \end{Bmatrix} = \begin{Bmatrix} \frac{\partial U}{\partial X} \\ \frac{U}{r} \\ \frac{\partial W}{\partial Z} \\ \frac{\partial U}{\partial Z} + \frac{\partial W}{\partial X} \end{Bmatrix} = \underline{\begin{bmatrix} \frac{\partial N_1}{\partial X} & 0 & \frac{\partial N_2}{\partial X} & 0 & \frac{\partial N_3}{\partial X} & 0 & \frac{\partial N_4}{\partial X} & 0 \\ \frac{N_1}{r} & 0 & \frac{N_2}{r} & 0 & \frac{N_3}{r} & 0 & \frac{N_4}{r} & 0 \\ 0 & \frac{\partial N_1}{\partial Z} & 0 & \frac{\partial N_2}{\partial Z} & 0 & \frac{\partial N_3}{\partial Z} & 0 & \frac{\partial N_4}{\partial Z} \\ \frac{\partial N_1}{\partial Z} & \frac{\partial N_1}{\partial X} & \frac{\partial N_2}{\partial Z} & \frac{\partial N_2}{\partial X} & \frac{\partial N_3}{\partial Z} & \frac{\partial N_3}{\partial X} & \frac{\partial N_4}{\partial Z} & \frac{\partial N_4}{\partial X} \end{bmatrix}} \begin{Bmatrix} U_1 \\ W_1 \\ U_2 \\ W_2 \\ U_3 \\ W_3 \\ U_4 \\ W_4 \end{Bmatrix}. \tag{5.50}$$

The derivative of the shape functions with respect to global coordinates using the inverse of the Jacobian matrix is given below.

**Jacobian matrix** [J]: The Jacobian matrix for this element (underlined in the equation below) relates the derivatives in the natural coordinate system to the derivatives in the global coordinate system:

$$\begin{Bmatrix} \frac{\partial N_i}{\partial \xi} \\ \frac{\partial N_i}{\partial \eta} \end{Bmatrix} = \underline{\begin{bmatrix} \frac{\partial X}{\partial \xi} & \frac{\partial Y}{\partial \xi} \\ \frac{\partial X}{\partial \eta} & \frac{\partial Y}{\partial \eta} \end{bmatrix}} \begin{Bmatrix} \frac{\partial N_i}{\partial X} \\ \frac{\partial N_i}{\partial Y} \end{Bmatrix}. \tag{5.51}$$

The derivative of global coordinates with respect to natural coordinates, at any point ($\xi$-$\eta$), is computed using shape function derivatives at that point and the node coordinates, e.g.:

$$\frac{\partial X}{\partial \xi} = \sum_{i=1}^{4} \frac{\partial N_i}{\partial \xi} X_i.$$

**Gauss integration rule:** $2 \times 2$ integration rule.

**Element Stiffness Matrix in global coordinate system:** The element stiffness matrix can now be computed using Gauss numerical integration rule as

$$\begin{aligned}[\mathrm{k}] &= \int_0^{2\pi}\int_{-1}^{+1}\int_{-1}^{+1}[\mathrm{B}(\xi,\eta)]^T[\mathrm{E}][\mathrm{B}(\xi,\eta)]J(\xi,\eta)r\,d\xi\,d\eta\,d\theta \\ &= 2\pi\sum_{i=1}^{2}\sum_{j=1}^{2}\left[\mathrm{B}(\xi_i,\eta_j)\right]^T[\mathrm{E}]\left[\mathrm{B}(\xi_i,\eta_j)\right]J(\xi_i,\eta_j)r(\xi_i,\eta_j)\mathrm{W_i W_j}.\end{aligned} \tag{5.52}$$

## 5.16 Stiffness Matrix for Beam Element

**Geometric features of a beam**: Consider a beam arbitrarily oriented in the global X–Y–Z coordinate system, as shown in Fig. 5.15. Beams are slender structures, which have one of their dimensions much larger than that in the other two directions. The cross sections of the beam, such as A-A shown in the figure, fall on the planes perpendicular to this longest direction. The cross section shall have a definite shape and geometric centroid. The line passing through the centroid of each section forms the *axis* of the beam. This axial direction (**x**) of a beam is also referred to as longitudinal direction. Two additional directions (**y** and **z**) identified (on the cross-sectional plane) which are mutually perpendicular to the axis of the beam are called the transverse directions. The coordinate system **x–y-z** forms the local coordinate system in which the deformation kinematics are enforced. By virtue of slender geometry, the deformation characteristic of the beam can be visualized to be a combination of axial, bending, shear, and torsional deformation in the local coordinate system.

A force acting along the longitudinal axis ($F_x$) of the beam causes extension or compression of the beam. A force acting along the transverse directions ($F_y$, $F_z$) causes shear deformation. An applied couple $C_y$ or a moment produced by the transverse force $F_z$ causes beam bending with its cross sections rotating about the **y**-axis. Similarly, an applied couple $C_z$ or a moment produced by transverse force $F_y$ causes beam bending with its cross sections rotating about the **z**-axis. These rotations of the beam cross sections about the **y-** and **z-**axes are designated as $\theta_y$ and $\theta_z$, respectively. An applied Couple $C_x$ about the longitudinal (**x**) axis causes twisting of the beam produced by rotation of the cross sections about the x-axis ($\theta_x$).

**Axial force ($N_x$)**: The longitudinal force applied along the centroid axis of the beam causes normal strain and normal stress, which are uniform at all points on

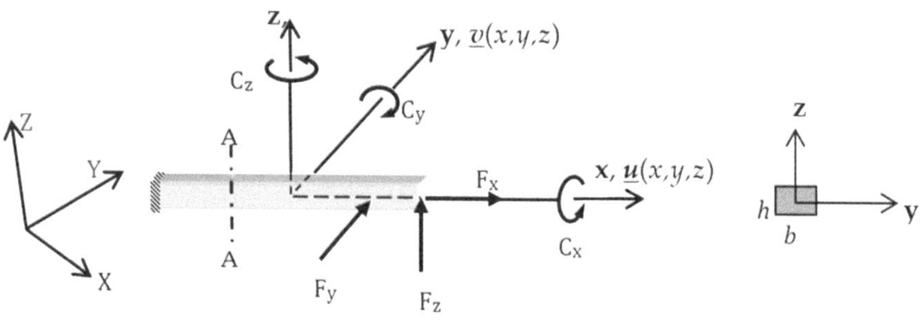


**Fig. 5.15** Geometry and Loads of a beam structure

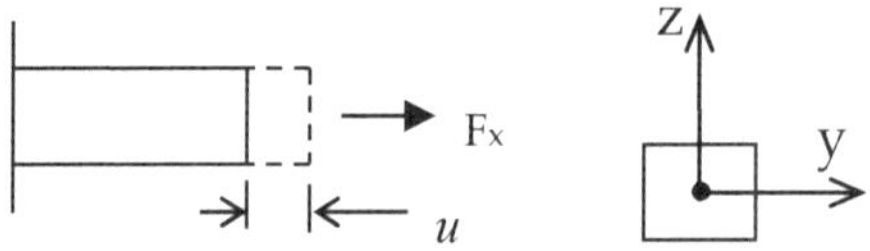


**Fig. 5.16** Displacement *u* of the cross section due to longitudinal force Fx

a cross section and hence cause a uniform axial displacement of the cross section. Hence, displacement at any point on a cross section due to the longitudinal force can be represented by $\underline{u}$(x,y,z) = $\underline{u}$(x,0,0) = *u*(x), the degree of freedom which denotes the value of axial displacement of the cross section. It may be noted that *u*(x) is the same for all the points on the cross section at x. The mathematical integration of the stress distribution over the cross-sectional area gives the internal stress resultant called the axial force ($N_x$) (Fig. 5.16).

**Note**: A force along the x-direction acting at a point not coinciding with the centroid C of the section has to be treated as a combination of **longitudinal force** (acting at the centroid) of the same magnitude and a **moment** of the magnitude equal to the product of the force and corresponding distance of the centroid axis from the point of application of the force, with appropriate sign convention.

**Bending moments ($M_y$ and $M_z$):** A +ve couple or moment about the y-axis causes +ve cross-sectional rotation $\theta_y$ about the y-axis and -ve transverse displacement $w_b$, as shown in Fig. 5.17. The section rotation results in +ve $\underline{u}$-displacements at points above the neutral plane (x–y plane where displacement will be zero) and –ve $\underline{u}$-displacement at points below the neutral plane. Hence, for small rotations, over a cross section, $\underline{u}$(x,z) = $z\theta_y(x)$, where *z* is measured from the neutral plane. The +ve couple or moment also causes transverse displacement of the section in the –ve z-direction. The transverse deflection of the beam due to bending is denoted as $w_b$. The cross sections are assumed to remain plain even after rotation and do not undergo any change in dimension. So, it can be appreciated that $w_b$(x,y,z) = $w_b$(x,0,0) = $w_b$. It can be seen that for a cross section at any x, as shown in Fig. 5.17, section rotation, $\theta_y(x)$ = -d$w_b$/dx = -(slope of beam deflection).

Following the same arguments, one can see that a + ve couple or moment (Cz) about the z-axis bends the beam in the x–y plane and the sections undergo + ve rotation about the z-axis, $\theta_z(x)$ and + ve transverse displacement $v_b$. The section rotation results in –ve $\underline{u}$ displacements at points on the + y half of the cross section

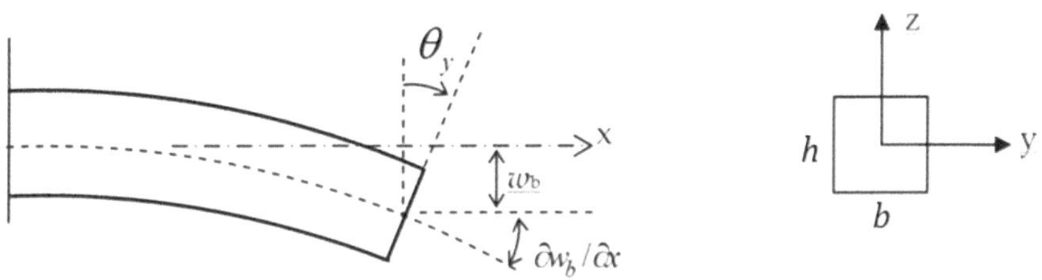


**Fig. 5.17** Bending and cross-section rotation due to couple or moment Cy

and + ve $\underline{u}$ displacement at points on the –y half of the cross section. Hence, for small rotations of $\theta_z$, $\underline{u}$(x,y) = -y$\theta_z(x)$. Also, section rotation $\theta_z(x) = dv_b/dx$.

The couple or moment about the y-axis causes a linear distribution of internal normal stress across the thickness in the z-direction of the beam, being compressive on one half and tensile on the other half, with zero stress along the neutral axis (centroid y-axis). The internal moment caused by this stress distribution about the y-axis is called the *bending moment* $M_y$.

Similarly, a couple or moment about the z-axis causes a combination of tensile and compressive linear distribution of internal normal stress across the thickness in the y-direction of the beam. The internal moment caused by this stress distribution about the z-axis is called the *bending moment* $M_z$.

**Shear forces ($Q_z$ and $Q_y$):** Forces applied on the beam in the transverse (i.e. perpendicular to the beam axis) directions (y and z) cause the beam to bend as well as shear. The resulting transverse deflection would comprise bending deflections ($w_b$, $v_b$) and shearing deflections ($w_s$, $v_s$). Since the thickness of the beam is a smaller dimension compared to the axial length, the transverse loads on the beam surface can be visualized to be acting on the centroid of the section. These forces ($F_y$ and $F_z$) cause transverse shear strains over the beam cross section. The figure below shows the shearing action of $F_z$. The shear strain $\gamma_{xz}$ will cause a displacement in the z-direction. As the shape of the cross section is assumed not to change, the transverse deflection due to shear $w_s$ will be the same at all points over a particular cross section. For the present discussion, let us assume that the shear strain and shear stress due to the transverse load $F_z$ are uniformly distributed over the cross section. Obviously, shear strain $\gamma_{xz} = dw_s/dx$, as shown in Fig. 5.18.

The total transverse deflection $w$ of the beam at any section along the axis of the beam is the sum of $w_b$ and $w_s$ for that section, i.e. $w = w_b + w_s$.

Hence, transverse shear strain, $\gamma_{xz} = dw_s/dx = dw/dx - dw_b/dx = dw/dx + \theta_y$.

Following similar arguments as above, the transverse shear strain $\gamma_{xy}$ due to the application of shear force $\mathbf{F_y}$ is given by $\gamma_{xy} = dv/dx - \theta_z$.

The mathematical integration of internal shear stress distributions $\tau_{xy}$ and $\tau_{xz}$, over the cross-sectional area of the beam, yields the internal shear stress resultants called the *shear forces* $Q_y$ and $Q_z$, respectively.

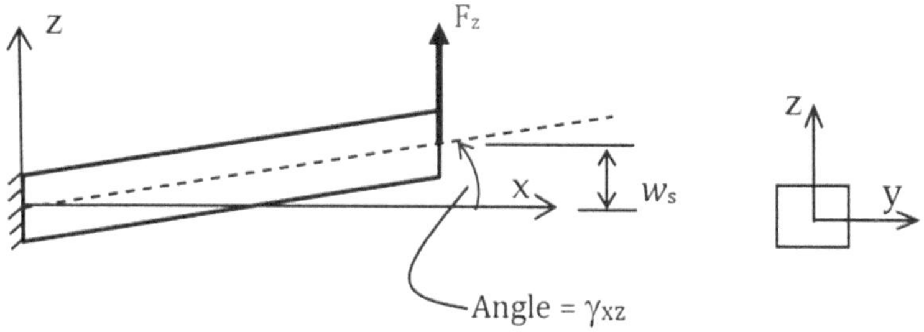


**Fig. 5.18** Shear deformation due to transverse load Fz

In the case of pure bending (bending without transverse shear, caused by a couple applied on the beam), the deformed plane and the deformed beam axis retain their perpendicularity. Interestingly, the presence of transverse strain spoils the perpendicularity between the plane of cross section and the beam axis, and it differs by an angle which is numerically equal to the transverse shear strain.

**Twisting Moment ($M_x$):** A torque applied on the beam causes cross-section rotation about the x-axis ($\theta_x$). For small rotations due to an applied positive torque (Cx), a point on the + ve y-axis of the cross section at a distance $y$ from the centroid undergoes a + ve $w$-displacement ($= y.\theta_x$) and will not have $v$ displacement. Similarly, a point on the + ve z-axis at a distance $z$ from the centroid undergoes a –ve $v$-displacement ($= -z.\theta_x$) and will not have $w$-displacement. Any arbitrary point ($y,z$) not on these axes on the cross section shall have both ***w*** and ***v*** displacements.

The displacement fields $\underline{u}$, $\underline{v}$, and $\underline{w}$ at any point ($x,y,z$) on a cross section (at $x$) can now be expressed in terms of the displacements ($u$, $v$, $w$) of the centroid of the cross section and rotations ($\theta_x$, $\theta_y$, $\theta_z$) of the cross section. All the centroid displacement fields and the section rotation fields are one-dimensional quantities, i.e. they are only functions of axial coordinate $x$ and do not vary over the cross section (y–z plane). These relations are called *displacement kinematics.*

This section describes the finite element formulation of a 2-node beam element which is arbitrarily oriented in the global coordinate system X–Y-Z and subjected to all the types of loads described above (Fig. 5.19).

**Global Coordinate system**: X–Y-Z.

**Local coordinate system**: (x–y-z) Local x-axis is defined along the axis of the element. Local $y$-axis and local $z$-axis are chosen to match with the axes about which the second moment of area $I_y$ and $I_z$ are specified.

**Natural coordinate system:** $\xi$, along the axis of the element, origin at the centroid. At $x = 0$, $\xi = -1$ and at $x = L$, $\xi = +1$.

**Relevant displacement fields in local coordinate system**: $u$(x),$v$(x), $w$(x), $\theta_x$(x),$\theta_y$(x), $\theta_z$(x).

**Global degrees of freedom:** $\{\mathbf{D}\} = \{U_1, V_1, W_1, \theta_{X1}, \theta_{Y1}, \theta_{Z1}, U_2, V_2, W_2, \theta_{X2}, \theta_{Y2}, \theta_{Z2}\}^T$.

**Local degrees of freedom: {d} =** $\{u_1, v_1, w_1, \theta_{x1}, \theta_{y1}, \theta_{z1}, u_2, v_2, w_2, \theta_{x2}, \theta_{y2}, \theta_{z2}\}^T$.

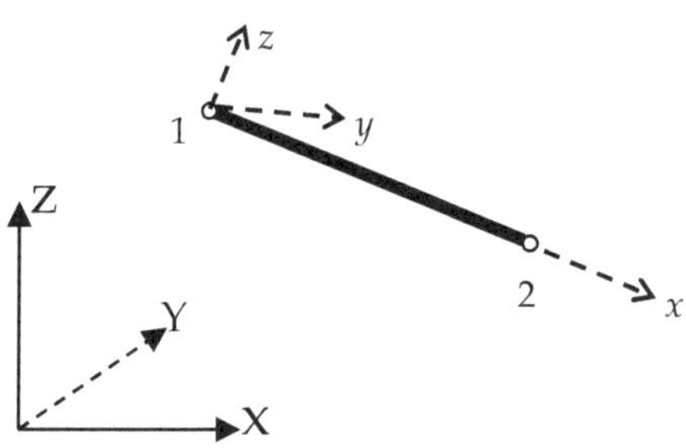


**Fig. 5.19** A 2-node beam finite element with global and local coordinate systems

$$\textbf{Shape functions}: \; N_1(\xi) = \frac{1}{2}(1-\xi); \quad N_2(\xi) = \frac{1}{2}(1+\xi). \tag{5.53}$$

**Displacement Field interpolation**: $u(x) = N_1(\xi)u_1 + N_2(\xi)u_2$. Similar interpolation is employed for the other displacement fields $v(x)$.

**Displacement Kinematics**: The three-dimensional displacement fields $\underline{u}$, $\underline{v}$, and $\underline{w}$ are expressed in terms of one-dimensional fields of centroid displacements and section rotations as

$$\begin{aligned} \underline{u}(x,y,z) &= u(x) + z\theta_y(x) - y\theta_z(x) \\ \underline{v}(x,y,z) &= v(x) - z\theta_x(x) \\ \underline{w}(x,y,z) &= w(x) + y\theta_x(x). \end{aligned} \tag{5.54}$$

**Strain and Stress components in local coordinate system**:

$$\varepsilon_{xx} = \frac{\partial \underline{u}}{\partial x} = \frac{\partial u}{\partial x} + z\frac{\partial \theta_y}{\partial x} - y\frac{\partial \theta_z}{\partial x}; \qquad \varepsilon_{yy} = \frac{\partial \underline{v}}{\partial y} = 0; \qquad \varepsilon_{zz} = \frac{\partial \underline{w}}{\partial x} = 0 \tag{5.55}$$

$$\gamma_{xy} = \frac{\partial \underline{u}}{\partial y} + \frac{\partial \underline{v}}{\partial x} = -\theta_z + \frac{\partial v}{\partial x} - z\frac{\partial \theta_x}{\partial x}; \quad \gamma_{yz} = \frac{\partial \underline{v}}{\partial z} + \frac{\partial \underline{w}}{\partial y} = 0;$$

$$\gamma_{xz} = \frac{\partial \underline{u}}{\partial z} + \frac{\partial \underline{w}}{\partial x} = \theta_y + \frac{\partial w}{\partial x} + y\frac{\partial \theta_x}{\partial x}$$

Stress resultants for a beam are the axial force and bending moments computed by integrating the effect of stress distribution over the cross section. They are useful for conveniently expressing the strain energy of the beam.

**The stress resultants**:

The Axial Force in terms of the area of cross section (A) is given by

$$\mathrm{N_x} = \int_A \sigma_x dA = \int_{-h/2}^{h/2}\int_{-b/2}^{b/2} E\left(\frac{\partial u}{\partial x} + z\frac{\partial \theta_y}{\partial x} - y\frac{\partial \theta_z}{\partial x}\right)dydz = EA\frac{\partial u}{\partial x}. \tag{5.56}$$

The Bending moments in terms of the second moments of cross-section area $I_z$ and $I_y$ about the local z- and y-axes:

$$\mathrm{M_y} = \int_A z\sigma_x dA = \int_{-h/2}^{h/2}\int_{-b/2}^{b/2} E\left(z\frac{\partial u}{\partial x} + z^2\frac{\partial \theta_y}{\partial x} - yz\frac{\partial \theta_z}{\partial x}\right)dydz = EI_y\frac{\partial \theta_y}{\partial x} \tag{5.57}$$

$$\mathrm{M_z} = \int_A y\sigma_x dA = \int_{-h/2}^{h/2}\int_{-b/2}^{b/2} E\left(y\frac{\partial u}{\partial x} + yz\frac{\partial \theta_y}{\partial x} - y^2\frac{\partial \theta_z}{\partial x}\right)dydz = EI_z\left(-\frac{\partial \theta_z}{\partial x}\right). \tag{5.58}$$

The Shear Forces:

$$Q_z = \int_A \tau_{xz} dA = \int_{-h/2}^{h/2} \int_{-b/2}^{b/2} G\left(\theta_y + \frac{\partial w}{\partial x} + y\frac{\partial \theta_x}{\partial x}\right) dydz = GA\left(\theta_y + \frac{\partial w}{\partial x}\right) \tag{5.59}$$

$$Q_y = \int_A \tau_{xy} dA = \int_{-h/2}^{h/2} \int_{-b/2}^{b/2} G\left(-\theta_z + \frac{\partial v}{\partial x} - z\frac{\partial \theta_x}{\partial x}\right) dydz = GA\left(-\theta_z + \frac{\partial v}{\partial x}\right). \tag{5.60}$$

The Twisting moment in terms of the polar moment of cross-sectional area $I_x$ about the x-axis

$$M_x = \int_A (y\tau_{xz} + z\tau_{xy}) dA = GI_x \frac{\partial \theta_x}{\partial x} \quad \text{where} \quad I_x = \int_A (y^2 + z^2) dA. \tag{5.61}$$

For a circular cross section of the beam, $I_x$ happens to be equal to the classical torsional constant for the section. For noncircular cross sections, where the shear distribution of the cross section is not 'smooth', the torsional constant slightly differ from the polar moment of area, which can be computed analytically. Hence, a correction factor, *k,* is generally multiplied by $I_x$ to take care of this difference. For rectangular cross sections, the value of *k* is typically taken as 5/6.

**Constitutive matrix [E]**: The constitutive matrix (underlined in the equation below) for the beam element which relates the strain vector to the stress resultant vector is given by

$$\begin{Bmatrix} N_x \\ M_y \\ M_z \\ Q_z \\ Q_y \\ M_x \end{Bmatrix} = \underline{\begin{bmatrix} EA & 0 & 0 & 0 & 0 & 0 \\ 0 & EI_y & 0 & 0 & 0 & 0 \\ 0 & 0 & EI_z & 0 & 0 & 0 \\ 0 & 0 & 0 & kGA & 0 & 0 \\ 0 & 0 & 0 & 0 & kGA & 0 \\ 0 & 0 & 0 & 0 & 0 & kGI_x \end{bmatrix}} \begin{Bmatrix} \frac{\partial u}{\partial x} \\ \frac{\partial \theta_y}{\partial x} \\ -\frac{\partial \theta_z}{\partial x} \\ \theta_y + \frac{\partial w}{\partial x} \\ -\theta_z + \frac{\partial v}{\partial x} \\ \frac{\partial \theta_x}{\partial x} \end{Bmatrix} \tag{5.62}$$

In Timoshenko beam theory, the effects of transverse shear strain/stress are accounted for in the element formulation. However, we assumed the transverse shear stress is distributed uniformly over the cross-sectional thickness and is equal to $Q_{xz}/A$. In fact, from the theory of elasticity, we know that this is not the reality. The shear strain distribution has to satisfy the traction-free top and bottom surface boundary condition, and the actual distribution takes a parabolic pattern across the thickness. In the Timoshenko beam theory, this difference is accounted for by introducing a factor *k* called the shear correction factor, in the constitutive matrix. The *shear correction factor* may be defined as the ratio of the effective area in bearing the shear strain to the actual area of the beam. The value of *k* depends on the actual shape of the cross section. *k* is taken to be equal to 5/6 for beams of rectangular cross sections.

**Elastic strain energy**: The elastic strain energy of the element, in terms of stress resultants and the strain components, can now be written as

$$
\begin{aligned}
\mathrm{U} = \int_0^L \Bigg[ & N_x\left(\frac{\partial u}{\partial x}\right) + M_y\left(\frac{\partial \theta_y}{\partial x}\right) + M_z\left(-\frac{\partial \theta_z}{\partial x}\right) + \\
& Q_z\left(\theta_y + \frac{\partial w}{\partial x}\right) + Q_y\left(-\theta_z + \frac{\partial v}{\partial x}\right) + M_x\left(\frac{\partial \theta_x}{\partial x}\right)\Bigg] dx.
\end{aligned}
\tag{5.63}
$$

**Strain–displacement matrix [B]:** The strain–displacement matrix [B] (shown underlined in the equation below) which relates the vector of strain field $\{\varepsilon\}$ to the node displacement vector {d} is given by

$$
\begin{Bmatrix} \frac{\partial u}{\partial x} \\ \frac{\partial \theta_y}{\partial x} \\ -\frac{\partial \theta_z}{\partial x} \\ \theta_y + \frac{\partial w}{\partial x} \\ -\theta_z + \frac{\partial v}{\partial x} \\ \frac{\partial \theta_x}{\partial x} \end{Bmatrix} = \underline{\begin{bmatrix}
\frac{\partial N_1}{\partial x} & 0 & 0 & 0 & 0 & 0 & \frac{\partial N_2}{\partial x} & 0 & 0 & 0 & 0 & 0 \\
0 & 0 & 0 & 0 & \frac{\partial N_1}{\partial x} & 0 & 0 & 0 & 0 & 0 & \frac{\partial N_2}{\partial x} & 0 \\
0 & 0 & 0 & 0 & 0 & -\frac{\partial N_1}{\partial x} & 0 & 0 & 0 & 0 & 0 & -\frac{\partial N_2}{\partial x} \\
0 & 0 & \frac{\partial N_1}{\partial x} & 0 & N_1 & 0 & 0 & 0 & \frac{\partial N_2}{\partial x} & 0 & N_2 & 0 \\
0 & \frac{\partial N_1}{\partial x} & 0 & 0 & 0 & -N_1 & 0 & \frac{\partial N_2}{\partial x} & 0 & 0 & 0 & -N_2 \\
0 & 0 & 0 & \frac{\partial N_1}{\partial x} & 0 & 0 & 0 & 0 & 0 & \frac{\partial N_2}{\partial x} & 0 & 0
\end{bmatrix}} \begin{Bmatrix} u_1 \\ v_1 \\ w_1 \\ \theta_{x1} \\ \theta_{y1} \\ \theta_{z1} \\ u_2 \\ v_2 \\ w_2 \\ \theta_{x2} \\ \theta_{y2} \\ \theta_{z2} \end{Bmatrix}
$$

$$
\text{where } \frac{\partial N_1}{\partial x} = \frac{\partial N_1}{\partial \xi} \cdot \frac{\partial \xi}{\partial x} = \frac{-1}{2} \cdot \frac{2}{L} = \frac{-1}{L} \quad \text{and} \quad \frac{\partial N_2}{\partial x} = \frac{1}{L}. \tag{5.64}
$$

**Jacobian matrix** [J]**:** The Jacobian matrix for this element is given by $[\mathrm{J}] = \left[\frac{\partial x}{\partial \xi}\right] = \left[\frac{L}{2}\right]$ where $L$ is the length of the beam element.

**Gauss integration rule**: 2-point scheme.

**Element Stiffness Matrix in local coordinate system**: The element stiffness matrix can now be computed using Gauss numerical integration rule as

$$
[\mathrm{k}] = \int_{-1}^{1} [\mathrm{B}]^T [\mathrm{E}][\mathrm{B}] J d\xi = \sum_{i=1}^{2} [\mathrm{B}(\xi_i)]^T [\mathrm{E}][\mathrm{B}(\xi_i)] J(\xi_i) \mathrm{W}_\mathrm{i}. \tag{5.65}
$$

*Relative Proportion of shear and bending energies in Timoshenko beams*

The above formulation is based on Timoshenko beam theory which assumes that plane cross sections remain plane even after the beam deformation, but are not necessarily normal to the beam axis. This theory is very important and relevant for thick beams, where the transverse shear deformation cannot be neglected. It is worthwhile to investigate the relative proportion of transverse shear and bending energies in a tip-loaded cantilever beam of length $l$ and thickness $h$. It can be shown that the ratio of transverse shear energy to bending energy is proportional to $(h/l)^2$. Hence, as the beam becomes thin ($h << l$), the shear energy stored in it becomes negligible compared to the bending energy and hence the effects of shear strain may be neglected. This leads to the well-known Euler–Bernoulli beam theory.

*Thin Beam (Euler–Bernoulli beam) theory*

Euler–Bernoulli beam formulation assumes that the plane cross section remains plane and normal to the beam axis, even after the beam deformation. This is more applicable for thin beams where shear deformation is very small in comparison with the bending deformation and can be ignored. It means that the beam slope in the local x–z plane, d$w$/d$x$, and the section rotation are numerically the same. Precisely, the section rotation $\theta_y = -\frac{\partial w}{\partial x}$ and the bending strain $\frac{\partial \theta_y}{\partial x} = -\frac{\partial^2 w}{\partial x^2}$. Similarly, on the local x–y plane, the slope d$v$/d$x$ and the section rotation are related as $\theta_z = \frac{\partial v}{\partial x}$ and the bending strain, $\frac{\partial \theta_z}{\partial x} = \frac{\partial^2 v}{\partial x^2}$.

## 5.17 Stiffness Matrix for Axisymmetric Shell Element

The formulation of a 2-node axisymmetric shell element is given here. An axisymmetric shell is formed by revolving a thin cross section in the R-Z plane (as shown in Fig. 5.20) about the Z-axis, by 360 degrees. The Z-axis is called the *axis of revolution.* R-axis is called *radial axis.*

The middle line of the section is called the ***meridian or meridional line***. A local x-axis is formed along this line, in the direction of the node. For the purpose of explanation, we consider here only the structures with a straight meridian. A typical point on this axisymmetric shell meridian represents a material ***circle (ring) of revolution*** whose radius is $r$, as shown in Fig. 5.20. At any point on the axisymmetric shell, a line normal to the meridian meets the axis of revolution at an angle $\psi$, called ***meridian angle***, as shown in Fig. 5.20.

For the applicability of the theory of axisymmetric shell, not only the geometry but also the applied load should be 'axisymmetric', so that the deformations are also axisymmetric. Therefore, a concentrated load drawn at a point on an axisymmetric

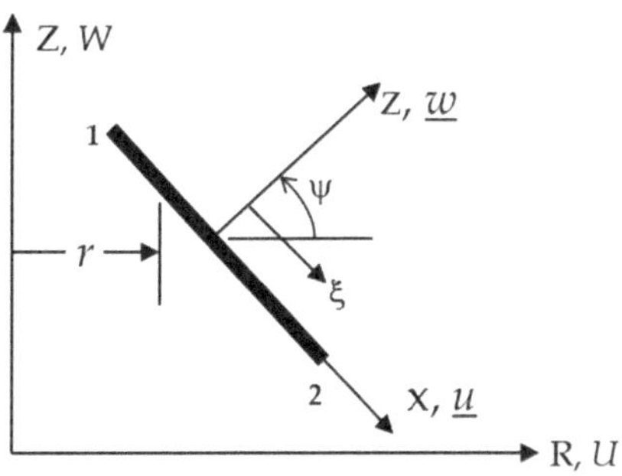


**Fig. 5.20** A 2-node Axisymmetric shell element

shell represents a line of load, distributed all around, along the circle of revolution passing through that point. Hence, displacements and rotations are restricted to the R-Z plane, and they do not vary in the circumferential direction. In other words, all the points lying on the same circle of revolution will experience the same strains and stresses.

The thin axisymmetric shell cross section which is revolved essentially appears like a beam with a centroid axis, *x*, and a transverse axis, *z*, as shown in Fig. 5.20. The *x–z* coordinate system is considered as the *Local coordinate system.* At any point on the meridian, the *y*-axis of the local coordinate system which is directed normal and into this paper is essentially tangent to the circle of rotation. Hence, this direction is also called *circumferential* direction or *hoop* direction or $\theta$-direction.

Like beam theory, we have already explained, for deriving the displacement kinematics for an axisymmetric shell, we have to represent the displacement fields ($\underline{u}$(x, z) and $\underline{w}$(x, z) in the x–z local coordinate system) in terms of their meridional values ($u$(x) and $w$(x)) and rotation ($\theta_y$) of the section-line. The section-line is perpendicular to the meridian. As evident from Fig. 5.20, the kinematics is going to be similar to a planar beam with a centroid axis as the meridional line and its thickness along the transverse direction, z.

This section describes the finite element formulation of stiffness matrix for a 2-node axisymmetric shell element which is arbitrarily oriented in the global coordinate system R-Z and subjected to axial force along x-direction ($N_x$), bending moment about ($M_y$), and shear force ($Q_z$) in the local coordinate system. Accordingly, the degrees of freedom in the local x–y-z coordinate system are $u$, $w$, and $\theta$.

**Global Coordinate system**: R–θ–Z.

**Local coordinate system**: (*x–z*) Local x-axis is defined along the axis of the element, in the direction from node-1 to node-2. Local z-axis is normal to the *x*-axis, such that the circumferential direction (θ) is directed into the paper, such that x–θ–z forms a right-handed local coordinate system.

**Natural coordinate system:** $\xi$, along the axis of the element, origin at the centroid. At $x = 0$, $\xi = -1$ and at $x = L$, $\xi = +1$.

**Relevant displacement fields in local coordinate system**: $u$(x), $w$(x), $\theta$(x).

**Global degrees of freedom:** $\{\mathbf{D}\} = \{U_1, W_1, \theta_1, U_2, W_2, \theta_2,\}^{\mathrm{T}}$.

**Local degrees of freedom:**$\{\mathbf{d}\} = \{u_1, w_1, \theta_1, u_2, w_2, \theta_2\}^{\mathrm{T}}$.

$$\textbf{Shape functions}: \quad N_1(\xi) = \frac{1}{2}(1-\xi); \quad N_2(\xi) = \frac{1}{2}(1+\xi). \tag{5.66}$$

**Displacement Field interpolation:** $u(x) = N_1(\xi)u_1 + N_2(\xi)u_2$. Similar interpolations are employed for $w$ and $\theta$ fields.

**Displacement Kinematics:** The two-dimensional displacement fields $\underline{u}(x,z)$, $\underline{w}(x,z)$ are expressed in terms of one-dimensional fields involving centroid displacements $u(x)$, $w(x)$ and section rotation $\theta(x)$ as

$$\begin{aligned} \underline{u}(x,z) &= u(x) + z\theta(x) \\ \underline{w}(x,z) &= w(x). \end{aligned} \tag{5.67}$$

**Strain components in local coordinate system**:

$$\varepsilon_x = \frac{\partial \underline{u}}{\partial x} = \frac{\partial u}{\partial x} + z\frac{\partial \theta}{\partial x} \tag{5.68}$$

$$\gamma_{xz} = \frac{\partial \underline{u}}{\partial z} + \frac{\partial \underline{w}}{\partial x} = \theta + \frac{\partial w}{\partial x}. \tag{5.69}$$

Apart from the above strain in the x–z plane, a normal strain called the *hoop strain* or *circumferential strain* also exists in the case of an axisymmetric shell. We shall call this strain $\varepsilon_{\theta}$. This strain is caused by the radial expansion/contraction of a *ring of revolution* at any material point on the structure.

From the concept of radial expansion or contraction of a ring, we know that the hoop or circumferential strain in the ring is $U/r$ where $U$ is the radial displacement (in global coordinate system), and $r$ is the distance of the point from the axis of revolution. In the case of an axisymmetric shell, the radial displacement ($U$) in the global coordinate system at any point has to be computed by adding the components (in the radial direction), of the local meridian ($u$) and local transverse displacement ($w$), as

$$U = \underline{w}Cos\psi + \underline{u}Sin\psi = wCos\psi + uSin\psi + z\theta Sin\psi. \tag{5.70}$$

Therefore, the hoop strain or circumferential strain is given by

$$\varepsilon_{\theta} = \frac{U}{r} = \frac{wCos\psi + uSin\psi}{r} + z\frac{\theta Sin\psi}{r}. \tag{5.71}$$

It is important to note that $\varepsilon_{\theta}$ consists of a part which is uniform over the thickness and another part which is linearly varying (function of z) across the thickness.

Stress resultants on the cross section formed by unit circumferential length and thickness of the shell ($h$) can be computed as force and moment by integrating the stress distribution over the cross section. They are useful for conveniently expressing the strain energy of the element.

**The role of plane stress relations**:

The normal stresses in the axisymmetric shell are meridian stress ($\sigma_x$) and the hoop stress ($\sigma_\theta$). As $\sigma_z$ is zero, it is appropriate to employ plane stress constitutive equations to relate the stresses and strains in the x-θ plane:

$$\begin{Bmatrix} \sigma_x \\ \sigma_\theta \\ \tau_{xz} \end{Bmatrix} = \frac{E}{1-\upsilon^2}\begin{bmatrix} 1 & \nu & 0 \\ \nu & 1 & 0 \\ 0 & 0 & \frac{(1-\nu)}{2} \end{bmatrix}\begin{Bmatrix} \varepsilon_x \\ \varepsilon_\theta \\ \gamma_{xz} \end{Bmatrix} \tag{5.72}$$

$$= \frac{E}{1-\upsilon^2}\begin{bmatrix} 1 & \nu & 0 \\ \nu & 1 & 0 \\ 0 & 0 & \frac{(1-\nu)}{2} \end{bmatrix}\begin{Bmatrix} \frac{\partial u}{\partial x} + z\frac{\partial \theta_y}{\partial x} \\ \frac{wCos\psi + uSin\psi}{r} + z\frac{\theta_y Sin\psi}{r} \\ \theta_y + \frac{\partial w}{\partial x} \end{Bmatrix}. \quad (5.73)$$

The transverse stress–strain relation is $\tau_{xz} = G\gamma_{xz}$.

**Elastic strain energy**:

The strain energy (U) of the axisymmetric shell is given by

$$\mathrm{U} = \frac{1}{2}\int_{\mathrm{V}} (\sigma_{\mathrm{x}}\varepsilon_{\mathrm{x}} + \sigma_\theta\varepsilon_\theta + \tau_{\mathrm{xz}}\gamma_{\mathrm{xz}})\mathrm{dv} \quad (5.74)$$

substituting for strains from Eqs. (5.68–5.71),

$$\mathrm{U} = \frac{1}{2}\int_0^{\mathrm{L}}\int_{-\mathrm{h}/2}^{\mathrm{h}/2}\left[\sigma_{\mathrm{x}}\left(\frac{\partial \mathrm{u}}{\partial \mathrm{x}}\right) + \mathrm{z}\sigma_{\mathrm{x}}\left(\frac{\partial \theta_{\mathrm{y}}}{\partial \mathrm{x}}\right)\right.$$
$$+\sigma_\theta\left(\frac{wCos\psi + uSin\psi}{r}\right) + z\sigma_\theta\left(\frac{\theta_y Sin\psi}{r}\right)$$
$$\left.+\tau_{xz}\left(\theta_y + \frac{\partial w}{\partial x}\right)\right]2\pi r dz dx$$

Since the deformations and strains and, hence, stresses do not vary along the circumferential direction, integration along the circumferential direction can be done by multiplying with 2πr, as

$$U = \frac{1}{2}\int_0^L\int_{-\frac{h}{2}}^{\frac{h}{2}}\left[\sigma_x\left(\frac{\partial u}{\partial x}\right) + z\sigma_x\left(\frac{\partial \theta_y}{\partial x}\right) + \sigma_\theta\left(\frac{wCos\psi + uSin\psi}{r}\right) + z\sigma_\theta\left(\frac{\theta_y Sin\psi}{r}\right) + \tau_{xz}\left(\theta_y + \frac{\partial w}{\partial x}\right)\right]$$
$$2\pi r dz dx \quad (5.75)$$

In the above equation, the quantities in curved brackets do not vary along the thickness direction. Therefore, integration of the above equation in thickness direction would lead to strain energy written in terms of the stress resultants viz. in-plane force, bending moment, and shear force per unit circumferential length:

$$U = \frac{1}{2}\int_0^L\left[N_x\left(\frac{\partial u}{\partial x}\right) + M_y\left(\frac{\partial \theta_y}{\partial x}\right) + N_\theta\left(\frac{wCos\psi + uSin\psi}{r}\right) + M_\theta\left(\frac{\theta_y Sin\psi}{r}\right) + Q_z\left(\theta_y + \frac{\partial w}{\partial x}\right)\right]$$
$$2\pi r dx \quad (5.76)$$

where $N_x = \int_{-h/2}^{h/2}\sigma_x dz$; $N_\theta = \int_{-h/2}^{h/2}\sigma_\theta dz$; $M_y = \int_{-h/2}^{h/2} z\sigma_x dz$; $M_\theta = \int_{-h/2}^{h/2} z\sigma_\theta dz$; $Q_z = \int_{-h/2}^{h/2}\gamma_{xz} dz$.

$N_x$ is called the meridional force; $N_\theta$ is called the hoop or circumferential force; $M_y$ is called meridional moment, $M_\theta$ is called the hoop or circumferential moment, and $Q_z$ is the transverse shear force. It should be noted that the above stress resultants are per unit circumferential length.

**Constitutive matrix [E]**: The constitutive matrix (underlined in the equation below) for axisymmetric shell element which relates the stress resultant vector to the strain vector is given by the relation:

$$\begin{Bmatrix} N_x \\ N_\theta \\ M_y \\ M_\theta \\ Q_z \end{Bmatrix} = \underline{\frac{Eh}{1-\nu^2}\begin{bmatrix} 1 & \nu & 0 & 0 & 0 \\ \nu & 1 & 0 & 0 & 0 \\ 0 & 0 & h^2/12 & \nu h^2/12 & 0 \\ 0 & 0 & \nu h^2/12 & h^2/12 & 0 \\ 0 & 0 & 0 & 0 & \frac{(1-\nu)}{2} \end{bmatrix}} \begin{Bmatrix} \frac{\partial u}{\partial x} \\ \frac{wCos\psi + uSin\psi}{r} \\ \frac{\partial \theta_y}{\partial x} \\ \frac{\theta_y Sin\psi}{r} \\ \theta_y + \frac{\partial w}{\partial x} \end{Bmatrix}. \tag{5.77}$$

**Strain–displacement matrix [B]**: The strain–displacement matrix [B] (shown underlined in the equation below) which relates the vector of the strain field $\{\varepsilon\}$ to the node displacement vector **{d}** is given by

$$\begin{Bmatrix} \frac{\partial u}{\partial x} \\ \frac{wCos\psi + uSin\psi}{r} \\ \frac{\partial \theta_y}{\partial x} \\ \frac{\theta_y Sin\psi}{r} \\ \theta_y + \frac{\partial w}{\partial x} \end{Bmatrix} = \begin{bmatrix} \frac{\partial N_1}{\partial x} & 0 & 0 & \frac{\partial N_2}{\partial x} & 0 & 0 \\ \frac{N_1 Sin\psi}{r} & \frac{N_1 Cos\psi}{r} & 0 & \frac{N_2 Sin\psi}{r} & \frac{N_2 Cos\psi}{r} & 0 \\ 0 & 0 & \frac{\partial N_1}{\partial x} & 0 & 0 & \frac{\partial N_2}{\partial x} \\ 0 & 0 & \frac{N_1 Sin\psi}{r} & 0 & 0 & \frac{N_2 Sin\psi}{r} \\ 0 & \frac{\partial N_1}{\partial x} & N_1 & 0 & \frac{\partial N_2}{\partial x} & N_2 \end{bmatrix} \begin{Bmatrix} u_1 \\ w_1 \\ \theta_1 \\ u_2 \\ w_2 \\ \theta_2 \end{Bmatrix} \tag{5.78}$$

$\frac{\partial N_1}{\partial x} = \frac{\partial N_1}{\partial \xi} \cdot \frac{\partial \xi}{\partial x} = \frac{-1}{2} \cdot \frac{2}{L} = \frac{-1}{L}$ and $\frac{\partial N_2}{\partial x} = \frac{1}{L}$.

**Jacobian matrix** [J]: The Jacobian for this element is given by $[J] = \left[\frac{\partial x}{\partial \xi}\right] = \left[\frac{L}{2}\right]$ where $L$ is the length of the element.

**Gauss integration rule**: 2-point rule.

**Element Stiffness Matrix in local coordinate system**: The element stiffness matrix can now be computed using Gauss numerical integration rule as

$$[k] = \int_0^{2\pi} \int_{-1}^{1} [B]^T [E][B] J \, d\xi \; r d\theta = 2\pi \sum_{i=1}^{2} [B(\xi_i)]^T [E][B(\xi_i)] J(\xi_i) \, r(\xi_i) W_i. \tag{5.79}$$

## 5.18 Stiffness Matrix for Plate element

This section considers the two-dimensional counterpart of beams, namely plates. A plate geometry is obtained by extending a long beam axis (x) in one of its transverse

(y) directions This results in a thin sheet of material having top and bottom surfaces (with +z and –z-axis as outward normals, respectively). While the beam had two dimensions much shorter than the large length dimension, the plate has only one dimension, called the thickness, much shorter than the other two large dimensions viz. length and breadth. Also, corresponding to the centroid axis in a beam, the plate has a mid-plane.

**Geometry**: Plates are characterized by a structure that is bounded by upper and lower surface planes that are separated by a distance $h$ as shown in Fig. 5.21. A mid-plane can be defined all along the plate, which spans the middle of thickness at all points. A local x–y coordinate system is placed on the mid-plane of the plate, so that the mid-plane corresponds to z = 0 at all locations. The x- and y-directions are called the "in-plane" directions. The z-axis at any point is normal to the x–y plane. In the present context, it is assumed that the thickness $h$ is a constant and material properties are homogeneous through the thickness. Consequently, the upper and lower surfaces correspond to $z = +h/2$ and $z = -h/2$, respectively.

In most practical applications of plate structures, they are subjected to distributed normal to the plate (z-direction) or concentrated loads normal to the external plate surface. These loads tend to produce bending and transverse shearing of the plate. They can also be subjected to in-plane tensile or compressive forces which produce in-plane normal stresses as well as in-plane shear stresses. Moments and transverse loads applied to the edge of the plate cause bending or transverse shearing of the plate. All the deformations produced by these loads, in general, can be represented by displacement fields $\underline{u}(x, y, z)$, $\underline{v}(x,y,z)$, and $\underline{w}(x,y,z)$.

The thickness at any point on the mid-plane is measured along the unique 'normal' which is perpendicular to the mid-plane, at that point. When extension, bending, and shear loads act on plates, the normal line undergoes displacements along x-direction (u), y-direction (v), and z-direction ($w$) as well as rotation about x-axis ($\theta_x$)

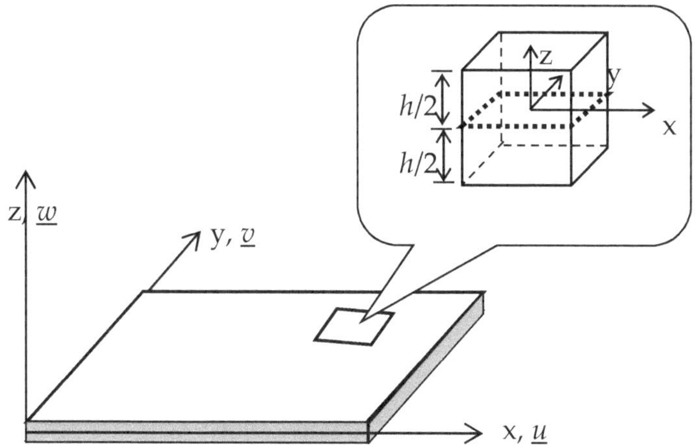


**Fig. 5.21** Geometry of a plate structure

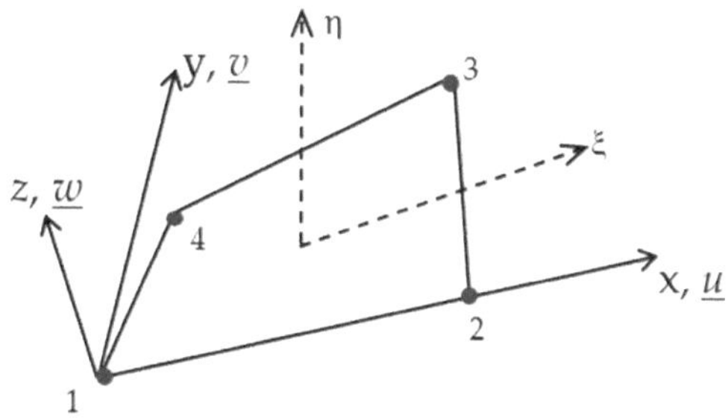


**Fig. 5.22** Geometry of a 4-node plate element lying on the $x$–$y$ plane of the local coordinate system

and rotation about y-axis ($\theta_y$).). These mid-plane displacements and rotations are functions of x and y (Fig. 5.22).

Plate elements are formulated to represent the mid-plane of a plate structure. In order to simplify the three-dimensional general displacement fields $\underline{u}$, $\underline{v}$, and $\underline{w}$ to a two-dimensional form suitable for the plate finite element, a few assumptions valid for *small deformation* and *small strain* situations are employed:

(1) A line normal to the mid-surface of the plate is inextensible (does not stretch or contract) during deformation, i.e. $\varepsilon_{zz} = 0$.
(2) When the plate bends, straight lines originally normal to the mid-plane rotate but remain straight. This implies linear distribution of $u$ and $v$ across the thickness, being zero at the mid-plane and maximum at the top/bottom surfaces.

### Deformation Kinematics

In accordance with the above assumptions, we can describe the three-dimensional displacement field in a two-dimensional form, in terms of displacement fields on the mid-plane: $u(x,y)$, $v(x,y)$, $w(x,y)$, and rotation of normal vector $\theta_x(x,y)$ and $\theta_y(x,y)$ which are independent of $z$:

$$\begin{aligned} \underline{u}(x, y, z) &= u(x, y) + z\theta_y(x, y) \\ \underline{v}(x, y, z) &= v(x, y) - z\theta_x(x, y) \\ \underline{w}(x, y, z) &= w(x, y). \end{aligned} \tag{5.80}$$

**Global degrees of freedom**: $\{D\} = \{U_1, V_1, W_1, \theta_{X1}, \theta_{Y1}, \theta_{Z1} \ldots\ldots U_4, V_4, W_4, \theta_{X4}, \theta_{Y4}, \theta_{Z4}\}^T$.

**Local degrees of freedom**: $\{D\} = \{u_1, v_1, w_1, \theta_{x1}, \theta_{y1} \ldots\ldots u_4, v_4, w_4, \theta_{x4}, \theta_{y4}\}^T$.

**Shape functions**: $N_i = \frac{1}{4}(1 + \xi\xi_i)(1 + \eta\eta_i)$. Substituting the natural coordinate of each of the nodes, we get

$$N_1 = \frac{1}{4}(1 - \xi)(1 - \eta); \quad N_2 = \frac{1}{4}(1 + \xi)(1 - \eta)$$
$$N_3 = \frac{1}{4}(1 + \xi)(1 + \eta); \quad N_4 = \frac{1}{4}(1 - \xi)(1 + \eta).$$

**Displacement Field interpolation**: $u(x,y) = \sum_{i=1}^{4} N_i(\xi,\eta)u_i$. Similarly, interpolation for the other fields $v$, $w$, $\theta_x$, and $\theta_y$.

**Strain–Displacement Relations**:

$$\text{In} - \text{plane strains}: \quad \varepsilon_{xx} = \frac{\partial \underline{u}}{\partial x} = \frac{\partial u}{\partial x} + z\frac{\partial \theta_y}{\partial x}; \quad \varepsilon_{yy} = \frac{\partial \underline{v}}{\partial y} = \frac{\partial v}{\partial y} - z\frac{\partial \theta_x}{\partial y}; \tag{5.81}$$

$$\gamma_{xy} = \frac{\partial \underline{u}}{\partial y} + \frac{\partial \underline{v}}{\partial x} = \frac{\partial u}{\partial y} + \frac{\partial v}{\partial x} + z\left(\frac{\partial \theta_y}{\partial y} - \frac{\partial \theta_x}{\partial x}\right). \tag{5.82}$$

Transverse shear strains:

$$\gamma_{xz} = \frac{\partial \underline{u}}{\partial z} + \frac{\partial \underline{w}}{\partial x} = \theta_y + \frac{\partial w}{\partial x} \tag{5.83}$$

$$\gamma_{yz} = \frac{\partial \underline{v}}{\partial z} + \frac{\partial \underline{w}}{\partial y} = -\theta_x + \frac{\partial w}{\partial y}. \tag{5.84}$$

All the above in-plane strains have a component which is independent of $z$ and a component which varies in the z-direction. Both the transverse shear strains are independent of $z$.

**Stress–Strain Relations**:

For in-plane stresses: (employing plane stress relations)

$$\sigma_x = \frac{E}{1-\nu^2}\left(\varepsilon_x + \nu\varepsilon_y\right); \quad \sigma_y = \frac{E}{1-\nu^2}\left(\varepsilon_y + \nu\varepsilon_x\right); \quad \tau_{xy} = G\gamma_{xy}. \tag{5.85}$$

$$\text{For transverse shear stresses}: \tau_{xz} = G\gamma_{xz}; \quad \tau_{yz} = G\gamma_{yz}. \tag{5.86}$$

**Elastic strain energy**:

$$U = \frac{1}{2}\int_V \left(\sigma_x\varepsilon_x + \sigma_y\varepsilon_y + \tau_{xy}\gamma_{xy} + \tau_{xz}\gamma_{xz} + \tau_{yz}\gamma_{yz}\right)dV. \tag{5.87}$$

Substituting for strains from Eqs. (5.81–5.84),

$$\begin{aligned} U = \int_x\int_y\int_{-h/2}^{+h/2} &\left[\sigma_x\left(\frac{\partial u}{\partial x}\right) + z\sigma_x\left(\frac{\partial \theta_y}{\partial x}\right) + \sigma_y\left(\frac{\partial v}{\partial y}\right)\right. \\ &-z\sigma_y\left(\frac{\partial \theta_x}{\partial y}\right) + \tau_{xy}\left(\frac{\partial u}{\partial y} + \frac{\partial v}{\partial x}\right) + z\tau_{xy}\left(\frac{\partial \theta_y}{\partial y} - \frac{\partial \theta_x}{\partial x}\right) \\ &\left.+\tau_{xz}\left(\theta_y + \frac{\partial w}{\partial x}\right) + \tau_{yz}\left(-\theta_x + \frac{\partial w}{\partial y}\right)\right]dxdydz. \end{aligned} \tag{5.88}$$

**Stress Resultants**

In the above equation, the quantities in curved brackets do not vary along the thickness direction. Therefore, integration of the above equation in thickness direction would lead to the strain energy written in terms of the stress resultants viz. in-plane forces, bending moments, and shear forces per unit edge length, as

$$\begin{aligned} U = \int_x \int_y \Bigg[ & N_x\left(\frac{\partial u}{\partial x}\right) + N_y\left(\frac{\partial v}{\partial y}\right) + N_{xy}\left(\frac{\partial u}{\partial y} + \frac{\partial v}{\partial x}\right) \\ & + M_x\left(-\frac{\partial \theta_x}{\partial y}\right) + M_y\left(\frac{\partial \theta_y}{\partial x}\right) + Q_{xz}\left(\theta_y + \frac{\partial w}{\partial x}\right) \\ & + Q_{yz}\left(-\theta_x + \frac{\partial w}{\partial y}\right) + M_{xy}\left(\frac{\partial \theta_y}{\partial y} - \frac{\partial \theta_x}{\partial x}\right)\Bigg] dxdy. \end{aligned} \quad (5.89)$$

Thus, the stress resultants per unit length (shown in Fig. 5.23) are defined as follows:

$$\text{In} - \text{plane Normal force, } N_x = \int_{-h/2}^{+h/2} \sigma_x dz = \frac{Eh}{1-\nu^2}\left(\frac{\partial u}{\partial x} + \nu \frac{\partial v}{\partial y}\right). \quad (5.90)$$

$$\text{In} - \text{plane Normal force, } N_y = \int_{-h/2}^{+h/2} \sigma_y dz = \frac{Eh}{1-\nu^2}\left(\frac{\partial v}{\partial y} + \nu \frac{\partial u}{\partial x}\right). \quad (5.91)$$

$$\text{In} - \text{plane Shear force, } N_{xy} = \int_{-h/2}^{+h/2} \tau_{xy} dz = Gh\left(\frac{\partial u}{\partial y} + \frac{\partial v}{\partial x}\right). \quad (5.92)$$

$$\text{Bending Moment about } x - \text{axis, } M_x = \int_{-h/2}^{+h/2} z\sigma_y dz = \frac{Eh^3}{12(1-\nu^2)}\left(-\frac{\partial \theta_x}{\partial y} + \nu \frac{\partial \theta_y}{\partial x}\right). \quad (5.93)$$

$$\text{Bending Moment about } y - \text{axis, } M_y = \int_{-h/2}^{+h/2} z\sigma_x dz = \frac{Eh^3}{12(1-\nu^2)}\left(\frac{\partial \theta_y}{\partial x} - \nu \frac{\partial \theta_x}{\partial y}\right). \quad (5.94)$$

$$\text{Twisting Moment, } M_{xy} = \int_{-h/2}^{+h/2} z\tau_{xy} dz = \frac{Gh^3}{12(1-\nu^2)}\left(\frac{\partial \theta_y}{\partial y} - \frac{\partial \theta_x}{\partial x}\right). \quad (5.95)$$

$$\text{Transverse Shear force, } Q_{xz} = \int_{-h/2}^{+h/2} \tau_{xz} dz = Gh\left(\theta_y + \frac{\partial w}{\partial x}\right). \quad (5.96)$$

$$\text{Transverse Shear force, } Q_{yz} = \int_{-h/2}^{+h/2} \tau_{yz} dz = Gh\left(\theta_x - \frac{\partial w}{\partial y}\right). \quad (5.97)$$

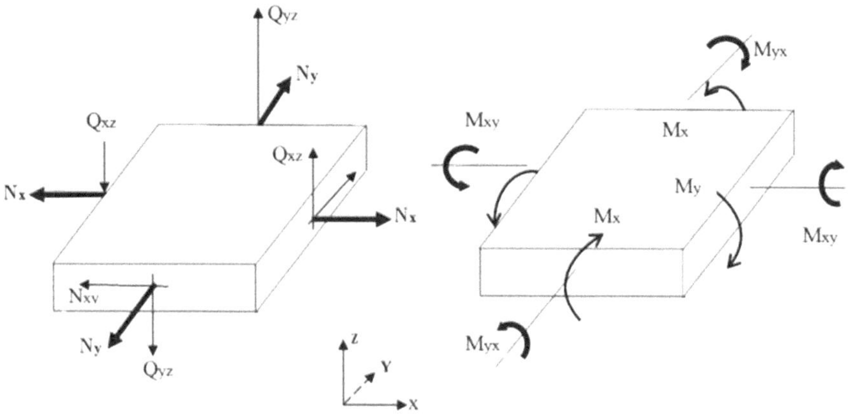


**Fig. 5.23** Stress resultants for a quadrilateral plate element

**Constitutive matrix [E]**: Using the above relations, the constitutive matrix (underlined in the equation below) for the plate element, which relates the stress resultants to the strain vector, can now be arranged as given below

$$\begin{Bmatrix} N_x \\ N_y \\ N_{xy} \\ M_x \\ M_y \\ M_{xy} \\ Q_{xz} \\ Q_{yz} \end{Bmatrix} = \underline{\frac{Eh}{1-\nu^2}\begin{bmatrix} 1 & \nu & 0 & 0 & 0 & 0 & 0 & 0 \\ \nu & 1 & 0 & 0 & 0 & 0 & 0 & 0 \\ 0 & 0 & (1-\nu)/2 & 0 & 0 & 0 & 0 & 0 \\ 0 & 0 & 0 & \nu h^2/12 & h^2/12 & 0 & 0 & 0 \\ 0 & 0 & 0 & h^2/12 & \nu h^2/12 & 0 & 0 & 0 \\ 0 & 0 & 0 & 0 & 0 & (1-\nu)h^2/24 & 0 & 0 \\ 0 & 0 & 0 & 0 & 0 & 0 & (1-\nu)/2 & 0 \\ 0 & 0 & 0 & 0 & 0 & 0 & 0 & (1-\nu)/2 \end{bmatrix}} \begin{Bmatrix} \frac{\partial u}{\partial x} \\ \frac{\partial v}{\partial y} \\ \frac{\partial v}{\partial x} + \frac{\partial u}{\partial y} \\ \frac{\partial \theta_y}{\partial x} \\ -\frac{\partial \theta_x}{\partial y} \\ \frac{\partial \theta_y}{\partial y} - \frac{\partial \theta_x}{\partial x} \\ \theta_y + \frac{\partial w}{\partial x} \\ -\theta_x + \frac{\partial w}{\partial y} \end{Bmatrix}. \tag{5.98}$$

**Strain–displacement matrix [B]:**

The strain–displacement matrix [B] (shown underlined in the equation below) which relates the vector of strain field $\{\varepsilon\}$ to the node displacement vector **{d}** is given by

$$\begin{Bmatrix} \frac{\partial u}{\partial x} \\ \frac{\partial v}{\partial y} \\ \frac{\partial v}{\partial x} + \frac{\partial u}{\partial y} \\ \frac{\partial \theta_y}{\partial x} \\ -\frac{\partial \theta_x}{\partial y} \\ \frac{\partial \theta_y}{\partial y} - \frac{\partial \theta_x}{\partial x} \\ \theta_y + \frac{\partial w}{\partial x} \\ -\theta_x + \frac{\partial w}{\partial y} \end{Bmatrix} = \underline{\begin{bmatrix} \frac{\partial N_1}{\partial x} & 0 & 0 & 0 & 0 & \dots & \frac{\partial N_4}{\partial x} & 0 & 0 & 0 & 0 \\ 0 & \frac{\partial N_1}{\partial y} & 0 & 0 & 0 & \dots & 0 & \frac{\partial N_4}{\partial y} & 0 & 0 & 0 \\ \frac{\partial N_1}{\partial y} & \frac{\partial N_1}{\partial x} & 0 & 0 & 0 & \dots & \frac{\partial N_4}{\partial y} & \frac{\partial N_4}{\partial x} & 0 & 0 & 0 \\ 0 & 0 & 0 & 0 & \frac{\partial N_1}{\partial x} & \dots & 0 & 0 & 0 & 0 & \frac{\partial N_4}{\partial x} \\ 0 & 0 & 0 & -\frac{\partial N_1}{\partial y} & 0 & \dots & 0 & 0 & 0 & -\frac{\partial N_4}{\partial y} & 0 \\ 0 & 0 & 0 & -\frac{\partial N_1}{\partial x} & \frac{\partial N_1}{\partial y} & \dots & 0 & 0 & 0 & -\frac{\partial N_4}{\partial x} & \frac{\partial N_4}{\partial y} \\ 0 & 0 & \frac{\partial N_1}{\partial x} & 0 & N_1 & \dots & 0 & 0 & \frac{\partial N_4}{\partial x} & 0 & N_4 \\ 0 & 0 & \frac{\partial N_1}{\partial y} & -N_1 & 0 & \dots & 0 & 0 & \frac{\partial N_4}{\partial y} & -N_4 & 0 \end{bmatrix}} \begin{Bmatrix} u_1 \\ v_1 \\ w_1 \\ \theta_{x1} \\ \theta_{y1} \\ u_2 \\ \cdot \\ \cdot \\ \cdot \\ \cdot \\ u_4 \\ v_4 \\ w_4 \\ \theta_{x4} \\ \theta_{y4} \end{Bmatrix}. \tag{5.99}$$

**Jacobian matrix** [J]**:** The Jacobian matrix for this element (underlined in the equation below) relates the derivatives in the natural coordinate system to the derivatives in the global coordinate system.

$$\begin{Bmatrix} \frac{\partial N_i}{\partial \xi} \\ \frac{\partial N_i}{\partial \eta} \end{Bmatrix} = \underline{\begin{bmatrix} \frac{\partial x}{\partial \xi} & \frac{\partial y}{\partial \xi} \\ \frac{\partial x}{\partial \eta} & \frac{\partial y}{\partial \eta} \end{bmatrix}} \begin{Bmatrix} \frac{\partial N_i}{\partial x} \\ \frac{\partial N_i}{\partial y} \end{Bmatrix}. \tag{5.100}$$

The derivatives in the Jacobian matrix here are with respect to the local coordinates *x* and *y*. However, we have only the global coordinates (*X*i, *Y*i, *Z*i) of the nodes in hand. Therefore, we have to employ a global-to-local transformation matrix given by

$$\begin{Bmatrix} x \\ y \\ z \end{Bmatrix} = \begin{bmatrix} l_1 & m_1 & n_1 \\ l_2 & m_2 & n_2 \\ l_3 & m_3 & n_3 \end{bmatrix} \begin{Bmatrix} X \\ Y \\ Z \end{Bmatrix} \quad e.g \quad \frac{\partial x}{\partial \xi} = l_1 \frac{\partial X}{\partial \xi} + m_1 \frac{\partial Y}{\partial \xi} + n_1 \frac{\partial Z}{\partial \xi} \tag{5.101}$$

$$\frac{\partial X}{\partial \xi} = \sum_{i=1}^{4} \frac{\partial N_i}{\partial \xi} X_i; \quad \frac{\partial Y}{\partial \xi} = \sum_{i=1}^{4} \frac{\partial N_i}{\partial \xi} Y_i; \quad \frac{\partial Z}{\partial \xi} = \sum_{i=1}^{4} \frac{\partial N_i}{\partial \xi} Z_i.$$

**Gauss integration rule**: 2 × 2 integration rule.

**Element Stiffness Matrix in local coordinate system**: The element stiffness matrix can now be computed using Gauss numerical integration rule as

$$[\mathrm{k}] = \int_{-1}^{+1}\int_{-1}^{+1} [\mathrm{B}(\xi,\eta,)]^T[\mathrm{E}][\mathrm{B}(\xi,\eta,)]J\,\mathrm{d}\xi\,\mathrm{d}\eta$$
$$= \sum_{i=1}^{2}\sum_{j=1}^{2}\left[\mathrm{B}(\xi_i,\eta_j\right]^T[\mathrm{E}]\left[\mathrm{B}\left(\xi_i,\eta_j\right)\right]J\left(\xi_i,\eta_j\right)\mathrm{W_i W_j}. \quad (5.102)$$

It may be noted that the rotational degree of freedom about the local z-axis ($\theta_z$) does not appear in the vector of node degrees of freedom, since the strains (Eqs. 5.81–5.84) defined in the local coordinate system x–y-z where the z-axis is normal to the plate does not involve this degree of freedom. However, in general plate structural elements, this degree of freedom will appear in the global vector of node degrees of freedom when the stiffness matrix computed in the local coordinate system is transformed to a global coordinate system using the transformation matrices explained in Sect. 5.10.

## 5.19 Worked Out Problems

**Problem 1** A triangular region in x–y plane has corner coordinates 1(0,0), 2(4,0), and 3(2,2) referred to a Cartesian coordinate system. Determine three shape functions $N_1, N_2, N_3$ which can be used for computing the value of a displacement field at any point (x,y) in this region.

The triangular region is shown in Fig. 5.24. Assume a polynomial to represent the displacement field in $x$ and $y$. Since there are three nodes, the polynomial should ideally have three terms: $u(x,y) = a_0 + a_1x + a_2y$. We have to find $\boldsymbol{N_1}, \boldsymbol{N_2}$, and $\boldsymbol{N_3}$ such that $u(x,y) = N_1u_1 + N_2u_2 + N_3u_3$. Substituting the node coordinate values in the assumed polynomial, we can write

$$u_1 = a_0; \qquad u_2 = a_0 + 4a_1; \qquad u_3 = a_0 + 2a_1 + 2a_2.$$

Solving for the coefficients, we get $a_0 = u_1$; $a_1 = \frac{1}{4}(u_2 - u_1)$; $a_2 = \frac{1}{4}(2u_3 - u_1 - u_2)$.

Substituting these values in the assumed polynomial for $u(x,y)$ and accumulating the common terms, we get $u(x,y) = \left(1 - \frac{x}{4} - \frac{y}{4}\right)u_1 + \left(\frac{x}{4} - \frac{y}{4}\right)u_2 + \left(\frac{y}{2}\right)u_3$.

Hence, the shape functions are $N_1 = \left(1 - \frac{x}{4} - \frac{y}{4}\right)$; $N_2 = \left(\frac{x}{4} - \frac{y}{4}\right)$; $N_3 = \left(\frac{y}{2}\right)$.

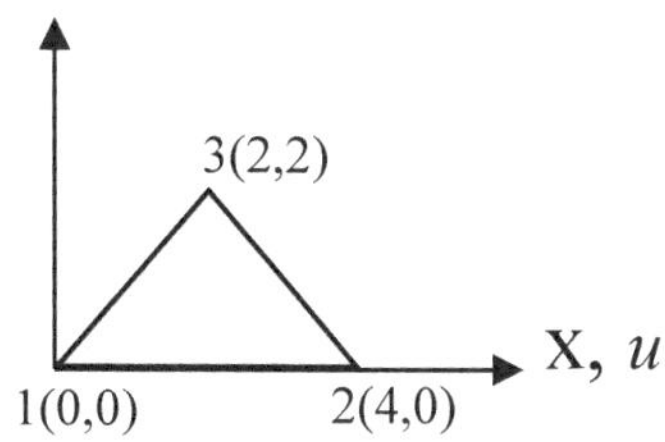


**Fig. 5.24** The triangular region with its corner coordinates

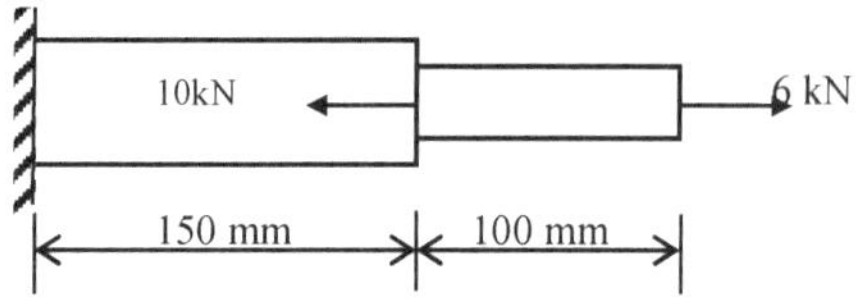


**Fig. 5.25** Geometry of the stepped bar

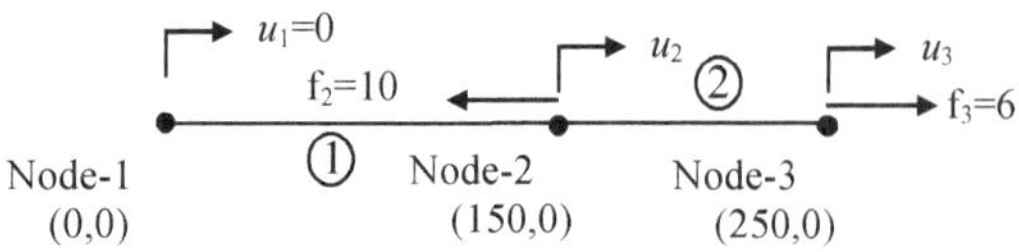


**Fig. 5.26** Finite element model of the stepped bar

**Problem 2** A stepped cylindrical bar, fixed at one end, is subjected to a longitudinal force of 10 kN and 6 kN, as shown in Fig. 5.25. Determine the stress and strain in both segments of the bar. Young's modulus of the material is 200 kN/mm$^2$. Diameters of the stepped segments are 20 mm and 10 mm.

Figure 5.26 shows the element-1 node connectivity is 1-2 and the element-2 node connectivity is 2-3. The cross-sectional area of the elements can be computed as $A_1 = 314.16$ mm$^2$ and $A_2 = 78.54$ mm$^2$.

The local element coordinate system of each element is along the line of node connection, which is also aligned with the global coordinate system. The stiffness matrix of element-1 and element-2 are given by Eq. 5.29 as

$$[\mathrm{k}_1] = \frac{EA_1}{L_1}\begin{bmatrix} 1 & -1 \\ -1 & 1 \end{bmatrix} = \frac{(2 \text{ x } 10^5)\text{x } 314.16}{150}\begin{bmatrix} 1 & -1 \\ -1 & 1 \end{bmatrix} = 10^5\begin{bmatrix} 4.188 & -4.188 \\ -4.188 & 4.188 \end{bmatrix}$$

$$[\mathrm{k}_2] = \frac{EA_2}{L_2}\begin{bmatrix} 1 & -1 \\ -1 & 1 \end{bmatrix} = \frac{(2 \text{ x } 10^5)\text{x } 78.54}{100}\begin{bmatrix} 1 & -1 \\ -1 & 1 \end{bmatrix} = 10^5\begin{bmatrix} 1.571 & -1.571 \\ -1.571 & 1.571 \end{bmatrix}.$$

The equations of $[k_1]$ correspond to the first and second equations of the global stiffness matrix. The equations of $[k_2]$ should add to the second and third equations of the global stiffness matrix. Hence, following the standard assembly procedures (Zienkiewicz 1977) of finite element equations, the global stiffness matrix $[K]$ can be formed by arranging terms of $[k_1]$ and $[k_2]$ into $[K]$, as shown below

$$[\mathrm{K}] = 10^5\begin{bmatrix} 4.188 & -4.188 & 0 \\ -4.188 & 4.188 + 1.571 & -1.571 \\ 0 & -1.571 & 1.571 \end{bmatrix}.$$

The global displacement vector is $\{0 \;\; u_2 \;\; u_3\}^{\mathrm{T}}$.

The global load vector is $\{f_1 \quad -10 \quad 6\}^{\mathrm{T}}$.

The prescribed boundary conditions can now be imposed in the global stiffness matrix by removing the row and columns of equations corresponding to the restrained degree of freedom ($u_1$). So, the first row and first column of $[K]$ are removed and the final equations to be solved can be written as

$$10^5\begin{bmatrix} 5.759 & -1.571 \\ -1.571 & 1.571 \end{bmatrix}\begin{Bmatrix} u_2 \\ u_3 \end{Bmatrix} = \begin{Bmatrix} -10 \\ 6 \end{Bmatrix}.$$

The unknown displacement can be solved as $u_2$=-9.55 × $10^{-3}$mm and $u_3$= 2.865× $10^{-2}$mm.

For the computation of strain in the elements, we can make use of the strain–displacement matrix, [B] (Eq. 5.28):

$$\varepsilon_1 = [\mathrm{B}]\begin{Bmatrix} u_1 \\ u_2 \end{Bmatrix} = \begin{bmatrix} \frac{-1}{L_1} & \frac{1}{L_1} \end{bmatrix}\begin{Bmatrix} u_1 \\ u_2 \end{Bmatrix} = \begin{bmatrix} \frac{-1}{150} & \frac{1}{150} \end{bmatrix}\begin{Bmatrix} 0 \\ = -9.55\text{x}10^{-3} \end{Bmatrix} = -6.36 \times 10^{-5}$$

$$\varepsilon_2 = [\mathrm{B}]\begin{Bmatrix} u_2 \\ u_3 \end{Bmatrix} = \begin{bmatrix} \frac{-1}{L_2} & \frac{1}{L_2} \end{bmatrix}\begin{Bmatrix} u_2 \\ u_3 \end{Bmatrix} = \begin{bmatrix} \frac{-1}{100} & \frac{1}{100} \end{bmatrix}\begin{Bmatrix} -9.55\text{x}10^{-3} \\ 2.865\text{x}10^{-2} \end{Bmatrix} = 3.82 \times 10^{-4}.$$

The stresses in the bar segments can be computed as

$$\sigma_1 = E\varepsilon_1 == 200\left(-6.36\text{x}10^{-5}\right) = -0.01272kN/mm^2$$

$$\sigma_2 = E\varepsilon_2 == 200\left(3.82\text{x}10^{-4}\right) = 0.0764kN/mm^2.$$

**Problem 3** A simple truss structure made of two rods of equal length (*L*) 1000 mm and cross-sectional area (*A*) 75 mm$^2$ are connected by pin joints at their ends and is subjected to a horizontal force of 10,000 N, as shown in Fig. 5.27. Find out the axial force in the rod elements. Young's Modulus (E) of the material is 2 × $10^5$ N/mm$^2$.

The truss structure can be modeled by two rod/link elements; element-1 with connectivity 1-2 and element-2 with connectivity 2-3.

The node coordinates are 1(0,0), 2(500,866.0254), and 3(1000,0). For this planar truss problem, the boundary conditions are $U_1$=$V_1$=$U_3$=$V_3$=0. As per the theory of rod finite elements, both elements would develop uniform axial stress along their length.

The equilibrium equation $[k^e][u^e]$=$[f^e]$ for both elment-1 and element-2 in the respective local coordinate system can be written in terms of their stiffness matrix and displacement vector and force vector in their local coordinate system, as

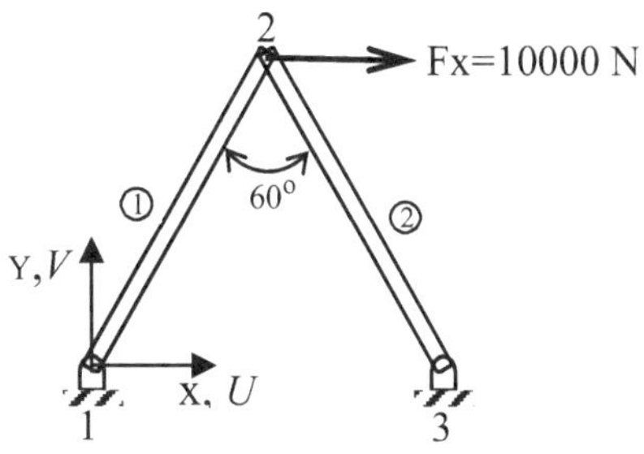


**Fig. 5.27** Geometry of the truss structure

$$\frac{EA}{L}\begin{bmatrix} 1 & 0 & -1 & 0 \\ 0 & 0 & 0 & 0 \\ -1 & 0 & 1 & 0 \\ 0 & 0 & 0 & 0 \end{bmatrix}\begin{Bmatrix} u_1 \\ v_1 \\ u_2 \\ v_2 \end{Bmatrix} = \begin{Bmatrix} f_{x1} \\ f_{y1} \\ f_{x2} \\ f_{y2} \end{Bmatrix}.$$

The above stiffness matrix in the local coordinate system shall be the same for both elements because both the rods are having same geometry and material properties. Now, for solving the structure in the common platform of the global coordinate system, these local stiffness matrices have to be transformed to the global coordinate system.

The node displacements $\{u_1, v_1, u_2, v_2,\}$ in the local coordinate system can be transformed to the node displacements in the global coordinate system $\{U_1, V_1, U_2, V_2\}$ by the transformation matrix, $T$, in the following relation, where $l = \cos\theta$ and $m = \sin\theta$ and $\theta$ is the orientation angle of the element with respect to +ve global X-axis:

$$\begin{Bmatrix} u_1 \\ v_1 \\ u_2 \\ v_2 \end{Bmatrix} = \begin{bmatrix} l & m & 0 & 0 \\ -m & l & 0 & 0 \\ 0 & 0 & l & m \\ 0 & 0 & -m & l \end{bmatrix}\begin{Bmatrix} U_1 \\ V_1 \\ U_2 \\ V_2 \end{Bmatrix}.$$

The transformation of the stiffness matrix from the local coordinate system $[k^e]$ to global coordinate system $[K^e]$ is obtained using $T$, as

$$[K^e] = T^T[k^e]T.$$

In terms of $l$ and $m$, the stiffness matrix of a rod element in the global coordinate system can now be written as

$$[K^e] = \frac{EA}{L}\begin{bmatrix} l^2 & lm & -l^2 & -lm \\ lm & m^2 & -lm & -m^2 \\ -l^2 & -lm & l^2 & lm \\ -lm & -m^2 & lm & m^2 \end{bmatrix}$$

For element-1, the angle $\theta$, which the node line 1–2 makes with the +ve global X-axis, is 60 degrees and for element-2, the node line 2–3 makes −60 degrees. Hence, for element −1, the values of $l$ and $m$ are $l_1 = 0.5$ and $m_1 = 0.866$ and for element-2, the values are $l_2 = 0.5$ and $m_2 = -0.866$. The transformed stiffness matrices for the elements $[K^e{}_1]$ and $[K^e{}_2]$ can now be assembled to form a 6x6 global stiffness matrix:

$$\begin{bmatrix} l_1^2 & l_1m_1 & -l_1^2 & -l_1m_1 & 0 & 0 \\ l_1m_1 & m_1^2 & -l_1m_1 & -m_1^2 & 0 & 0 \\ -l_1^2 & -l_1m_1 & l_1^2+l_2^2 & l_1m_1+l_2m_2 & -l_2^2 & -l_2m_2 \\ -l_1m_1 & -m_1^2 & l_1m_1+l_2m_2 & m_1^2+m_2^2 & -l_2m_2 & -m_2^2 \\ 0 & 0 & -l_1^2 & -l_2m_2 & l_2^2 & l_2m_2 \\ 0 & 0 & -l_2m_2 & -m_2^2 & l_2m_2 & m_2^2 \end{bmatrix} \begin{Bmatrix} U_1 \\ V_1 \\ U_2 \\ V_2 \\ U_3 \\ V_3 \end{Bmatrix} = \begin{Bmatrix} F_1^x \\ F_1^y \\ F_2^x \\ F_2^y \\ F_3^x \\ F_3^y \end{Bmatrix}.$$

As nodes 1 and 3 are fully restrained, the rows and columns corresponding to $\{U_1, V_1, U_3, V_3\}$ are eliminated from the global stiffness matrix, retaining the equation corresponding to $\{U_2, V_2\}$. The final set of the equation to be solved is

$$\frac{EA}{L}\begin{bmatrix} l_1^2+l_2^2 & l_1m_1+l_2m_2 \\ l_1m_1+l_2m_2 & m_1^2+m_2^2 \end{bmatrix}\begin{Bmatrix} U_2 \\ V_2 \end{Bmatrix} = \begin{Bmatrix} F_2^x \\ F_2^y \end{Bmatrix}.$$

Substituting the values, the equations become

$$\frac{2\text{x}10^5}{1000}\begin{bmatrix} 0.5 & 0 \\ 0 & 1.5 \end{bmatrix}\begin{Bmatrix} U_2 \\ V_2 \end{Bmatrix} = \begin{Bmatrix} 10000 \\ 0 \end{Bmatrix}.$$

The solution is $U_2 = 1.3333$ and $V_2 = 0$.
Axial force in element-1 can be computed as

$$\begin{aligned} N_1 &= \frac{AE}{L}(u_2 - u_1) = \frac{AE}{L}[(l_1U_2 + m_1V_2) - (l_1U_1 + m_1V_1)] \\ &= \frac{75\text{x}2\text{x}10^5}{1000}(0.5 \text{ x } 1.33333 - 0.0) = +10000N \text{ (Tension)}. \end{aligned}$$

Axial force in element-2 can be computed as

$$\begin{aligned} N_2 &= \frac{AE}{L}(u_3 - u_2) = \frac{AE}{L}[(l_2U_3 + m_2V_3) - (l_2U_2 + m_2V_2)] \\ &= \frac{75\text{x}2\text{x}10^5}{1000}(0 - 0.5 \text{ x } 1.33333) = -10000N \text{ (Compression)}. \end{aligned}$$

**Problem-4:** A rectangular plane stress element x–y plane has corner coordinates: 1(0,0); 2(4,0); 3(4,2); 4(0,2). The x-direction displacement field u(x,y) existing in this region has corner values $u_1 = 0.1$, $u_2 = 0.3$, $u_3 = -0.3$, $u_4 = 0.0$. Using the appropriate shape functions for interpolation, determine the extensional strain $\varepsilon_x$ at the centroid of the element (Fig. 5.28).

A natural element coordinate system ξ-η is formed at the centroid of the element, so that, for interpolation of the displacement field, we can use the standard shape functions given below

$$N_1 = \frac{1}{4}(1-\xi)(1-\eta);\ N_2 = \frac{1}{4}(1+\xi)(1-\eta);\ N_3 = \frac{1}{4}(1+\xi)(1+\eta);\ N_4 = \frac{1}{4}(1-\xi)(1+\eta).$$

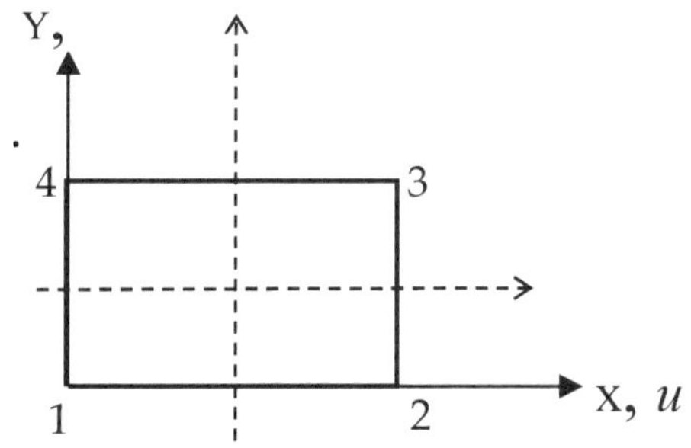


**Fig. 5.28** Geometry of rectangular plane stress element

At the centroid ($\xi = 0, \eta = 0$): $\frac{\partial N_1}{\partial \xi} = -\frac{1}{4}$; $\frac{\partial N_2}{\partial \xi} = \frac{1}{4}$; $\frac{\partial N_3}{\partial \xi} = \frac{1}{4}$; $\frac{\partial N_4}{\partial \xi} = -\frac{1}{4}$.

Since the natural coordinate axes are orthogonal (by virtue of the rectangular geometry), the Jacobian matrix for the element can be easily formed by observation: $[J] = \begin{bmatrix} \frac{\partial x}{\partial \xi} & \frac{\partial y}{\partial \xi} \\ \frac{\partial x}{\partial \eta} & \frac{\partial y}{\partial \eta} \end{bmatrix} = \begin{bmatrix} 2 & 0 \\ 0 & 1 \end{bmatrix}$. Hence, $\frac{\partial N_i}{\partial x} = \frac{\partial N_i}{\partial \xi}\frac{\partial \xi}{\partial x} = \frac{1}{2}\frac{\partial N_i}{\partial \xi}$ and $\frac{\partial N_i}{\partial y} = \frac{\partial N_i}{\partial \eta}\frac{\partial \eta}{\partial y} = \frac{\partial N_i}{\partial \eta}$.

Therefore, at the centroid, the extensional strain in the x-direction can be computed as

$$\begin{aligned} \varepsilon_x &= \frac{\partial u}{\partial x} = \sum \frac{\partial N_i}{\partial x} u_i \\ &= \frac{1}{2} \sum \frac{\partial N_i}{\partial \xi} u_i \\ &= \frac{1}{2}\left[\left(-\frac{1}{4}\right)(0.1) + \left(\frac{1}{4}\right)(0.3) + \left(\frac{1}{4}\right)(-0.3) + \left(-\frac{1}{4}\right)(0)\right] \\ &= -0.0125. \end{aligned}$$

## 5.20 Exercise Problems

5.20.1 The free end of a solid cantilever beam of circular cross section of diameter 5 cm and length 50 cm is attached with a rigid plate of diameter 10 cm. Three concentrated forces $F_A$, $F_B$, and $F_C$ are acting on the rigid plate, as shown in the figure. Determine the bending moment, axial force, and shear force at the fixed end.

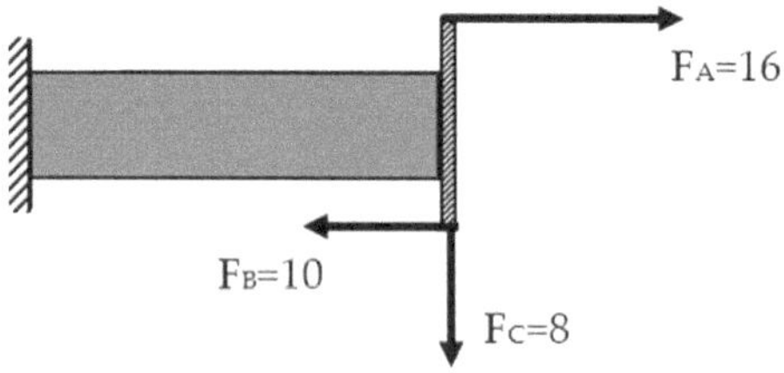


5.20.2 A system of two springs of stiffness $k_1$ and $k_2$ in series is loaded by a force F. By applying *principle of minimum potential energy*, derive expressions for the displacement $u_1$ and $u_2$.

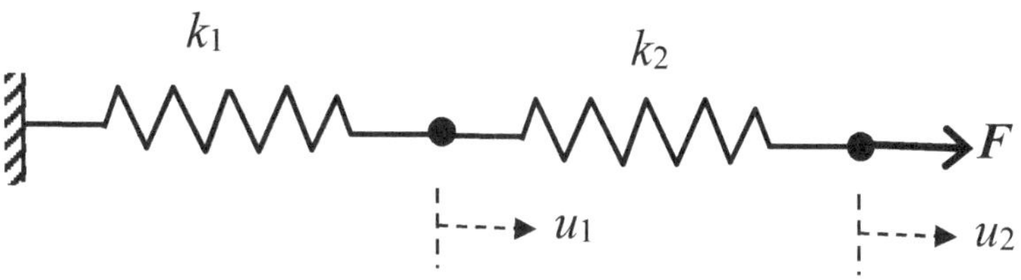


5.20.3 Using the concept of the principle of minimum potential energy,

(a) Construct the stiffness matrix of a system of springs in series, as shown in the figure.
(b) Solve for displacements if horizontal node forces applied are $F_2 = 10$, $F_3 = -5$;

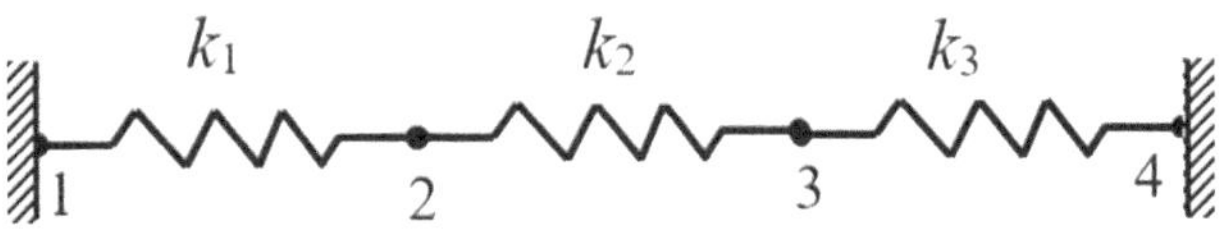


and $k_1 = 15$, $k_2 = 20$, $k_3 = 25$, in consistent units.

5.20.4 A system of four springs of stiffness $k_1$, $k_2$, $k_3$, $k_4$ with three translational degrees of freedom ($U_1$, $U_2$, $U_3$) is loaded by

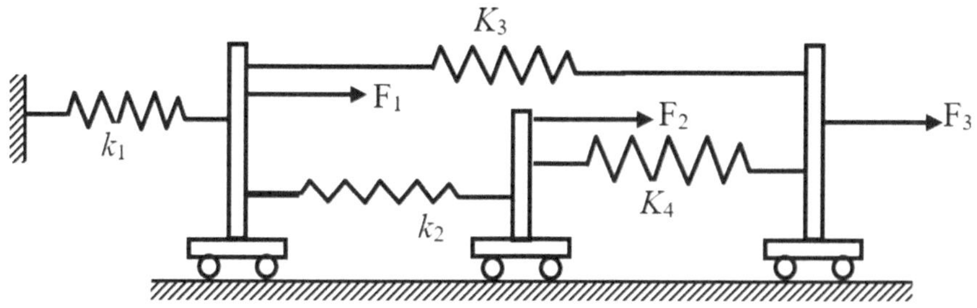


forces $F_1$, $F_2$, $F_3$. By applying *principle of minimum potential energy*, derive the stiffness matrix for the system.

5.20.5 Using 2 × 2 Gaussian Quadrature (numerical integration), obtain the value of the following integral. Also, compare with the analytical integration solution:

$$\int_{-1}^{1}\int_{-1}^{1}(\xi^3 - 1)(\eta - 1)^2 d\xi d\eta.$$

5.20.6 Find the analytical solution of the following integral expression in $x$:

$$\int_{-2}^{3}(x^3 + 2x)dx.$$

Using isoparametric mapping, rewrite the above integral in terms of a natural coordinate $\xi$ and evaluate the intergral using Gaussian Quadrature. Compare the solutions.

5.20.7 A structural designer is trying to interpret stress analysis results for a beam. Shown here is a small segment (length = 15) of the long beam along the X-axis, with the normal stress distribution on the left and right cross sections. Determine **(a)** axial force, **(b)** bending moment, and **(c)** shear force in the beam. (All quantities are given in consistent units.)

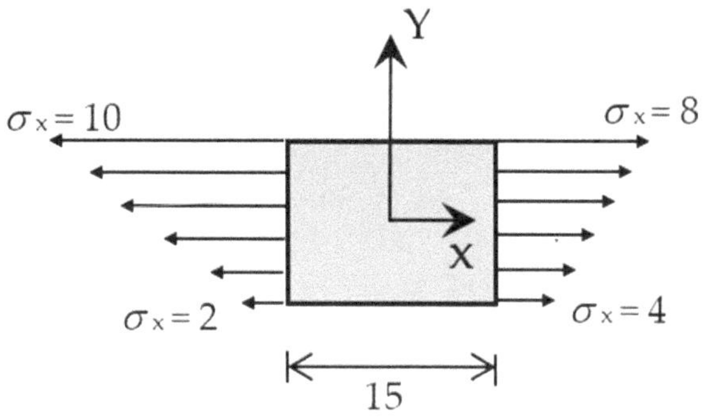


5.20.8 Starting from an assume quadratic polynomial, derive shape functions suitable for interpolation of field variables of a 3-node isoparametric Timoshenko beam finite element.

5.20.9 Imagine a local coordinate system x–y–z, in which a 2-node Timoshenko beam element with its axis aligned along the y-axis. This means that the beam cross section is in the local x–z plane. In such a configuration, $\underline{v}(x, y, z)$ represents the axial displacement field, and $\underline{u}(x, y, z)$ and $\underline{w}(x, y, z)$ represent the transverse displacement fields. Using appropriate sign conventions, form the displacement kinematic relations for this beam, which express the displacement fields $\underline{u}(x, y, z)$, $\underline{v}(x, y, z)$, and $\underline{w}(x, y, z)$ in terms of centroid displacements $u(x)$, $v(x)$, and $w(x)$ and section rotations $\theta_x$, $\theta_y$, and $\theta_z$.

## Bibliography

Bathe KJ, Wilson EL (1976) Numerical methods in finite element analysis. Prentice-Hall Inc., Englewood Cliffs, NJ

Zienkiewicz OC (1977) The finite element method. McGraw-Hill Book Company Limited, London

# Chapter 6
# Boundary Conditions and Loads

**R. Marimuthu**

*In all things of nature there is something of marvelous*—Aristotle

## 6.1 Introduction

The structural configurations are supported at appropriate locations called boundary conditions, to withstand mechanical and thermal loads that arise during its service period, without any failure. Accordingly, boundary conditions need to be specified on finite element nodes to represent the physics of the problem. In static analysis, the application of boundary conditions is essential to prevent rigid body motion and hence to avoid singularity of the assembled global stiffness matrix. The following loads are discussed in this chapter: (1) Point loads applied on nodes of finite element mesh, (2) Traction loads applied on edges in the case of two-dimensional elements, and on surfaces in the case of three-dimensional elements, (3) Body forces that arise due to gravity, magnetic force, and centrifugal loading, and (4) Thermal loads that occur due to change in temperature of the body.

R. Marimuthu (✉)
DH, FSDD/SDAG/STR, Thiruvananthapuram, Kerala 695022, India
e-mail: r_marimuthu@vssc.gov.in



S. U. Nair and S. Somanath (eds.), *Introduction to Finite Element Analysis*,
https://doi.org/10.1007/978-981-19-7989-7_6

## 6.2 Types of Boundary Conditions

In general, as discussed in previous chapter, there are three types of boundary conditions, namely: Dirichlet (essential), Neumann (natural), and Robin (combination of the above two) boundary conditions. Displacement-based finite element formulations use only essential boundary conditions which are either translations and/or rotations. Constraints are of two types, viz., (1) Single Point Constraints (SPC) and (2) Multi Point Constraints (MPC). In SPC, specific values (zero or non-zero) are prescribed for translations/rotations at selected degrees of freedom. MPCs are relations between different translations/rotations degrees of freedom that are written in the form of equations. Typically, MPC equations result from arresting translations/rotations degrees of freedom with respect to an inclined edge or surface. This is explained in example 5 of Sect. 6.5 where MPCs are effectively used for representing structural parts sliding on an inclined edge. The boundary conditions can be imposed on a finite element model either by (i) penalty method, or (ii) elimination method. The penalty method, which is easy to implement in general purpose software for SPC and MPC, is described and demonstrated in the following sections.

## 6.3 Single Point Constraints (SPC)

SPC is a restriction imposed on a particular degree of freedom to either have zero or non-zero value. Let the ith degree of freedom $d_i$ in the finite element model be specified with a fixed value $\alpha_i$. This can be written as

$$d_i = \alpha_i \; or \; d_i - \alpha_i = 0. \tag{6.1}$$

The above equation can be imposed as a constraint in the total potential of the system as

$$\pi = \frac{1}{2}\{\mathrm{D}\}^{\mathrm{T}}[\mathrm{K}]\{\mathrm{D}\} - \{\mathrm{D}\}^{\mathrm{T}}\{\mathrm{F}\} + \frac{1}{2}\mathrm{C}(d_i - \alpha_i)^2. \tag{6.2}$$

In the above equation, [K] is assembled global stiffness matrix, {D} and {F} are global displacement vector and applied global load vector, respectively. C is a penalty value that is used to enforce the constraint imposed by Eq. (6.1). In general, C is assumed to be at least six orders higher than the maximum diagonal value appearing in the global stiffness matrix. Minimization of total potential (Eq. 6.2) results in

$$\delta\pi = \delta\{\mathrm{D}\}^{\mathrm{T}}[\mathrm{K}]\{\mathrm{D}\} - \delta\{\mathrm{D}\}^{\mathrm{T}}\{\mathrm{F}\} + \delta d_i \mathrm{C} d_i - \delta d_i \mathrm{C}\alpha_i = 0. \tag{6.3}$$

Let $[\mathrm{K}]_{\mathrm{spc}}$ be an initially null stiffness matrix and $\{\mathrm{F}\}_{\mathrm{spc}}$ be an initially null force vector. The stiffness corresponding to SPC mentioned in 6.1 is formed by placing C at $i$th diagonal position of $[\mathrm{K}]_{\mathrm{spc}}$. The corresponding load due to that SPC is formed by placing $\mathrm{C}\alpha_i$ at $i$th position of $\{\mathrm{F}\}_{\mathrm{spc}}$. Then $d_i$ and $\delta d_i$ can be represented in the global displacement vector as $\{\mathrm{D}\}$ and $\delta\{\mathrm{D}\}$, respectively.

Then Eq. (6.3) can be written as

$$\delta\pi = \delta\{\mathrm{D}\}^{\mathrm{T}}[\mathrm{K}]\{\mathrm{D}\} - \delta\{\mathrm{D}\}^{\mathrm{T}}\{\mathrm{F}\} + \delta\{\mathrm{D}\}^{\mathrm{T}}[\mathrm{K}]_{\mathrm{spc}}\{\mathrm{D}\} - \delta\{\mathrm{D}\}^{\mathrm{T}}\{\mathrm{F}\}_{\mathrm{spc}} = 0 \tag{6.4}$$

where

$$[\mathrm{K}]_{spc} = \begin{bmatrix} 0\,0 \ldots 0 \ldots 0 \ldots 0 \\ 0\,0 \ldots 0 \ldots 0 \ldots 0 \\ \vdots\,\vdots \ldots \vdots \ldots \vdots \ldots 0 \\ 0\,0 \ldots \mathrm{C} \ldots 0 \ldots 0 \\ \vdots\,\vdots \ldots \vdots \ldots \vdots \ldots \vdots \\ 0\,0 \ldots 0 \ldots 0 \ldots 0 \\ \vdots\,\vdots \ldots \vdots \ldots \vdots \ldots \vdots \\ 0\,0 \ldots 0 \ldots 0 \ldots 0 \end{bmatrix} \text{ and } \{\mathrm{F}\}_{spc} = \begin{Bmatrix} 0 \\ 0 \\ \vdots \\ C\alpha_i \\ \vdots \\ 0 \\ \vdots \\ 0 \end{Bmatrix}. \tag{6.5}$$

Equation (6.4) can be rewritten as

$$\delta\pi = \delta\{\mathrm{D}\}^{\mathrm{T}}\big(([\mathrm{K}] + [\mathrm{K}]_{\mathrm{spc}})\{\mathrm{D}\} - (\{\mathrm{F}\} + \{\mathrm{F}\}_{\mathrm{spc}})\big) = 0$$

Since variation in displacement is arbitrary, $\delta\{D\}^T \neq 0$ leads to the equilibrium equation:

$$\big([\mathrm{K}] + [\mathrm{K}]_{spc}\big)\{\mathrm{D}\} = \{\mathrm{F}\} + \{\mathrm{F}\}_{spc}. \tag{6.6}$$

The above equation can be written as

$$([\mathrm{K}]_{net})\{\mathrm{D}\} = \{\mathrm{F}\}_{net} \tag{6.7}$$

where

$[\mathrm{K}]_{net} = [\mathrm{K}] + [\mathrm{K}]_{spc}$ and $\{\mathrm{F}\}_{net} = \{\mathrm{F}\} + \{\mathrm{F}\}_{spc}$.

if $\alpha_i = 0$, as in the case of a restrained degree of freedom, $\{F\}_{spc}$ will be a null vector.

## 6.4 Multi Point Constraints (MPC)

MPC are equations that relate to one or more degrees of freedom and are expressed in the form of equations. The following equation is a result of relating three degrees of freedom i, j, and k.

$$\alpha_i d_i + \alpha_j d_j + \alpha_k d_k = \alpha_0. \tag{6.8}$$

where $\alpha_i, \alpha_j, \alpha_k,$ and $\alpha_0$ are MPC constants.

The above equation can be imposed as a constraint in the total potential of the system as

$$\pi = \frac{1}{2}\{\mathrm{D}\}^{\mathrm{T}}[\mathrm{K}]\{\mathrm{D}\} - \{\mathrm{D}\}^{\mathrm{T}}\{\mathrm{F}\} + \frac{1}{2}C\left(\alpha_i d_i + \alpha_j d_j + \alpha_k d_k - \alpha_0\right)^2. \tag{6.9}$$

Following the same procedure described in Sect. 6.3, for SPC, the system of equations for enforcing MPC is

$$\left([\mathrm{K}] + [\mathrm{K}]_{mpc}\right)\{\mathrm{D}\} = \{\mathrm{F}\} + \{\mathrm{F}\}_{mpc}. \tag{6.10}$$

The above equation can be written as

$$([\mathrm{K}]_{net})\{\mathrm{D}\} = \{\mathrm{F}\}_{net} \tag{6.11}$$

where

$[\mathrm{K}]_{net} = [\mathrm{K}] + [\mathrm{K}]_{spc}$ and $\{\mathrm{F}\}_{net} = \{\mathrm{F}\} + \{\mathrm{F}\}_{mpc}$

$$[\mathrm{K}]_{mpc} = \begin{bmatrix} 0 \cdots & 0 & \cdots & 0 & \cdots & 0 & \cdots 0 \\ \vdots\ \vdots & \vdots & \vdots & \vdots & \vdots & \vdots & \vdots\ \vdots \\ 0 \cdots & C\alpha_i^2 & \cdots & C\alpha_i\alpha_j & \cdots & C\alpha_i\alpha_k & \cdots 0 \\ \vdots\ \vdots & \vdots & \vdots & \vdots & \vdots & \vdots & \vdots\ \vdots \\ 0 \cdots & C\alpha_i\alpha_j & \cdots & C\alpha_j^2 & \cdots & C\alpha_j\alpha_k & \cdots 0 \\ \vdots\ \vdots & \vdots & \vdots & \vdots & \vdots & \vdots & \vdots\ \vdots \\ 0 \cdots & C\alpha_i\alpha_k & \cdots & C\alpha_j\alpha_k & \cdots & C\alpha_k^2 & \cdots 0 \\ \vdots\ \vdots & \vdots & \vdots & \vdots & \vdots & \vdots & \vdots\ \vdots \\ 0 \cdots & 0 & \cdots & 0 & \cdots & 0 & \cdots 0 \end{bmatrix} \text{ and } \{\mathrm{F}\}_{mpc} = \begin{Bmatrix} 0 \\ \vdots \\ C\alpha_i\alpha_0 \\ \vdots \\ C\alpha_j\alpha_0 \\ \vdots \\ C\alpha_k\alpha_0 \\ \vdots \\ 0 \end{Bmatrix}.$$

The reaction force vector $\{R\}$ corresponding to SPC and MPC can be computed as

$$\{R\} = [K]\{D\} - \{F\}. \tag{6.12}$$

## 6.5 Numerical Examples of Boundary Conditions

The concepts of imposing SPC and MPC are explained with a straight bar of length '$l$'. The bar is discretized using 3 elements of unequal lengths $l_1$, $l_2$, and $l_3$; Young's modulus of $E_1, E_2$, and $E_3$; and the area of cross sections A1, A2, and A3 are assigned to elements 1, 2, and 3, respectively. Figure 6.1 shows the discretized bar with node 1 arrested and point load P applied at node 4.

The stiffness matrix of a bar element can be written as

$$[\mathrm{K}] = \frac{AE}{l}\begin{bmatrix} 1 & -1 \\ -1 & 1 \end{bmatrix}.$$

Therefore, the stiffness matrix for elements 1, 2, and 3 can be written as

$$[\mathrm{K1}] = \frac{A_1E_1}{l_1}\begin{bmatrix} 1 & -1 \\ -1 & 1 \end{bmatrix};\ [\mathrm{K2}] = \frac{A_2E_2}{l_2}\begin{bmatrix} 1 & -1 \\ -1 & 1 \end{bmatrix};\ [\mathrm{K3}] = \frac{A_3E_3}{l_3}\begin{bmatrix} 1 & -1 \\ -1 & 1 \end{bmatrix}. \tag{6.13}$$

**Assembly of finite elements**: From the above three individual element stiffness matrices, the global stiffness matrix for the entire bar is assembled as shown in Fig. 6.2. Since the bar has four nodes with one degree of freedom each, the size of the global stiffness matrix is 4 × 4. Figure 6.2 depicts the contribution of each element to the global stiffness matrix [K]. It can be seen that the stiffness contribution from element-1 is given as K1(1,1), K1(1,2), K1(2,1), and K1(2,2). Similarly, the stiffness contributions from [K2] and [K3] are also shown. As shown in Fig. 6.1, node 2 is common for elements 1 and 2, and hence at node 2, there will be stiffness contributions from both elements 1 and 2. Similarly, at node 3, there will be stiffness contributions from both elements 2 and 3. There is no stiffness contribution at node 3 from element-1 and hence zero is added to the global stiffness matrix in cell (1, 3). Similarly, element 2 has no contribution to node 4 and hence zero is added to the global stiffness matrix in cell (1, 4).

If stiffness values of elements from Eq. (6.13) are added as shown in Fig. 6.2, the global stiffness matrix of the bar can be written as

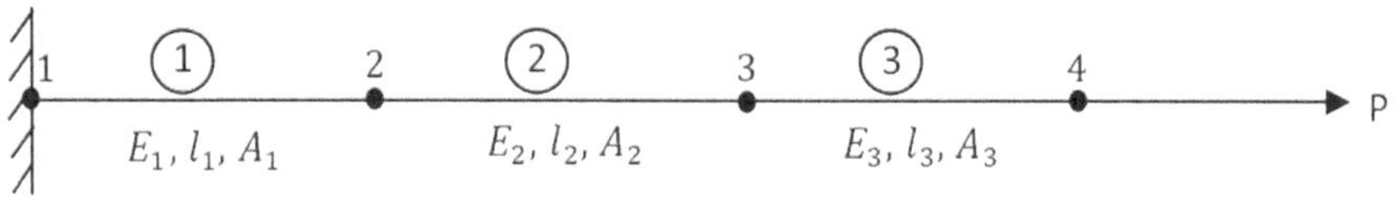


**Fig. 6.1** Axial bar with finite element discretization and point load

**Fig. 6.2** Contribution of each element to the global stiffness matrix

| Node | 1 | 2 | 3 | 4 |
|---|---|---|---|---|
| 1 | K1(1,1) | K1(1,2) | 0 | 0 |
| 2 | K1(2,1) | K1(2,2) + K2(1,1) | K2(1,2) | 0 |
| 3 | 0 | K2(2,1) | K2(2,2) + K3(1,1) | K3(1,2) |
| 4 | 0 | 0 | K3(2,1) | K3(2,2) |

$$[\mathrm{K}] = \begin{bmatrix} \frac{A_1E_1}{l_1} & -\frac{A_1E_1}{l_1} & 0 & 0 \\ -\frac{A_1E_1}{l_1} & \frac{A_1E_1}{l_1} + \frac{A_2E_2}{l_2} & -\frac{A_2E_2}{l_2} & 0 \\ 0 & -\frac{A_2E_2}{l_2} & \frac{A_2E_2}{l_2} + \frac{A_3E_3}{l_3} & -\frac{A_3E_3}{l_3} \\ 0 & 0 & -\frac{A_3E_3}{l_3} & \frac{A_3E_3}{l_3} \end{bmatrix}. \tag{6.14}$$

The global force vector can be written as

$$\{\mathrm{F}\}^{\mathrm{T}} = \{0, 0, 0, \mathrm{P}\}. \tag{6.15}$$

Let us assume the global displacement vector as

$$\{\mathrm{D}\}^{\mathrm{T}} = \{\mathrm{d}_1, \mathrm{d}_2, \mathrm{d}_3, \mathrm{d}_4\} \tag{6.16}$$

where, $\mathrm{d}_1$, $\mathrm{d}_2$, $\mathrm{d}_3$, and $\mathrm{d}_4$ are displacements at nodes 1, 2, 3, and 4, respectively.

In the section that follows, numerical studies are conducted to demonstrate the concepts explained for SPC and MPC. For simplicity, a combination of geometric and material properties has been chosen such that $\frac{A_1E_1}{l_1} = 1$, $\frac{A_2E_2}{l_2} = 2$, and $\frac{A_3E_3}{l_3} =$ 3 units and the point load applied at node 4 is 2 units. With these numerical values, the assembled stiffness matrix and load vector are written as

$$[K] = \begin{bmatrix} 1 & -1 & 0 & 0 \\ -1 & 3 & -2 & 0 \\ 0 & -2 & 5 & -3 \\ 0 & 0 & -3 & 3 \end{bmatrix} \tag{6.17}$$

and

$$\{F\}^T = \{0\ 0\ 0\ 2\}. \tag{6.18}$$

The net stiffness matrix is obtained by adding the global stiffness matrix and constrained stiffness matrix of SPC/MPC. Similarly, the net force is obtained by adding the applied force vector and constrained force vector due to SPC/MPC.

**Example 1** Obtain the displacement and reaction force vector for thc problem shown in Fig. 6.1.

Since the boundary condition is specified as SPC, this problem can be solved using the penalty method;

$[K]_{spc}$ matrix and $\{F\}_{spc}^{T}$ are given as

$$[\mathrm{K}]_{\mathrm{spc}} = \begin{bmatrix} 10^6 & 0\,0\,0 \\ 0 & 0\,0\,0 \\ 0 & 0\,0\,0 \\ 0 & 0\,0\,0 \end{bmatrix} \tag{6.19}$$

$$\{\mathrm{F}\}_{spc}^{T} = \{0\,0\,0\,0\}. \tag{6.20}$$

The net stiffness matrix is obtained by adding $[K]$ and $[K]_{spc}$ and net force vector is obtained by adding $\{F\}$ and $\{F\}_{spc}^{T}$

$$[K]_{Net} = \begin{bmatrix} 1+10^6 & -1 & 0 & 0 \\ -1 & 3 & -2 & 0 \\ 0 & -2 & 5 & -3 \\ 0 & 0 & -3 & 3 \end{bmatrix} \text{a}\,\{\mathrm{F}\}_{Net}^{T} = \{0\,0\,0\,2\}. \tag{6.21}$$

By solving the equation

$$([\mathrm{K}]_{net})\{\mathrm{D}\} = \{\mathrm{F}\}_{net},$$

the displacement vector $\{D\}$ is obtained as

$$\{D\}^{T} = \{0\ 2\ 3\ 3.6667\}. \tag{6.22}$$

From (6.12), the reaction force vector $\{R\}$ is obtained as

$$\{R\}^{T} = \{-2\,0\,0\,0\}. \tag{6.23}$$

**Example 2** Obtain the displacement and reaction force vector for the problem shown in Fig. 6.1 with a specified displacement of 5 units at node 3.

Since the boundary condition is specified as SPC, this problem can be solved using the penalty method;

$[K]_{spc}$ matrix and $\{F\}_{spc}^{T}$ are given as

$$[K]_{spc} = \begin{bmatrix} 10^6 & 0 & 0 & 0 \\ 0 & 0 & 0 & 0 \\ 0 & 0 & 10^6 & 0 \\ 0 & 0 & 0 & 0 \end{bmatrix} \tag{6.24}$$

$$\{\mathrm{F}\}_{spc}^T = \left\{ 0\ 0\ 5 \times 10^6\ 0 \right\}. \tag{6.25}$$

The net stiffness matrix is obtained by adding $[K]$ and $[K]_{spc}$ and net force vector is obtained by adding $\{\mathrm{F}\}$ and $\{\mathrm{F}\}_{spc}^T$

$$[K]_{Net} = \begin{bmatrix} 1+10^6 & -1 & 0 & 0 \\ -1 & 3 & -2 & 0 \\ 0 & -2 & 5+10^6 & -3 \\ 0 & 0 & -3 & 3 \end{bmatrix} \text{ and } \{\mathrm{F}\}_{Net}^T = \left\{ 0\ 0\ 5 \times 10^6\ 2 \right\}. \tag{6.26}$$

By solving the equation

$$([\mathrm{K}]_{net})\{\mathrm{D}\} = \{\mathrm{F}\}_{net},$$

the displacement vector $\{D\}$ is obtained as

$$\{D\}^T = \left\{ 0\ 3.3333\ 5.0\ 5.6667 \right\}. \tag{6.27}$$

From (6.12), the reaction force vector $\{R\}$ is obtained as

$$\{R\}^T = \left\{ -3.3333\ 0\ 1.3333\ 0 \right\}. \tag{6.28}$$

**Example 3** Obtain the displacement and reaction force vector for the problem shown in Fig. 6.1 with the same displacement at node 2 and node 3.

Since the boundary condition is specified as MPC, this problem can be solved using the penalty method;

$[K]_{mpc}$ matrix and $\{\mathrm{F}\}_{mpc}^T$ are given as

$$[K]_{mpc} = \begin{bmatrix} 10^6 & 0 & 0 & 0 \\ 0 & 10^6 & -10^6 & 0 \\ 0 & -10^6 & 10^6 & 0 \\ 0 & 0 & 0 & 0 \end{bmatrix} \tag{6.29}$$

$$\{\mathrm{F}\}_{mpc}^T = \left\{ 0\ 0\ 0\ 0 \right\}. \tag{6.30}$$

The net stiffness matrix is obtained by adding $[K]$ and $[K]_{mpc}$ and net force vector is obtained by adding $\{\mathrm{F}\}$ and $\{\mathrm{F}\}_{mpc}^T$

$$[K]_{Net} = \begin{bmatrix} 1+10^6 & -1 & 0 & 0 \\ -1 & 3+10^6 & -2-10^6 & 0 \\ 0 & -2-10^6 & 5+10^6 & -3 \\ 0 & 0 & -3 & 3 \end{bmatrix} \text{ and } \{\mathrm{F}\}_{Net}^T = \left\{ 0\ 0\ 0\ 2 \right\}. \quad (6.31)$$

By solving the equation

$$([\mathrm{K}]_{net})\{\mathrm{D}\} = \{\mathrm{F}\}_{net},$$

the displacement vector $\{D\}$ is obtained as

$$\{D\}^T = \left\{ 0\ 2\ 2\ 2.6667 \right\}. \quad (6.32)$$

From (6.12), the reaction force vector $\{R\}$ is obtained as

$$\{R\}^T = \left\{ -2\ 2\ -2\ 0 \right\}. \quad (6.33)$$

**Example 4** Obtain the displacement and reaction force vector for the problem shown in Fig. 6.1 such that the addition of displacement at node 2 and node 3 is equal to 4 units.

Since the boundary condition is specified as MPC, this problem can be solved using the penalty method;
$[K]_{mpc}$ matrix and $\{\mathrm{F}\}_{mpc}^T$ are given as

$$[K]_{mpc} = \begin{bmatrix} 10^6 & 0 & 0 & 0 \\ 0 & 10^6 & 10^6 & 0 \\ 0 & 10^6 & 10^6 & 0 \\ 0 & 0 & 0 & 0 \end{bmatrix} \quad (6.34)$$

$$\{\mathrm{F}\}_{mpc}^T = \left\{ 0\ 4\times10^6\ 4\times10^6\ 0 \right\}. \quad (6.35)$$

The net stiffness matrix is obtained by adding $[K]$ and $[K]_{spc}$ and net force vector is obtained by adding $\{\mathrm{F}\}$ and $\{\mathrm{F}\}_{mpc}^T\{\mathrm{F}\}_{mpc}^T$

$$[K]_{Net} = \begin{bmatrix} 1+10^6 & -1 & 0 & 0 \\ -1 & 3+10^6 & -2+10^6 & 0 \\ 0 & -2+10^6 & 5+10^6 & -3 \\ 0 & 0 & -3 & 3 \end{bmatrix} \text{ and}$$

$$\{\mathrm{F}\}_{Net}^T = \left\{ 0\ 4\times10^6\ 4\times10^6\ 2 \right\}.$$

By solving the equation

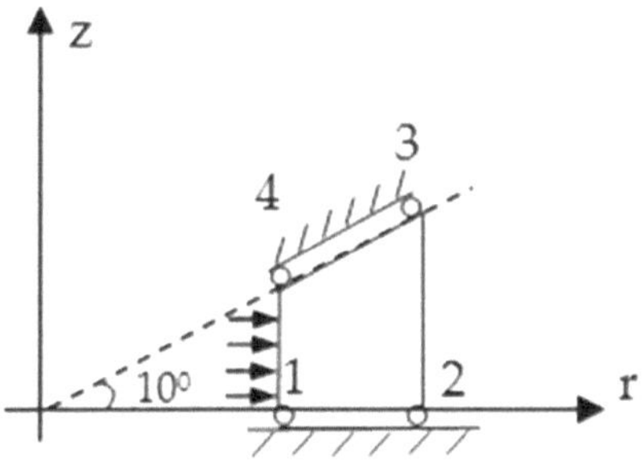


| Node | X | Y |
|---|---|---|
| 1 | 0.2 | 0 |
| 2 | 0.4 | 0 |
| 3 | 0.394 | 0.06946 |
| 4 | 0.197 | 0.03473 |

**Fig. 6.3** Inclined boundary condition applied on an edge

$$([\mathrm{K}]_{net})\{\mathrm{D}\} = \{\mathrm{F}\}_{net},$$

the displacement vector $\{D\}$ is obtained as

$$\{D\}^T = \left\{0\ 1.556\ 2.444\ 3.1111\right\}. \tag{6.36}$$

From (6.12), the reaction force vector $\{R\}$ is obtained as

$$\{R\}^T = \left\{-1,556\ -0.2222\ -0.2222\ 0\right\}. \tag{6.37}$$

**Example 5** Figure 6.3 shows a plane strain element subjected to pressure load at the edge connecting nodes 1 and 4. Displacement normal to the edge connecting nodes 4 and 3 of length 'L' is suppressed. Form the MPC equation to achieve such a constraint. Element connectivity is (1 2 3 4).

To achieve such a constraint, the tangent vector of the edge connecting nodes 3 and 4 has to be calculated. The cross product of the tangent vector with the displacement vector at the nodes that are lying on that inclined edge is performed and is equated to zero.

The vector to the edge connecting nodes 3 and 4 is written as

$$\vec{\mathrm{t}} = \frac{\mathrm{X3} - \mathrm{X4}}{\mathrm{L}}\vec{\mathrm{i}} + \frac{\mathrm{Y3} - \mathrm{Y4}}{\mathrm{L}}\vec{\mathrm{j}} \tag{6.38}$$

where

$$\mathrm{L} = \sqrt{(\mathrm{X3} - \mathrm{X4})^2 + (\mathrm{Y3} - \mathrm{Y4})^2} = 0.2 \tag{6.39}$$

$$\vec{\mathrm{t}} = 0.985\,\vec{\mathrm{i}} + 0.17365\,\vec{\mathrm{j}}. \tag{6.40}$$

The displacement vector for node 3 is

$$\vec{d_3} = u_3 \vec{i} + v_3 \vec{j} \tag{6.41}$$

$$\vec{d_3} \times \vec{t} = 0 = \begin{vmatrix} i & j & k \\ u_3 & v_3 & 0 \\ 0.985 & 0.17365 & 0 \end{vmatrix} \tag{6.42}$$

$$= \vec{i}(0) - \vec{j}(0) + \vec{k}(0.17365u_3 - 0.985v_3) = 0. \tag{6.43}$$

Hence, the MPC equation for node 3 is

$$0.17365u_3 - 0.985v_3 = 0. \tag{6.44}$$

Similarly, the MPC equation for node 4 is

$$0.17365u_4 - 0.985v_4 = 0. \tag{6.45}$$

## 6.6 Load Vectors

The global equilibrium equation to be solved after the application of all boundary conditions can be written as

$$\left([K] + [K]_{mpc} + [K]_{spc}\right)\{D\} = \{F\} + \{F\}_{mpc} + \{F\}_{spc}. \tag{6.46}$$

This section explains the computation of the global applied load vector {F}, which is the sum of the global applied element load vector $\{F^e\}$ and global applied point load vector $\{F^p\}$

$$\{F\} = \left\{F^e\right\} + \left\{F^p\right\}. \tag{6.47}$$

$\{F^p\}$ is formed by directly assigning the applied point load components in the global coordinate system.

$\{F^e\}$ is formed by assembling individual element load vectors $\{f_i\}$

$$\left\{F^e\right\} = \sum_{i=1}^{nel}\{f_i\}. \tag{6.48}$$

The element load vector {f} is the summation of load vectors due to the applied body force ($\left\{f^B\right\}$), applied surface force ($\left\{f^S\right\}$), and thermal load $\left(\left\{f^T\right\}\right)$

$$\{f\} = \left\{f^B\right\} + \left\{f^S\right\} + \left\{f^T\right\}. \tag{6.49}$$

The element body force $\{f^B\}$ and traction force vectors $\{f^S\}$ can be computed as

$$\{f^B\} = \int_V [N]^T \{b\} dv \tag{6.50}$$

and

$$\{f^S\} = \int_S [N]^T \{t\} ds \tag{6.51}$$

where {b} and {t} are vectors of body force and traction force components, respectively.

The thermal load vector $\{f^T\}$ arises from the thermal strains due to the thermal expansion of the structure. It is computed as

$$\{f^T\} = \int_V [B]^T [E] \{\varepsilon_0\} dv \tag{6.52}$$

where [B] is the strain displacement matrix, [E] is the constitutive matrix, and $\{\varepsilon_0\}$ is the vector of thermal strains. For a general three-dimensional problem, $\{\varepsilon_0\}$ takes the following form:

$$\{\varepsilon_0\}^T = \{ E\alpha\Delta T \ E\alpha\Delta T \ E\alpha\Delta T \ 0 \ 0 \ 0 \} \tag{6.53}$$

where $E$ is Young's modulus of the material, $\alpha$ is the coefficient of thermal expansion, and $\Delta T$ is the temperature change. The sections that follow explains the computation of $\{f^B\}$, $\{f^S\}$, and $\{f^T\}$, specific to a set of most commonly used finite elements and demonstration using numerical examples.

Gauss integration rule is followed to numerically integrate the expressions. For a polynomial of order $2n - 1$, the number of Gauss points required to exactly integrate is '$n$'. In the section that follows, appropriate Gauss points are selected for computing the load vector.

### 6.6.1 Load Vector for Rod/Truss/Link Element

A rod element can be oriented in any direction in a three-dimensional coordinate system. Load vector for this element is computed first in the local coordinate system ($\xi$,, aligned with the axis of the rod element) and then transformed to a global coordinate system, using an appropriate transformation matrix [T] formed for each element. Let the coordinates of the nodes of an element be ($X1$, $Y1$, $Z1$) and ($X2$, $Y2$, $Z2$). Let $A$ be the area of a cross section of the bar.

Length of the element,

$$L = \sqrt{(X2 - X1)^2 + (Y2 - Y1)^2 + (Z2 - Z1)^2}. \tag{6.54}$$

Shape function matrix,

$$[\mathrm{N}] = \left[\frac{1-\xi}{2}\ \frac{1+\xi}{2}\right]. \tag{6.55}$$

Strain-displacement matrix,

$$[\mathrm{B}] = \left[-\frac{1}{L}\ \frac{1}{L}\right]. \tag{6.56}$$

Determinant of the Jacobian,

$$J = \left|\frac{\partial x}{\partial \xi}\right| \tag{6.57}$$

Vector of applied body force = {b}, the axial component of body force.

Vector of applied traction = {t}, the axial component of the traction, input as force per unit length of the element.

Body force vector,

$$\{\mathrm{f}^{\mathrm{B}}\} = \int_{\mathrm{V}} [\mathrm{N}]^{\mathrm{T}}\{\mathrm{b}\}\mathrm{dv} = A\left[\int_{-1}^{1} \{N(\xi)\}^{\mathrm{T}} b(\xi) J d\xi\right]. \tag{6.58}$$

Traction load vector,

$$\{\mathrm{f}^{\mathrm{S}}\} = \int_{\mathrm{L}} [\mathrm{N}]^{\mathrm{T}}\{\mathrm{t}\}\mathrm{d}L = \left[\int_{-1}^{1} \{N(\xi)\}^{\mathrm{T}} t(\xi) J d\xi\right]. \tag{6.59}$$

Thermal load vector,

$$\{\mathrm{f}^{\mathrm{T}}\} = \int_{\mathrm{V}} [\mathrm{B}]^{\mathrm{T}}[\mathrm{E}]\{\varepsilon_0\}\mathrm{dv} = A\left[\int_{-1}^{1} [\mathrm{B}]^{\mathrm{T}} E\alpha\Delta T J d\xi\right]. \tag{6.60}$$

For a linear variation of traction, the value is given by

$$t(\xi) = N_1 t_1 + N_2 t_2. \tag{6.61}$$

Transformation vector,

$$[\mathrm{T}] = \begin{bmatrix} \frac{X2-X1}{L} & \frac{Y2-Y1}{L} & \frac{Z2-Z1}{L} & 0 & 0 & 0 \\ 0 & 0 & 0 & \frac{X2-X1}{L} & \frac{Y2-Y1}{L} & \frac{Z2-Z1}{L} \end{bmatrix}.$$

*Gauss integration rule*: A 1-point is used for numerical integration of the load vector if the load intensity components or the temperature change are constant over the element. A 2-point integration is used if they are linearly varying over the length.

Load vector in the global coordinate system $\{\mathrm{f}\} = [\mathrm{T}]^{\mathrm{T}}\left(\left\{\mathrm{f}^{\mathrm{B}}\right\} + \left\{\mathrm{f}^{\mathrm{S}}\right\} + \left\{\mathrm{f}^{\mathrm{T}}\right\}\right)$.

### 6.6.2 Load Vector for 3D Solid Element

Being a continuum element, the stiffness matrix and load vectors for three-dimensional solid elements are formulated directly in the global X-Y-Z coordinate. The load vector computation for the same element is shown below.

The shape function matrix

$$[\mathrm{N}] = \begin{bmatrix} N_1 & 0 & 0 & N_2 & 0 & 0 \ldots & N_8 & 0 & 0 \\ 0 & N_1 & 0 & 0 & N_2 & 0 \ldots & 0 & N_8 & 0 \\ 0 & 0 & N_1 & 0 & 0 & N_2 \ldots & 0 & 0 & N_8 \end{bmatrix}, \tag{6.62}$$

where $N_i = \frac{1}{8}(1+\xi\xi_i)(1+\eta\eta_i)(1+\zeta\zeta_i)$, for the $i$th node.

Constitutive Matrix, [E] is given as

$$[\mathrm{E}] = \frac{E}{(1+\upsilon)(1-2\upsilon)}\begin{bmatrix} 1-\upsilon & \upsilon & \upsilon & 0 & 0 & 0 \\ \upsilon & 1-\upsilon & \upsilon & 0 & 0 & 0 \\ \upsilon & \upsilon & 1-\upsilon & 0 & 0 & 0 \\ 0 & 0 & 0 & \frac{1-2\upsilon}{2} & 0 & 0 \\ 0 & 0 & 0 & 0 & \frac{1-2\upsilon}{2} & 0 \\ 0 & 0 & 0 & 0 & 0 & \frac{1-2\upsilon}{2} \end{bmatrix}.$$

Strain-displacement matrix, [B] is given as

$$[B] = \begin{bmatrix} \frac{\partial N_1}{\partial X} & 0 & 0 & \frac{\partial N_2}{\partial X} & 0 & 0 & \cdots\cdots & \frac{\partial N_{81}}{\partial X} & 0 & 0 \\ 0 & \frac{\partial N_1}{\partial Y} & 0 & 0 & \frac{\partial N_2}{\partial Y} & 0 & \cdots\cdots & 0 & \frac{\partial N_8}{\partial Y} & 0 \\ 0 & 0 & \frac{\partial N_1}{\partial Y} & 0 & 0 & \frac{\partial N_2}{\partial Y} & \cdots\cdots & 0 & 0 & \frac{\partial N_8}{\partial Y} \\ \frac{\partial N_1}{\partial Y} & \frac{\partial N_1}{\partial X} & 0 & \frac{\partial N_2}{\partial Y} & \frac{\partial N_2}{\partial X} & 0 & \cdots\cdots & \frac{\partial N_8}{\partial Y} & \frac{\partial N_8}{\partial X} & 0 \\ 0 & \frac{\partial N_1}{\partial Z} & \frac{\partial N_1}{\partial Y} & 0 & \frac{\partial N_2}{\partial Z} & \frac{\partial N_2}{\partial Y} & \cdots\cdots & 0 & \frac{\partial N_8}{\partial Z} & \frac{\partial N_8}{\partial Y} \\ \frac{\partial N_1}{\partial Z} & 0 & \frac{\partial N_1}{\partial X} & \frac{\partial N_2}{\partial Z} & 0 & \frac{\partial N_2}{\partial X} & \cdots\cdots & \frac{\partial N_8}{\partial Z} & 0 & \frac{\partial N_8}{\partial X} \end{bmatrix}.$$

Vector of applied body force, $\{\mathrm{b}\}^{\mathrm{T}} = \{b_X, b_Y, b_Z\}$.

Vector of applied traction, $\{t\}^T = \{t_X, t_Y, t_Z\}$, input as force per unit area on the element face(s).

Determinant of Jacobian, $J = \begin{vmatrix} \frac{\partial X}{\partial \xi} & \frac{\partial Y}{\partial \xi} & \frac{\partial Y}{\partial \xi} \\ \frac{\partial X}{\partial \eta} & \frac{\partial Y}{\partial \eta} & \frac{\partial Z}{\partial \eta} \\ \frac{\partial X}{\partial \zeta} & \frac{\partial Y}{\partial \zeta} & \frac{\partial Z}{\partial \zeta} \end{vmatrix}$.

Body force vector,

$$\{f^B\} = \int_V [N]^T \{b\} dv = \int_V [N]^T \begin{Bmatrix} b_X \\ b_Y \\ b_Z \end{Bmatrix} dv = \int_{-1}^{1}\int_{-1}^{1}\int_{-1}^{1} [N]^T \begin{Bmatrix} b_X \\ b_Y \\ b_Z \end{Bmatrix} J d\xi d\eta d\zeta. \tag{6.63}$$

Traction load vector,

$$\{f^S\} = \int_s [N]^T \{t\} ds = \int_s [N]^T \begin{Bmatrix} t_X \\ t_Y \\ t_Z \end{Bmatrix} ds = \int_{-1}^{1}\int_{-1}^{1} [N]^T \begin{Bmatrix} t_X \\ t_Y \\ t_Z \end{Bmatrix} J d\xi d\eta. \tag{6.64}$$

Thermal load vector

$$\{f^T\} = \int_V [B]^T [E] \{\varepsilon_0\} dv = \int_V [B]^T [E] \begin{Bmatrix} \alpha \Delta T \\ \alpha \Delta T \\ \alpha \Delta T \\ 0 \\ 0 \\ 0 \end{Bmatrix} dv$$

$$= \int_{-1}^{1}\int_{-1}^{1}\int_{-1}^{1} [B]^T [E] \begin{Bmatrix} \alpha \Delta T \\ \alpha \Delta T \\ \alpha \Delta T \\ 0 \\ 0 \\ 0 \end{Bmatrix} = J\, d\xi d\eta d\zeta. \tag{6.65}$$

*Gauss integration rule*: A 2 × 2 × 2 order is used for numerical integration of body force vector and thermal load vector. A 2 × 2 order (on the loaded quadrilateral face) is used for the surface load vector.

### 6.6.3 Load Vector for Planar Element

The plane stress and plane strain elements are directly formulated in the global coordinate system. The load vector computation in the global X-Y coordinate system is shown below.

The shape function Matrix is computed at 2 × 2 Gauss points for body force and 2 Gauss points for traction load.

The shape function matrix,

$$[\mathrm{N}] = \begin{bmatrix} N_1 & 0 & N_2 & 0 & N_3 & 0 & N_4 & 0 \\ 0 & N_1 & 0 & N_2 & 0 & N_3 & 0 & N_4 \end{bmatrix}, \tag{6.66}$$

where $N_i = \frac{1}{4}(1+\xi\xi_i)(1+\eta\eta_i)$, for the $i$th node.

Constitutive Matrix, [E] for plane stress element is

$$[\mathrm{E}] = \frac{E}{1-\upsilon^2}\begin{bmatrix} 1 & \upsilon & 0 \\ \upsilon & 1 & 0 \\ 0 & 0 & (1-\upsilon)/2 \end{bmatrix},$$

and for plane strain element is

$$[\mathrm{E}] = \frac{E}{(1+\upsilon)(1-2\upsilon)}\begin{bmatrix} 1-\upsilon & \upsilon & 0 \\ \upsilon & 1-\upsilon & 0 \\ 0 & 0 & \frac{1-2\upsilon}{2} \end{bmatrix}.$$

Strain–displacement matrix [B] is given as

$$[B] = \begin{bmatrix} \frac{\partial N_1}{\partial X} & 0 & \frac{\partial N_2}{\partial X} & 0 & \frac{\partial N_3}{\partial X} & 0 & \frac{\partial N_4}{\partial X} & 0 \\ 0 & \frac{\partial N_1}{\partial Y} & 0 & \frac{\partial N_2}{\partial Y} & 0 & \frac{\partial N_3}{\partial Y} & 0 & \frac{\partial N_4}{\partial Y} \\ \frac{\partial N_1}{\partial Y} & \frac{\partial N_1}{\partial X} & \frac{\partial N_2}{\partial Y} & \frac{\partial N_2}{\partial X} & \frac{\partial N_3}{\partial Y} & \frac{\partial N_3}{\partial X} & \frac{\partial N_4}{\partial Y} & \frac{\partial N_4}{\partial X} \end{bmatrix}.$$

Vector of applied body force, $\{\mathrm{b}\}^{\mathrm{T}} = \{b_X, b_Y\}$.

Vector of applied traction, $\{\mathrm{t}\}^{\mathrm{T}} = \{t_X, t_Y\}$, input as force per unit length on the edge(s).

Determinant of Jacobian, $J = \begin{vmatrix} \frac{\partial X}{\partial \xi} & \frac{\partial Y}{\partial \xi} \\ \frac{\partial X}{\partial \eta} & \frac{\partial Y}{\partial \eta} \end{vmatrix}$.

Body force vector

$$\{\mathrm{f^B}\} = \int_{\mathrm{V}} [\mathrm{N}]^{\mathrm{T}}\{\mathrm{b}\}\mathrm{dv} = \int_{\mathrm{A}} h[\mathrm{N}]^{\mathrm{T}}\begin{Bmatrix} \mathrm{b_X} \\ \mathrm{b_Y} \end{Bmatrix}\mathrm{d}A = \int_{-1}^{1}\int_{-1}^{1} h[\mathrm{N}]^{\mathrm{T}}\begin{Bmatrix} b_X \\ b_Y \end{Bmatrix} J\,\mathrm{d}\xi\,\mathrm{d}\eta. \tag{6.67}$$

Traction (on edges) load vector

$$\{f^S\} = \int_L [N]^T\{t\}dL = \int_L [N]^T \begin{Bmatrix} t_X \\ t_Y \end{Bmatrix} dL = \int_{-1}^{1} [N]^T \begin{Bmatrix} t_X \\ t_Y \end{Bmatrix} J\,d\xi. \tag{6.68}$$

Thermal load vector

$$\{f^T\} = \int_V [B]^T[E]\{\varepsilon_0\}dv = \int_A h[B]^T[E] \begin{Bmatrix} \alpha\Delta T \\ \alpha\Delta T \\ 0 \end{Bmatrix} dA$$

$$= \int_{-1}^{1}\int_{-1}^{1} h[B]^T[E] \begin{Bmatrix} \alpha\Delta T \\ \alpha\Delta T \\ 0 \end{Bmatrix} J\,d\xi\,d\eta. \tag{6.69}$$

*Gauss integration rule*: A 2 × 2 order is used for numerical integration of body force vector and thermal load vector. A 2-point integration (on the loaded quadrilateral face) is used for the traction load vector on the edges of the element. In the above equations, $h$ refers to the actual thickness of the structure at the Gauss point, for plane stress element. For plane strain element, $h$ is assumed as unity. '$L$' refers to the length of the edge on which traction acts.

### 6.6.4 Load Vector for Axi-Symmetric Solid Element

Axi-symmetric solid elements are directly formulated in the global coordinate system. The computation of the load vector in the global R-Z coordinate system is shown below. It should be noted that the R-direction matches with the conventional global X-direction.

The shape function matrix,

$$[N] = \begin{bmatrix} N_1 & 0 & N_2 & 0 & N_3 & 0 & N_4 & 0 \\ 0 & N_1 & 0 & N_2 & 0 & N_3 & 0 & N_4 \end{bmatrix} \tag{6.70}$$

where $N_i = \frac{1}{4}(1 + \xi\xi_i)(1 + \eta\eta_i)$, for the $i$th node.

Constitutive Matrix, [E], is given as

$$[E] = \frac{E}{(1+\upsilon)(1-2\upsilon)} \begin{bmatrix} 1-\upsilon & \upsilon & \upsilon & 0 \\ \upsilon & 1-\upsilon & \upsilon & 0 \\ \upsilon & \upsilon & 1-\upsilon & 0 \\ 0 & 0 & 0 & \frac{1-2\upsilon}{2} \end{bmatrix}.$$

Strain–displacement matrix, [B], is given as

$$[B] = \begin{bmatrix} \frac{\partial N_1}{\partial X} & 0 & \frac{\partial N_2}{\partial X} & 0 & \frac{\partial N_3}{\partial X} & 0 & \frac{\partial N_4}{\partial X} & 0 \\ \frac{N_1}{r} & 0 & \frac{N_2}{r} & 0 & \frac{N_3}{r} & 0 & \frac{N_4}{r} & 0 \\ 0 & \frac{\partial N_1}{\partial X} & 0 & \frac{\partial N_2}{\partial X} & 0 & \frac{\partial N_3}{\partial X} & 0 & \frac{\partial N_4}{\partial X} \\ \frac{\partial N_1}{\partial Z} & \frac{\partial N_1}{\partial X} & \frac{\partial N_2}{\partial Z} & \frac{\partial N_2}{\partial X} & \frac{\partial N_3}{\partial Z} & \frac{\partial N_3}{\partial X} & \frac{\partial N_4}{\partial Z} & \frac{\partial N_4}{\partial X} \end{bmatrix}.$$

Vector of applied body force, $\{b\}^T = \{ b_R \; b_Z \} = \{ \rho r \omega^2 \; \rho g \}$,

where $\rho$—Density of the material; $r$—Gauss point radius; $\omega$—angular velocity; g—acceleration due to gravity.

Vector of applied traction, $\{\mathrm{t}\}^{\mathrm{T}} = \{t_R, t_Z\}$.

Determinant of Jacobian, $J = \begin{vmatrix} \frac{\partial R}{\partial \xi} & \frac{\partial Z}{\partial \xi} \\ \frac{\partial R}{\partial \eta} & \frac{\partial Z}{\partial \eta} \end{vmatrix}$.

Body force vector is given as

$$\{\mathrm{f^B}\} = \int_{\mathrm{v}} [\mathrm{N}]^{\mathrm{T}}\{\mathrm{b}\}\mathrm{dv} = 2\pi \int_A [\mathrm{N}]^{\mathrm{T}} \begin{Bmatrix} \mathrm{b_R} \\ \mathrm{b_Z} \end{Bmatrix} r\mathrm{dA} = 2\pi \int_{-1}^{1}\int_{-1}^{1} [\mathrm{N}]^{\mathrm{T}} \begin{Bmatrix} \mathrm{b}_R \\ \mathrm{b}_Z \end{Bmatrix} r\,J\,\mathrm{d}\xi\,\mathrm{d}\eta. \tag{6.71}$$

Traction load vector is given as

$$\{\mathrm{f^S}\} = \int_{\mathrm{L}} [\mathrm{N}]^{\mathrm{T}}\{\mathrm{t}\}\mathrm{dL} = 2\pi \int_L [\mathrm{N}]^{\mathrm{T}} \begin{Bmatrix} t_R \\ t_Z \end{Bmatrix} r\,\mathrm{d}L = 2\pi \int_{-1}^{1} [\mathrm{N}]^{\mathrm{T}} \begin{Bmatrix} t_R \\ t_z \end{Bmatrix} r\,J\,\mathrm{d}\xi. \tag{6.72}$$

Thermal load vector is given as

$$\{\mathrm{f^T}\} = \int_{\mathrm{v}} [\mathrm{B}]^{\mathrm{T}}[\mathrm{E}]\{\varepsilon_0\}\mathrm{dv} = 2\pi \int_{\mathrm{A}} [\mathrm{B}]^{\mathrm{T}}[\mathrm{E}] \begin{Bmatrix} \alpha\Delta T \\ \alpha\Delta T \\ \alpha\Delta T \\ 0 \end{Bmatrix} r\mathrm{d}A$$

$$= 2\pi \int_{-1}^{1}\int_{-1}^{1} [\mathrm{B}]^{\mathrm{T}}[\mathrm{E}] \begin{Bmatrix} \alpha\Delta T \\ \alpha\Delta T \\ \alpha\Delta T \\ 0 \end{Bmatrix} r\,J\,\mathrm{d}\xi\,\mathrm{d}\eta. \tag{6.73}$$

*Gauss integration rule*: A 2 × 2 order is used for numerical integration of body force vector and thermal load vector. A 2-point integration (on the loaded edge) is used for the traction load vector on the edges of the element. In the above equations, $r$ refers to the distance of the Gauss point from the axis of the revolution, which we chose as the global Z-axis. Hence $r$ is also the global X-coordinate of the concerned Gauss point, which can be computed from the X-coordinate of the nodes, using the shape functions. $L$ refers to the length of the edge on which traction acts.

### 6.6.5 Load Vector for Beam Element

A beam element can be oriented in any direction in a three-dimensional coordinate system. Each 2-node beam element is associated with a local x-y-z coordinate system. The local x-axis is defined along the axis of the beam connecting the end nodes. The local y and z axes are formed on the cross section of the beam such that it matches with the axes about which the second moment area of the cross section $I_y$ and $I_z$ are specified. The body load ($b_x$, $b_y$, *and* $b_z$) and traction ($t_x$, $t_y$, *and* $t_z$) components are specified in this local coordinate system, for computing the element load vector in the local coordinate system. It is then transformed into the global coordinate system, using an appropriate transformation matrix [T] formed for each element. Let the coordinates of the nodes of the element be (*X1, Y1, Z1*) and (*X2, Y2, Z2*). Let *A* be the area of the cross section of the beam.

Length of the element, $L = \sqrt{(X2 - X1)^2 + (Y2 - Y1)^2 + (Z2 - Z1)^2}$.

Direction cosine of the local x, y, and z axes: ($l_1$, $m_1$, $n_1$), ($l_2$, $m_2$, $n_2$), and ($l_3$, $m_3$, $n_3$).

Shape function matrix, $[\mathrm{N}] = \begin{bmatrix} \mathrm{N}_1 & 0 & 0 & \mathrm{N}_2 & 0 & 0 \\ 0 & \mathrm{N}_1 & 0 & 0 & \mathrm{N}_2 & 0 \\ 0 & 0 & \mathrm{N}_1 & 0 & 0 & \mathrm{N}_2 \end{bmatrix}$,

where $N_1 = \frac{1-\xi}{2}$, $N_2 = \frac{1+\xi}{2}$. This shape function matrix is formed by considering only the translational displacement fields (u, v, w) and the corresponding degrees of freedom at the nodes. The section rotation fields are not considered.

Determinant of the Jacobian, $= \left|\frac{\partial x}{\partial \xi}\right|$.

Vector of applied body force $= \{\mathrm{b}\} = \left\{ b_x \; b_y \; b_z \right\}^T$ in the local coordinate system, input as force per unit volume of the body.

Vector of applied traction $= \{\mathrm{t}\} = \left\{ t_x \; t_y \; t_z \right\}^T$ in the local coordinate system, input as traction per unit length of the beam.

Body force vector,

$$\left\{\mathrm{f^B}\right\} = \int\limits_{\mathrm{V}} [\mathrm{N}]^{\mathrm{T}}\{\mathrm{b}\}\mathrm{dv} = A\left[\int\limits_{-1}^{1} \{N(\xi)\}^{\mathrm{T}} b(\xi) J d\xi\right]. \tag{6.74}$$

Traction load vector,

$$\left\{\mathrm{f^S}\right\} = \int\limits_{\mathrm{L}} [\mathrm{N}]^{\mathrm{T}}\{\mathrm{t}\}\mathrm{d}L = \left[\int\limits_{-1}^{1} \{N(\xi)\}^{\mathrm{T}} t(\xi) J d\xi\right]. \tag{6.75}$$

For a linear variation of traction value is given by $t(\xi) = N_1(\xi)t_1 + N_2(\xi)t_2$.

*Gauss integration rule*: A 2-point is used for numerical integration of both body load and traction vectors.

$$\text{Transformation vector, [T]} = \begin{bmatrix} l_1 & m_1 & n_1 & 0 & 0 & 0 \\ l_2 & m_2 & n_2 & 0 & 0 & 0 \\ l_3 & m_3 & n_3 & 0 & 0 & 0 \\ 0 & 0 & 0 & l_1 & m_1 & n_1 \\ 0 & 0 & 0 & l_2 & m_2 & n_2 \\ 0 & 0 & 0 & l_3 & m_3 & n_3 \end{bmatrix}.$$

Element load vector in the global coordinate system $\{\mathrm{f}\} = [\mathrm{T}]^{\mathrm{T}}\left(\left\{\mathrm{f}^{\mathrm{B}}\right\} + \left\{\mathrm{f}^{\mathrm{S}}\right\}\right)$.

The computed load vector, after transformation, shall have 6 components corresponding to forces at each translational degree of freedom. They have to be appropriately placed in the element load vector of size 12, where nodal force components appear. At the positions in the global assembled load vector corresponding to the rotational degrees of freedom, concentrated moments can be applied directly.

### 6.6.6 Load Vector for Axi-Symmetric Shell Element

An axi-symmetric shell element can be oriented in any direction in the x-z plane. Each 2-node axi-symmetric shell element is associated with a local x-z coordinate system. The local x-axis is defined along the element connecting nodes from 1 to 2. The local z-axis is formed normal to the element such that the positive hoop direction (y-axis) is into the paper in the right-handed x-y-z system. The body load $(b_x, b_z)$ and traction $(t_x, t_z)$ components are specified in this local coordinate system, for computing the element load vector in the local coordinate system. It is then transformed into the global coordinate system, using an appropriate transformation matrix [T] formed for each element. Let the coordinates of the nodes of the element be (*X1, Z1*) and (*X2, Z2*). Let $h$ be the thickness of the shell in the local z-direction.

Direction cosine of the local x and z axes: $(l_1, m_1, n_1)$, $(l_2, m_2, n_2)$

Length of the element, $L = \sqrt{(X2 - X1)^2 + (Z2 - Z1)^2}$.

Direction cosine of the local x, and z axes: $(l_1, m_1, n_1)$, $(l_2, m_2, n_2)$

$$\text{Shape function matrix, [N]} = \begin{bmatrix} N_1 & 0 & N_2 & 0 \\ 0 & N_1 & 0 & N_2 \end{bmatrix}$$

where $N_1 = \frac{1-\xi}{2}$, $N_2 = \frac{1+\xi}{2}$. This shape function matrix is formed by considering only the translational displacement fields (u, w) and the corresponding degrees of freedom at the nodes. The section rotation field is considered.

Determinant of the Jacobian, $J = \left|\frac{\partial x}{\partial \xi}\right|$.

Vector of applied body force $= \{\mathrm{b}\} = \{b_x b_z\}^T$, in the local coordinate system, input as force per unit volume of the body.

Vector of applied traction $= \{\mathrm{t}\} = \{t_x t_z\}^T$, in the local coordinate system, input as traction per unit length of the element.

Body force vector,

$$\{f^B\} = \int_V [N]^T\{b\}dv = 2\pi \int_{-1}^{1} \{N(\xi)\}^T\{b(\xi)\}rhJd\xi. \tag{6.76}$$

Traction load vector,

$$\{f^S\} = \int_L [N]^T\{t\}dL = 2\pi \int_{-1}^{1} \{N(\xi)\}^T\{t(\xi)\}rJd\xi. \tag{6.77}$$

For a linear variation of traction value is given by $t(\xi) = N_1(\xi)t_1 + N_2(\xi)t_2$.

*Gauss integration rule*: A 2-point is used for numerical integration of both body load and traction vectors.

$$\text{Transformation vector, } [T] = \begin{bmatrix} l_1 & m_1 & 0 & 0 \\ l_2 & m_2 & 0 & 0 \\ 0 & 0 & l_1 & m_1 \\ 0 & 0 & l_2 & m_2 \end{bmatrix}.$$

Element load vector in the global coordinate system $\{f\} = [T]^T(\{f^B\} + \{f^S\})$.

The computed load vector, after transformation, shall have 4 components corresponding to forces at each translational degree of freedom. They have to be appropriately placed in the element load vector of size 6, where nodal force components appear. At the positions in the global assembled load vector corresponding to the rotational degrees of freedom, concentrated moments can be applied directly.

### 6.6.7 Load Vector for Plate Element

A plate element can be oriented in any direction in a three-dimensional coordinate system. For a 4-noded flat plate element, the displacement kinematics is defined in a local coordinate system, whose z-axis is normal to the mid-plane. The local x- and y-axes can be easily assigned on the mid-plane, to form a rectangular coordinate system. The body load ($b_x$, $b_y$, $b_z$) and traction ($t_x$, $t_y$, $t_z$) components are specified in this local coordinate system, for computing the element load vector in the local coordinate system. It is then transformed to the global coordinate system, using an appropriate transformation matrix [T] formed for each element. In case of a 4-node flat plate element, the procedure given below converts the distributed body load field and the traction field acting on the plate element to their work equivalent forces acting on the corresponding nodes. Since there are three translational degrees of freedom at each node, a vector of size 12 is generated in the local coordinate system.

Direction cosine of the local x, y, and z axes: $(l_1, m_1, n_1)$, $(l_2, m_2, n_2)$, and $(l_3, m_3, n_3)$.

Shape function matrix, $[N] = \begin{bmatrix} N_1 & 0 & 0 & N_2 & 0 & 0 & N_3 & 0 & 0 & N_4 & 0 & 0 \\ 0 & N_1 & 0 & 0 & N_2 & 0 & 0 & N_3 & 0 & 0 & N_4 & 0 \\ 0 & 0 & N_1 & 0 & 0 & N_2 & 0 & 0 & N_3 & 0 & 0 & N_4 \end{bmatrix} [N]$.

Where $N_i = \frac{1}{4}(1 + \xi\xi_i)(1 + \eta\eta_i)$. This shape function matrix is formed by considering only the translational displacement fields (u, v, w) and the corresponding degrees of freedom at the nodes. The section rotation fields are not considered.

Determinant of the Jacobian, $J = \begin{vmatrix} \frac{\partial x}{\partial \xi} & \frac{\partial y}{\partial \xi} \\ \frac{\partial x}{\partial \eta} & \frac{\partial y}{\partial \eta} \end{vmatrix}$.

Vector of applied body force $= \{\mathrm{b}\} = \left\{ b_x \ b_y \ b_z \right\}^T$, in the local coordinate system, input as force per unit volume of the body.

Vector of applied traction $= \{\mathrm{t}\} = \left\{ t_x \ t_y \ t_z \right\}^T$, in the local coordinate system, input as traction per area of the plate.

Body force vector,

$$\left\{\mathrm{f^B}\right\} = \int_{\mathrm{V}} [\mathrm{N}]^{\mathrm{T}}\{\mathrm{b}\}\mathrm{dv} = \int_{-1}^{1}\int_{-1}^{1} \{N(\xi, \eta)\}^{\mathrm{T}} b(\xi, \eta) h J d\xi d\eta. \tag{6.78}$$

Traction load vector acting on the face of the element

$$\left\{\mathrm{f^S}\right\} = \int_{\mathrm{A}} [\mathrm{N}]^{\mathrm{T}}\{\mathrm{t}\}\mathrm{d}A = \int_{-1}^{1} \{N(\xi, \eta)\}^{\mathrm{T}} J d\xi d\eta. \tag{6.79}$$

For a linear variation of traction over the surface of the plate, the required value at the Gauss point is given by $t(\xi) = N_1(\xi)t_1 + N_{2(\xi)}t_2 + N_3(\xi)t_3 + N_4(\xi)t_4$.

Transformation vector, $[\mathrm{T}] = \begin{bmatrix} [\mathrm{M}] & [0] & [0] & [0] \\ [0] & [\mathrm{M}] & [0] & [0] \\ [0] & [0] & [\mathrm{M}] & [0] \\ [0] & [0] & [0] & [\mathrm{M}] \end{bmatrix}$.

Where $[\mathrm{M}] = \begin{bmatrix} l_1 & m_1 & n_1 \\ l_2 & m_2 & n_2 \\ l_3 & m_3 & n_3 \end{bmatrix}$ and [0] is a null matrix of $3 \times 3$.

*Gauss integration rule*: A $2 \times 2$ order for numerical integration of both body load and traction vectors.

Element load vector in the global coordinate system $\{\mathrm{f}\} = [\mathrm{T}]^{\mathrm{T}}\left(\left\{\mathrm{f^B}\right\} + \left\{\mathrm{f^S}\right\}\right)$.

The computed load vector, after transformation, shall have 12 components corresponding to forces at each translational degree of freedom. They have to be appropriately placed in the element load vector of size 24, where nodal force components appear. At the positions in the global assembled load vector corresponding to the rotational degrees of freedom, concentrated moments can be applied directly.

## 6.7 Numerical Examples of Load Vector

In this section, load vector computation for various elements defined above will be demonstrated through numerical examples. The following properties are assumed for computations:

$A = 0.01\ \text{m}^2; \alpha = 1.0 \times 10^{-6}/^{\circ}\text{C}; E = 2.0 \times 10^{10}\ \text{N/m}^2; \nu = 0.3, \rho = 7800\ \text{kg/m}^3$; $g = 9.81\ \text{m/s}^2$; $\Delta t = 30$.

**Example 6** Figure 6.4 shows a plane truss with the point load at node 2. Compute the elemental load vector in the local system, transform them into the global system, and assemble them in a global load vector for (a) Point load, (b) Thermal load, and (c) Gravity load.

Element connectivity for Element 1 and 2 are given as (1 2) and (2 3), respectively. Length of the Element one

$$l_1 = \sqrt{(X2 - X1)^2 + (Y2 - Y1)^2} = \sqrt{(0.15)^2 + (0.4)^2} = 0.4272. \tag{6.80}$$

Length of the element two is

$$l_2 = \sqrt{(X3 - X2)^2 + (Y3 - Y2)^2} = \sqrt{(0.35)^2 + (0.8)^2} = 0.8732. \tag{6.81}$$

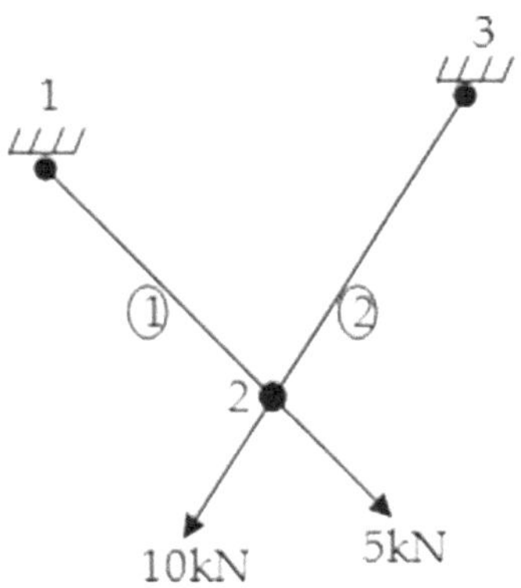


| Node | X | Y |
|---|---|---|
| 1 | 0.2 | 0.6 |
| 2 | 0.35 | 0.2 |
| 3 | 0.7 | 1 |

**Fig. 6.4** Plane truss with point load

Transformation matrix for Element one is

$$[\mathrm{T}_1] = \begin{bmatrix} \frac{\mathrm{X2-X1}}{l_1} & \frac{\mathrm{Y2-Y1}}{l_1} & 0 & 0 \\ 0 & 0 & \frac{\mathrm{X2-X1}}{l_1} & \frac{\mathrm{Y2-Y1}}{l_1} \end{bmatrix} \tag{6.82}$$

$$[\mathrm{T}_1] = \begin{bmatrix} \frac{0.15}{0.4272} & \frac{0.4}{0.4272} & 0 & 0 \\ 0 & 0 & \frac{0.15}{0.4272} & \frac{0.4}{0.4272} \end{bmatrix} \tag{6.83}$$

$$[\mathrm{T}_1] = \begin{bmatrix} 0.351124 & 0.93633 & 0 & 0 \\ 0 & 0 & 0.351124 & 0.93633 \end{bmatrix}. \tag{6.84}$$

Transformation matrix for Element two is

$$[\mathrm{T}_2] = \begin{bmatrix} \frac{\mathrm{X3-X2}}{l_2} & \frac{\mathrm{Y3-Y2}}{l_2} & 0 & 0 \\ 0 & 0 & \frac{\mathrm{X3-X2}}{l_2} & \frac{\mathrm{Y3-Y2}}{l_2} \end{bmatrix} \tag{6.85}$$

$$[T_2] = \begin{bmatrix} 0.4008 & 0.9162 & 0 & 0 \\ 0 & 0 & 0.4008 & 0.9162 \end{bmatrix}. \tag{6.86}$$

(a) Load vector for Point load

Point load vector for Element 1 and 2 in the local x-coordinate system is

$$\{\mathrm{f}_1\}^T = \left\{ 0\ 5 \times 10^3 \right\} \tag{6.87}$$

$$\{\mathrm{f}_2\}^T = \left\{ 10 \times 10^3\ 0 \right\}. \tag{6.88}$$

Point load vector for Element 1 and 2 in the global X-Y coordinate system is

$$\{\mathrm{f}_1\}_\mathrm{g}^\mathrm{T} = [\mathrm{T}_1]^\mathrm{T}\{f_1\}^T \tag{6.89}$$

$$\{\mathrm{f}_1\}_\mathrm{g}^\mathrm{T} = \left\{ 0\ 0\ 1755.62\ 4681.65 \right\} \tag{6.90}$$

$$\{\mathrm{f}_2\}_\mathrm{g}^\mathrm{T} = [\mathrm{T}_2]^\mathrm{T}\{f_2\}^T \tag{6.91}$$

$$\{\mathrm{f}_2\}_\mathrm{g} = \left\{ 4000.8\ 9162\ 0\ 0 \right\} \tag{6.92}$$

$$\{\mathrm{f}\}_\mathrm{g}^\mathrm{T} = \{\mathrm{f}_1\}_\mathrm{g}^\mathrm{T} + \{\mathrm{f}_2\}_\mathrm{g}^\mathrm{T} \tag{6.93}$$

$$\{\mathrm{f}\}_\mathrm{g}^\mathrm{T} = \left\{ 0\ 0\ (1755.62 + 4000.8)\ (4681.65 + 9162)\ 0\ 0 \right\} \tag{6.94}$$

$$\{f\}_g^T = \{0\ 0\ 5756.42\ 13843.65\ 0\ 0\}. \tag{6.95}$$

(b) Load vector due to thermal expansion

Thermal load vector for Element one in the local x-coordinate system is

$$\{f_1^T\} = \begin{Bmatrix} -AE\alpha\Delta T \\ AE\alpha\Delta T \end{Bmatrix} = \begin{Bmatrix} -0.01 \times 2.0 \times 10^{10} \times 1.0 \times 10^{-6} \times 30 \\ +0.01 \times 2.0x10^{10} \times 1.0 \times 10^{-6} \times 30 \end{Bmatrix} \tag{6.96}$$

$$\{f_1^T\} = \begin{Bmatrix} -6 \times 10^3 \\ 6 \times 10^3 \end{Bmatrix}. \tag{6.97}$$

Thermal load vector for Element 2in the local x-coordinate system is same as Element 1

$$\{f_1^T\} = \begin{Bmatrix} -6 \times 10^3 \\ 6 \times 10^3 \end{Bmatrix}. \tag{6.98}$$

Global load vector due to thermal load for element (1) is given as

$$\{f_1^T\}_g^T = [T_1]^T\{f_1^T\} \tag{6.99}$$

$$\{f_1^T\}_g^T = \{-2106.744\ 5617.98\ -2106.744\ 5617.98\} \tag{6.100}$$

$$\{f_2^T\}_g^T = [T_2]^T\{f_2^T\} \tag{6.101}$$

$$\{f_2\}_g^T = \{-2404.8\ 5497.2\ -2404.8\ 5497.2\} \tag{6.102}$$

$$\{f\}_g^T = \{f_2^T\}_g^T + \{f_2^T\}_g^T \tag{6.103}$$

$$\{f^T\}_g^T = \{-2106.744\ \ 5417.98\ \ -(2106.744 + 2404.8)\ \ (5617.98 + 5497.2)\ \ \ -2404.8\ \ 5497.2\} \tag{6.104}$$

$$\{f^T\}_g^T = \{-2106.744\ 5617.98\ -4511.544\ 11115.18\ -2404.8\ 5497.2\}. \tag{6.105}$$

(c) Load vector due to gravity

Gravity load in the local x-direction for Element one and two is given as

$$\{f_1^B\} = \begin{Bmatrix} \rho\, gAl_1 \\ \rho\, gAl_1 \end{Bmatrix} = \begin{Bmatrix} 7800 \times 9.81 \times 0.01 \times 0.4272 \\ 7800 \times 9.81 \times 0.01 \times 0.4272 \end{Bmatrix} \tag{6.106}$$

$$\{f_1^B\} = \begin{Bmatrix} 326.884896 \\ 326.884896 \end{Bmatrix} \tag{6.107}$$

$$\left\{ f_2^B \right\} = \begin{Bmatrix} \rho\, gAl_2 \\ \rho\, gAl_2 \end{Bmatrix} = \begin{Bmatrix} 668.155716 \\ 668.155716 \end{Bmatrix}. \tag{6.108}$$

Load vector in global X-Y-direction is given as

$$\left\{ f_1^B \right\}_g^T = [T_1]^T \left\{ f_1^B \right\} \tag{6.109}$$

$$\left\{ f_1^B \right\}_g^T = \left\{ 114.777132\ 326.894269\ 114.777132\ 326.894269 \right\} \tag{6.110}$$

$$\left\{ f_2^B \right\}_g^T = [T_2]^T \left\{ f_2^B \right\} \tag{6.111}$$

$$\left\{ f_2^B \right\}_g^T = \left\{ 267.796595\ 612.163772\ 267.796595\ 612.163722 \right\} \tag{6.112}$$

$$\{f\}_g^T = \left\{ f_1^B \right\}_g^T + \left\{ f_2^B \right\}_g^T \tag{6.113}$$

$$\left\{ f^B \right\}_g^T = \left\{ 114.777132\ 326.894259\ \begin{matrix} 114.777132+ \\ 267.796595 \end{matrix}\ \begin{matrix} (326.89425+ \\ 612.163722) \end{matrix}\ 267.796595\ 612.163722 \right\} \tag{6.114}$$

$$\left\{ f^B \right\}_g^T = \left\{ 114.7771\ 326.8943\ 382.5679\ 939.0580\ 267.7966\ 612.1637 \right\}. \tag{6.115}$$

**Example 7** Compute the load vector for the planar structural configuration shown in Fig. 6.5. Compute the load vector for (a) pressure, (b) gravity, and (c) thermal loads. Element connectivity is (1 2 3 4).

(a) Pressure Load

Compute the normal to the edge connecting nodes 3 and 4. Vector tangent to the edge is

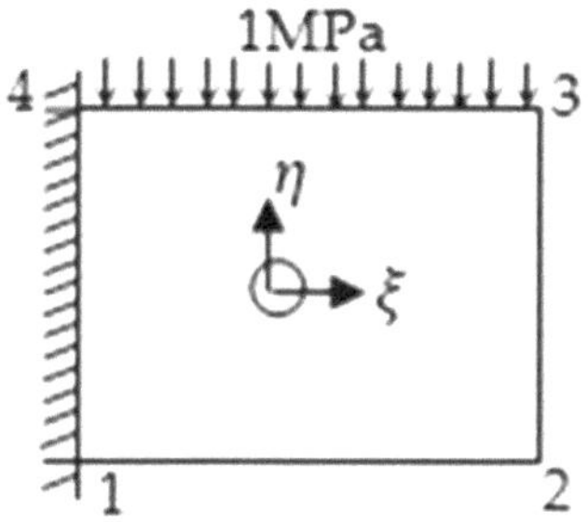


| Node | X | Y |
|---|---|---|
| 1 | 0 | 0 |
| 2 | 1 | 0 |
| 3 | 1 | 1 |
| 4 | 0 | 1 |

**Fig. 6.5** Planar structure

$$\overrightarrow{t_1} = 1\,\overrightarrow{i} + 0\,\overrightarrow{j} + 0\,\overrightarrow{k} \tag{6.116}$$

The vector normal to $\overrightarrow{t_1}$ is obtained by taking a cross product with the unit vector in Z direction and is given as

$$\overrightarrow{t_2} = 0\,\overrightarrow{i} + 0\,\overrightarrow{j} + 1\,\overrightarrow{k} \tag{6.117}$$

$$\hat{n} = \overrightarrow{t_1} \times \overrightarrow{t_2} = \begin{vmatrix} i & j & k \\ 1 & 0 & 0 \\ 0 & 0 & 1 \end{vmatrix} = i(0) - j(1) + k(0). \tag{6.118}$$

Hence, the direction cosines normal to the edge is, $l_1 = 0; m_1 = -1$.

Determinant of Jacobian connecting the edges 4 and 3 is given as

$$J = \left|\frac{\partial X}{\partial \xi}\right| = \left(\frac{1}{2}\right).$$

Load vector due to pressure load is obtained as

$$\{f^S\} = h\int_{-1}^{1} [N]^T \begin{Bmatrix} 0 \\ -1 \times 10^6 \end{Bmatrix} \left(\frac{1}{2}\right) d\varepsilon \tag{6.119}$$

where $[N]$ is the shape function matrix corresponding to the edge connecting nodes 3 and 4. This can be obtained by putting $\eta = +1$

$$[N] = \begin{bmatrix} 0\,0\,0\,0 & \frac{1}{2}(1+\xi) & 0 & \frac{1}{2}(1-\xi) & 0 \\ 0\,0\,0\,0 & 0 & \frac{1}{2}(1+\xi) & 0 & \frac{1}{2}(1-\xi) \end{bmatrix} \tag{6.120}$$

$$\{F^S\}^T = \frac{h}{2}\{0\;0\;0\;0\;0\;-10^6\;0\;-10^6\}. \tag{6.121}$$

Load vector due to pressure can be obtained by substituting, h = thickness for plane stress and unity for plane strain.

(b) Gravity Load Vector

Gravity direction is assumed opposite to Y-direction, hence

$$\{b_x\; b_y\}^T = \{0\; -9.81 \times 7800\} = \{0\; -76518\} \tag{6.122}$$

Load vector due to gravity load is given as

$$\{f^B\} = h\int_{-1}^{1}\int_{-1}^{1} [N]^T \begin{Bmatrix} 0 \\ -76518 \end{Bmatrix} J\, d\xi\, d\eta \tag{6.123}$$

Shape function for four node quadrilateral element

$$N_1 = \frac{1}{4}(1-\xi)(1-\eta);\ N_2 = \frac{1}{4}(1+\xi)(1-\eta) \tag{6.124}$$

$$N_3 = \frac{1}{4}(1+\xi)(1+\eta);\ N_4 = \frac{1}{4}(1-\xi)(1+\eta) \tag{6.125}$$

$$\frac{\partial N_1}{\partial \xi} = -\frac{1}{4}(1-\eta);\ \frac{\partial N_2}{\partial \xi} = \frac{1}{4}(1-\eta) \tag{6.126}$$

$$\frac{\partial N_3}{\partial \xi} = \frac{1}{4}(1+\eta);\ \frac{\partial N_4}{\partial \xi} = -\frac{1}{4}(1+\eta) \tag{6.127}$$

$$\frac{\partial N_1}{\partial \eta} = -\frac{1}{4}(1-\xi);\ \frac{\partial N_2}{\partial \eta} = -\frac{1}{4}(1+\xi) \tag{6.128}$$

$$\frac{\partial N_3}{\partial \eta} = \frac{1}{4}(1+\xi);\ \frac{\partial N_4}{\partial \eta} = \frac{1}{4}(1-\xi) \tag{6.129}$$

$$[J] = \begin{bmatrix} \frac{\partial X}{\partial \xi} & \frac{\partial X}{\partial \eta} \\ \frac{\partial Y}{\partial \xi} & \frac{\partial Y}{\partial \eta} \end{bmatrix} \tag{6.130}$$

$$x = N_i X_i;\ \frac{\partial X}{\partial \xi} = \frac{\partial N_i}{\partial \xi} X_i = \frac{1}{2};\ \frac{\partial Y}{\partial \xi} = \frac{\partial N_i}{\partial \xi}.Y_i = 0 \tag{6.131}$$

$$y = N_i Y_i;\ \frac{\partial X}{\partial \eta} = \frac{\partial N_i}{\partial \eta} X_i = 0;\ \frac{\partial Y}{\partial \eta} = \frac{\partial N_i}{\partial \eta} Y_i = \frac{1}{2} \tag{6.132}$$

$$[J] = \begin{bmatrix} \frac{1}{2} & 0 \\ 0 & \frac{1}{2} \end{bmatrix};\ J = \frac{1}{4}. \tag{6.133}$$

This Jacobian computation is applicable to the calculation of the thermal load vector also

$$= h\sum_{i=1}^{2}\sum_{j=1}^{2}[N]^T\begin{Bmatrix} 0 \\ -76518 \end{Bmatrix}\left(\frac{1}{4}\right)W_i W_j \tag{6.134}$$

$2 \times 2$ Gaussian quadrature is used for integration and h is thickness for plane stress and unity for plane strain.

(c) Thermal load vector

$$\{f^T\} = h\int_{-1}^{1}\int_{-1}^{1}[B]^T[E]\begin{Bmatrix} 3\times 10^{-5} \\ 3\times 10^{-5} \\ 0 \end{Bmatrix} J\, d\xi\, d\eta \tag{6.135}$$

where thickness for plane stress and unity for plane strain. Thermal load vector for the plane strain element is given below assuming h as the unity

$$\{f^T\} = h\sum_{i=1}^{2}\sum_{j=1}^{2}[B]^T[E]\begin{Bmatrix} 1.3\times 3\times 10^{-5} \\ 1.3\times 3\times 10^{-5} \\ 0 \end{Bmatrix} JW_iW_j. \tag{6.136}$$

Appropriate strain displacement matrix and material property matrix are provided in Chap. 5.

**Example 8** Figure 6.6 shows thick cylinder fixed at the bottom and subjected to an internal pressure of 1MPa. Compute the load vectors for (a) pressure, (b) gravity, and (c) thermal loads. Element connectivity is (1, 2, 3, 4).

(a) Load vector due to pressure

$$\{F^S\} = 2\pi\int_{-1}^{1}[N_1]^T\begin{Bmatrix} l_1 p \\ m_1 p \end{Bmatrix} Jrd\eta. \tag{6.137}$$

Vector normal to the edge connecting nodes 1 and 4 is

$$\hat{n} = 1i + 0j.$$

Determinant of Jacobian connecting the edge 1 and 4 is given as

$$J = \left|\frac{\partial X}{\partial \xi}\right| = \left(\frac{1}{2}\right). \tag{6.138}$$

Direction cosines connecting the edge 1 and 4 is $\{l_1, m_1\} = \{1, 0\}$.

Shape function matrix [N] can be obtained by putting = −1 in the shape function matrix results in

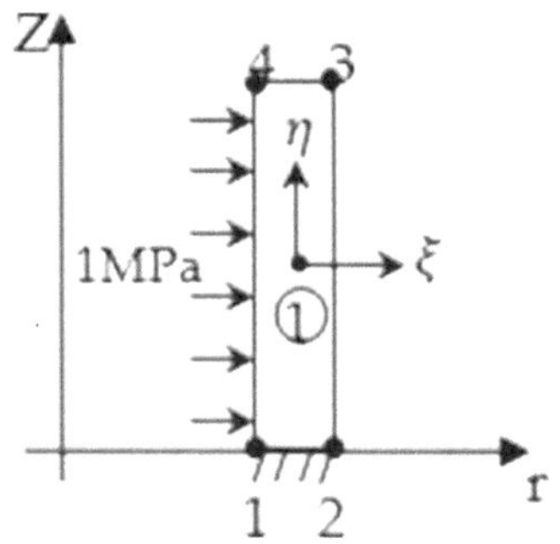


| Node | r | z |
|---|---|---|
| 1 | 0.2 | 0 |
| 2 | 0.3 | 0 |
| 3 | 0.3 | 1 |
| 4 | 0.2 | 1 |

**Fig. 6.6** Thick cylinder with pressure applied at inner radius

$$[\mathrm{N}] = \begin{bmatrix} \frac{1-\eta}{2} & 0 & 0\,0\,0\,0 & \frac{1+\eta}{2} & 0 \\ 0 & \frac{1-\eta}{2} & 0\,0\,0\,0 & 0 & \frac{1+\eta}{2} \end{bmatrix}. \tag{6.139}$$

Load vector due to pressure is given as

$$\left\{\mathrm{f}^{\mathrm{S}}\right\} = 2\pi \int_{-1}^{1} [\mathrm{N}_{\mathrm{l}}]^{\mathrm{T}} \left\{ \begin{matrix} 10^6 \\ 0 \end{matrix} \right\} \left(\frac{1}{2}\right)(0.2)\mathrm{d}\eta \tag{6.140}$$

$$\left\{\mathrm{f}^{\mathrm{S}}\right\}^{\mathrm{T}} = 0.2\pi\left\{ 10^6\ 0\ 0\ 0\ 0\ 0\ 10^6\ 0 \right\}. \tag{6.141}$$

(b) Load vector due to gravity

$$\left\{\mathrm{f}^{\mathrm{B}}\right\} = 2\pi \int_{-1}^{1}\int_{-1}^{1} [\mathrm{N}]^{\mathrm{T}} \begin{bmatrix} 0 \\ \rho g \end{bmatrix} J\,\mathrm{r}\mathrm{d}\xi\,\mathrm{d}\eta. \tag{6.142}$$

Shape function and its derivative with respect to $\xi$ and $\eta$ are provided in the previous example

$$[\mathrm{J}] = \begin{bmatrix} \frac{\partial \mathrm{r}}{\partial \xi} & \frac{\partial \mathrm{r}}{\partial \eta} \\ \frac{\partial \mathrm{z}}{\partial \xi} & \frac{\partial \mathrm{z}}{\partial \eta} \end{bmatrix} \tag{6.143}$$

$$\frac{\partial \mathrm{r}}{\partial \xi} = \frac{\partial \mathrm{N_i}}{\partial \xi}\mathrm{r_i} = 0.05;\ \frac{\partial \mathrm{r}}{\partial \eta} = 0 \tag{6.144}$$

$$\frac{\partial \mathrm{z}}{\partial \xi} = \frac{\partial \mathrm{N_i}}{\partial \xi}\mathrm{Z_i} = 0;\ \frac{\partial \mathrm{z}}{\partial \eta} = \frac{1}{2} \tag{6.145}$$

$$[\mathrm{J}] = \begin{bmatrix} 0.05 & 0 \\ 0 & \frac{1}{2} \end{bmatrix};\ J = 0.025 \tag{6.146}$$

$$\left\{\mathrm{f}^{\mathrm{B}}\right\} = 2\pi \sum_{\mathrm{i}=1}^{2}\sum_{\mathrm{j}=1}^{2} [\mathrm{N}]^{\mathrm{T}} \left\{ \begin{matrix} 0 \\ -76518 \end{matrix} \right\} (0.025)\mathrm{r}\ \mathrm{W_i W_j}. \tag{6.147}$$

(c) Load vector due to thermal expansion

$$\left\{\mathrm{f}^{\mathrm{T}}\right\} = 2\pi \int_{-1}^{1}\int_{-1}^{1} [\mathrm{B}]^{\mathrm{T}}[\mathrm{E}] \left\{ \begin{matrix} 3\times 10^{-5} \\ 3\times 10^{-5} \\ 3\times 10^{-5} \\ 0 \end{matrix} \right\} J\,\mathrm{r}\mathrm{d}\xi\,\mathrm{d}\eta \tag{6.148}$$

$$\{f^T\} = 2\pi \sum_{i=1}^{2}\sum_{j=1}^{2}[B]^T[E]\begin{Bmatrix} 3\times 10^{-5} \\ 3\times 10^{-5} \\ 3\times 10^{-5} \\ 0 \end{Bmatrix}(0.025)rW_iW_j. \tag{6.149}$$

## 6.8 Exercises

The following properties are assumed for numerical computation of the load vector: $A = 0.01\ m^2$; $\alpha = 24 \times 10^{-6}/^oC$; $E = 7 \times 10^{10}\ N/m^2$; $\nu = 0.3$, $\rho = 2700\ kg/m^3$; $g = 9.81\ m/s^2$; $\Delta t = 30$. Axi-symmetric structure undergoes uniform angular velocity of 2500 rpm.

6.8.1 An axial bar of length 3 m is discretized with 3 elements as shown in Fig. 6.7. It is fixed at both ends and a 1MN load is applied at node 3. Find the displacements at nodes 1, 2, 3, and 4 and the reaction at the fixed ends by Penalty method.

6.8.2 An axial bar of length 3 m is discretized with four elements as shown in Fig. 6.8. It is fixed at one end and constrained to have displacements as follows (a) $d_2$–$d_3 = 0$ (b) $d_3 = 2.5$.

6.8.3 An axial bar shown in Fig. 6.9 is fixed at node 1 and load is applied at node 6. Nodes 3 and 4 are at the same coordinate and are not connected. Solve the problem for two cases (a) use the tying option with the constraint $d_3$-$d_4 = 0$ and (b) Merge the nodes 3 and 4 and make the bar continuous with 4 elements having five nodes having the load at node 5 and solve the problem. Check for displacements for both the cases.

6.8.4 Figure 6.10 shows a planar element with a pressure load of 1Mpa on one of the edges. Compute the elemental load vector for (a) pressure, (b) gravity, and (c) thermal loads. Assume thickness 'h' as 0.125 for the plane stress element and unity for the plane strain element. Element connectivity is (1 2 3 4).

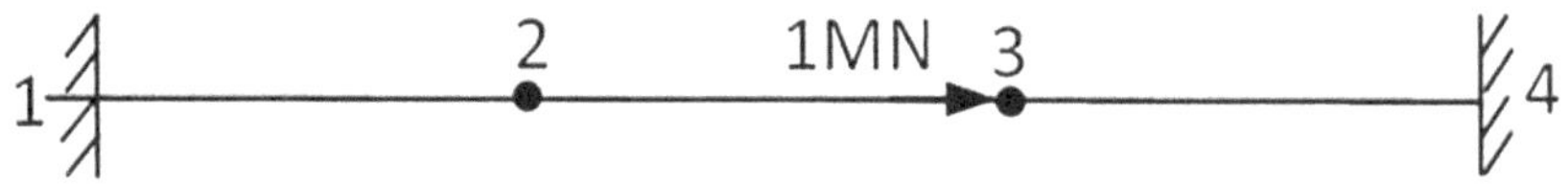


**Fig. 6.7** Axial bar having constraints at both ends with load at node 3

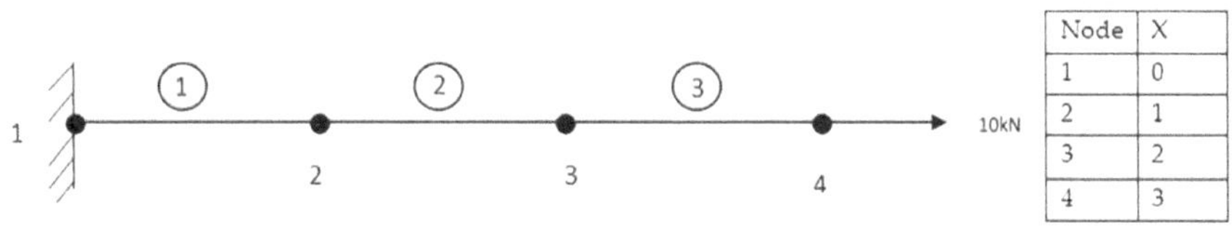


**Fig. 6.8** Axial bar with point load at the node 4

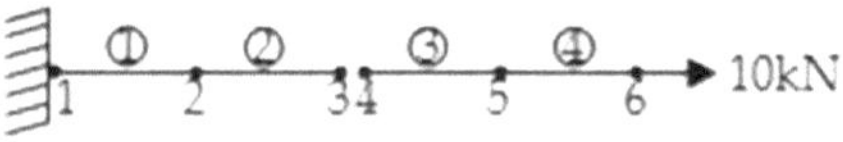


| Node | 1 | 2 | 3 | 4 | 5 | 6 |
|---|---|---|---|---|---|---|
| X | 0 | 1 | 2 | 2 | 3 | 4 |

**Fig. 6.9** Rod element with tying option

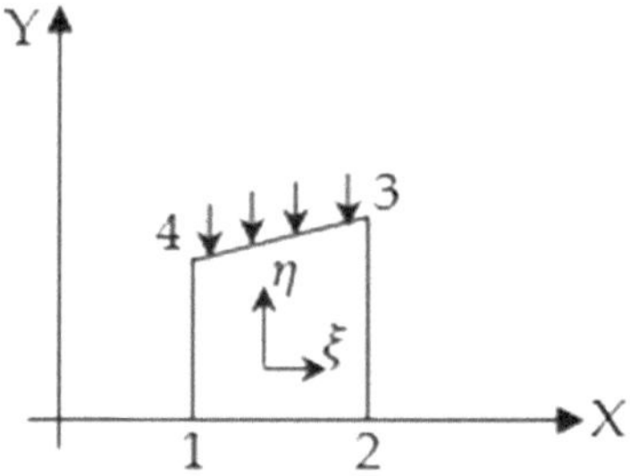


| Node | X | Y |
|---|---|---|
| 1 | 5 | 0 |
| 2 | 10 | 0 |
| 3 | 10 | 8 |
| 4 | 5 | 5 |

**Fig. 6.10** Planar element with pressure load

6.8.5 An axi-symmetric element shown in Fig. 6.11 is subjected to a pressure of 1 Mpa at the edge connecting nodes 1 and 4. Compute the load vector due to (a) pressure, (b) thermal, and (c) gravity loads. To exploit the symmetry of the structure, only a 20° sector is modeled. Formulate the MPC equation to suppress displacements normal to the edge connecting nodes 4 and 3.

Element connectivity is (1 2 3 4).

6.8.6 A thick cylinder with the internal pressure of 1Mpa as shown in Fig. 6.12 is modeled with an axi-symmetric element. Compute the load vector due to (a) Pressure, (b) Thermal loads, and (c) Centrifugal load.

6.8.7 A tripod shown in Fig. 6.13 is subjected to point loads and pressure loads of 1Mpa in the local coordinate system of Element 2. Compute the load vector for (a) point, (b) pressure, (c) gravity, and (d) thermal loads. Element connectivity (1, 4), (2, 4), and (3, 4).

6.8.8 Compute the pressure load vector for the axi-symmetric shell element as shown in Fig. 6.14 with the pressure of 1 Mpa.

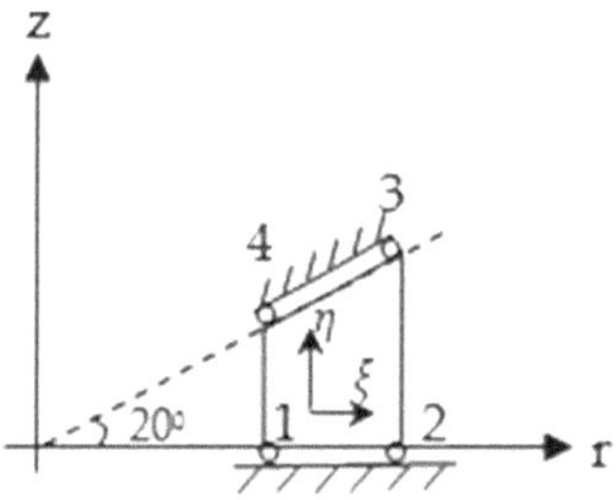


| Node | r | z |
|---|---|---|
| 1 | 0.3 | 0 |
| 2 | 0.5 | 0 |
| 3 | 0.46985 | 0.17100 |
| 4 | 0.28191 | 0.10261 |

**Fig. 6.11** Axi-symmetric element with inclined boundary condition

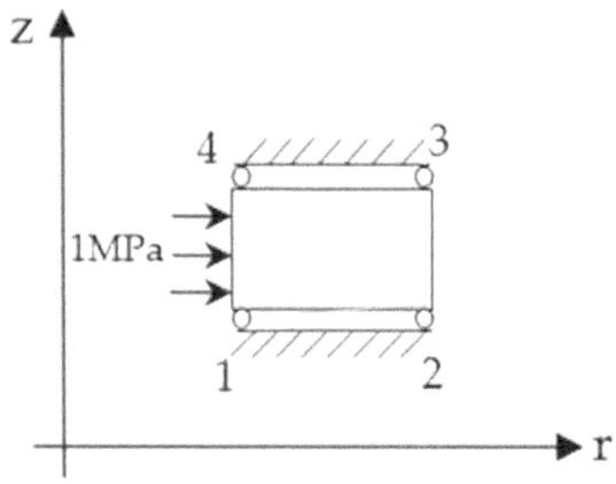


| Node | r | z |
|---|---|---|
| 1 | 0.2 | 2 |
| 2 | 0.3 | 2 |
| 3 | 0.3 | 3 |
| 4 | 0.2 | 3 |

**Fig. 6.12** Thick cylinder subjected to internal pressure

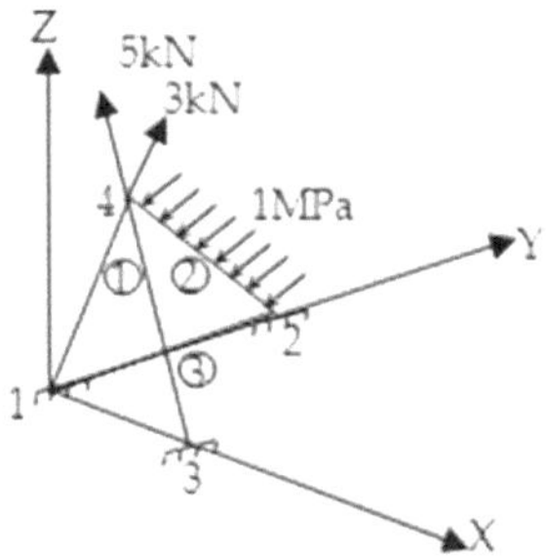


| Node | X | Y | Z |
|---|---|---|---|
| 1 | 0 | 0 | 0 |
| 2 | 0 | 3 | 0 |
| 3 | 3 | 0 | 0 |
| 4 | 2 | 2 | 3 |

**Fig. 6.13** Tripod with point load in local system

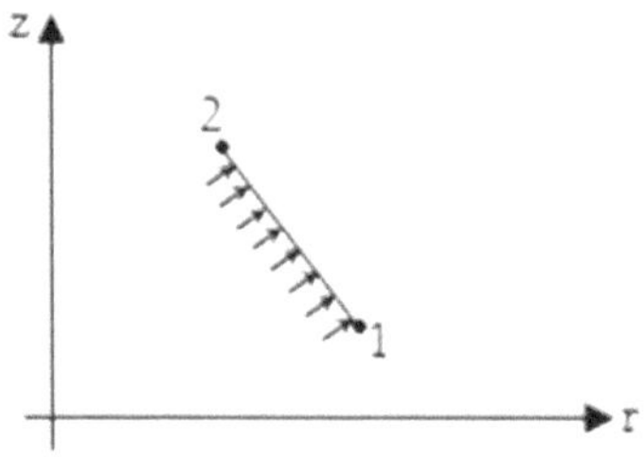


| Node | R | z |
|---|---|---|
| 1 | 3 | 1 |
| 2 | 2 | 3 |

**Fig. 6.14** Axi-symmetric shell with pressure load

6.8.9 Figure 6.15 shows 8 node brick elements. Form the MPC equation to have displacement normal to the inclined elemental face having nodes (3 4 8 7) (Hint: Form the vector normal to the face and equate it to zero after taking the dot product with displacement vector for the nodes in the inclined face).

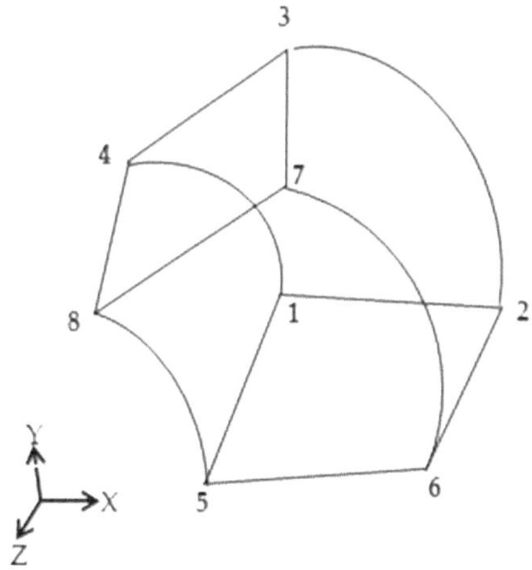


| Node | X | Y | Z |
|---|---|---|---|
| 1 | 10 | 0 | 0 |
| 2 | 20 | 0 | 0 |
| 3 | 9.8481 | 1.7365 | 0 |
| 4 | 19.6962 | 3.473 | 0 |
| 5 | 10 | 0 | 10 |
| 6 | 20 | 0 | 20 |
| 7 | 10 | 9.8481 | 1.7365 |
| 8 | 20 | 19.6962 | 3.473 |

**Fig. 6.15** MPC equations for inclined surface

## Bibliography

Cook RD, Malkus DS, Plesha ME, Witt RJ (2007) Concepts and applications of finite element analysis. John Wiley & Sons (Asia) Pte Ltd. ISBN-13 978-81-265-1336-9

Tirupathi R, Chandrapatla, Ashok D, Belegundu. Introduction to finite elements in engineering, 4th edn. Prentice Hall, Upper Saddle River, New Jersey 07458 ISBN:13:978-0132162746

Zienkiewicz OC, Taylor RL (2000) The finite element method, vol 1, The basis, 5th edn. Butterworth-Heinemann, ISBN 0 7506 5049 4

# Chapter 7
# Finite Element Formulation for Heat Transfer

**Philip George**

*Imagination is more important than knowledge. Knowledge is limited. Imagination encircles the world*
*—Albert Einstein*

## 7.1 Introduction

Since finite element method (FEM) is a general mathematical approach for solving partial differential equations (PDEs), its application extend to various field phenomena, which are governed by PDEs such as elasticity, heat transfer, fluid dynamics, electro-magnetics, etc. This chapter explores the application of FEM in heat transfer. Beginning with a brief overview of the fundamentals of heat transfer, the chapter describes how to model a typical heat transfer problem using FEM. Explanations made are from an engineering point of view through simple cases. The

P. George (✉)
ADAD/AHTG/AERO, Veli, VSSC, Thiruvananthapuram 695022, India
e-mail: philip_george@vssc.gov.in; pgcet1@yahoo.com

 
S. U. Nair and S. Somanath (eds.), *Introduction to Finite Element Analysis*,
https://doi.org/10.1007/978-981-19-7989-7_7

equivalence between a problem in elasticity and heat transfer is brought out wherever possible.

## 7.2 Heat Transfer Essentials

Thermodynamics defines heat as an energy interaction between the system and its surroundings driven by the temperature difference between them. By virtue of the Second Law of Thermodynamics, heat is transferred from a higher-temperature region to a lower-temperature region. Thermodynamics deals with energy interactions with no emphasis on the time taken for such interactions, whereas heat transfer deals with the time rate of heat interaction between the system and its surroundings. The three modes of heat transfer are:

- Conduction
- Convection
- Radiation.

### 7.2.1 *Conduction*

Conduction is the mode of heat transfer within a continuous medium because of the temperature gradient existing within the medium. Mechanisms of conductive heat transfer include the lattice vibrations or phonon transport (dominant in non-metals) as well as the transfer of free electrons (dominant in metals). The fundamental law in conductive heat transfer is Fourier's law of conduction, which states that the total heat transfer is directly proportional to the area of cross section and the temperature gradient along the direction of heat transfer

$$Q \propto A\frac{\partial T}{\partial x}, \tag{7.1}$$

where

Q is the heat transferred in Watts (W);

A is the area of heat transfer in $m^2$; and

$\frac{\partial T}{\partial x}$ is the temperature gradient in K/m.

Introducing thermal conductivity $k$ as the proportionality constant in the above equation and by virtue of the Second Law of Thermodynamics that heat flows in the direction of the negative temperature gradient, Fourier's law of conduction can be written as

$$Q = -kA\frac{\partial T}{\partial x}, \tag{7.2}$$

where

$k$ is the thermal conductivity of the medium in W/mK, which is a characteristic property of the material.

Or in terms of heat flux, the Fourier's law becomes

$$q = -k\frac{\partial T}{\partial x}, \tag{7.3}$$

where

$q$ is the heat flux in W/m$^2$.

Heat flux $q_i$ being a vector with a magnitude and direction, the vector form of Fourier's law is

$$q_i = -\left(k_{i1}\frac{\partial T}{\partial x_1} + k_{i2}\frac{\partial T}{\partial x_2} + k_{i3}\frac{\partial T}{\partial x_3}\right). \tag{7.4}$$

By Einstein's summation convention (a notational convention for simplifying expressions including summations of vectors, matrices, and general tensors)

$$q_i = -k_{ij}\frac{\partial T}{\partial x_j}, \tag{7.5}$$

where

$q_i$ is the heat flux in $i$ direction;

$\frac{\partial T}{\partial x_j}$ is the temperature gradient in $j$ direction; and

$k_{ij}$ is the proportionality constant relating $q_i$ and $\frac{\partial T}{\partial x_j}$.

Here thermal conductivity $k_{ij}$ is a tensor. For a general three-dimensional problem, Fourier's law can also be expressed in the matrix form as

$$\begin{bmatrix} q_1 \\ q_2 \\ q_3 \end{bmatrix} = -\begin{bmatrix} k_{11} & k_{12} & k_{13} \\ k_{21} & k_{22} & k_{23} \\ k_{31} & k_{32} & k_{33} \end{bmatrix}\begin{bmatrix} \frac{\partial T}{\partial x_1} \\ \frac{\partial T}{\partial x_2} \\ \frac{\partial T}{\partial x_3} \end{bmatrix}. \tag{7.6}$$

This holds for all materials including anisotropic materials such as wood. If the off-diagonal terms of the thermal conductivity matrix are zero, then the material is said to be orthotropic

$$\begin{bmatrix} q_1 \\ q_2 \\ q_3 \end{bmatrix} = -\begin{bmatrix} k_1 & 0 & 0 \\ 0 & k_2 & 0 \\ 0 & 0 & k_3 \end{bmatrix}\begin{bmatrix} \frac{\partial T}{\partial x_1} \\ \frac{\partial T}{\partial x_2} \\ \frac{\partial T}{\partial x_3} \end{bmatrix}. \tag{7.7}$$

To describe Fourier's Law for orthotropic materials, we need three proportionality constants $k_1$, $k_2$, and $k_3$ along the principal directions. Examples of orthotropic materials are composites such as Carbon Fiber Reinforced Polymer (CFRP), where conductivity varies along and across the fiber directions. If the magnitude of $k_1$, $k_2$, and $k_3$ are the same, then the material is isotropic:

$$\begin{bmatrix} q_1 \\ q_2 \\ q_3 \end{bmatrix} = -[k]\begin{bmatrix} \frac{\partial T}{\partial x_1} \\ \frac{\partial T}{\partial x_2} \\ \frac{\partial T}{\partial x_3} \end{bmatrix} \text{ or } \vec{q} = -k\vec{\nabla}T. \tag{7.8}$$

Since there is only one proportionality constant to describe Fourier's Law for isotropic materials, it can be stated that thermal conductivity is independent of direction for these classes of materials. Most of the metals exhibit isotropic properties. For the sake of brevity, we will restrict ourselves to isotropic materials where $k$ is constant in any direction, but can vary with temperature.

### 7.2.2 *Convection*

Convection is the mode of heat transfer that is characterized by the presence of fluid flow over a surface. Heat is transferred to/from the surface by means of conduction adjacent to the surface where fluid velocities are zero (no slip conditions) and by means of advection or bulk motion of fluid away from the surface where fluid velocities exist. A general expression for quantifying the convective heat transfer is given by Newton's law of cooling, which is defined as

$$Q_{\text{conv}} = hA(T_s - T_f), \tag{7.9}$$

where

$Q_{\text{conv}}$ is heat transferred in *W*;

$h$ is the convective heat transfer coefficient in $W/m^2\ K$;

$T_s$ is the surface temperature in *K*;

$T_f$ is the fluid temperature in *K*.

Heat transfer coefficient $h$ is a strong function of flow parameters like fluid velocity, density, viscosity, etc. Empirical correlations derived from experiments or boundary layer theory are used to quantify the value of $h$. For more details on the estimation of the convective heat transfer coefficient, one can refer to Incropera and Dewitt (2007). Typical values of $h$ for different types of flow are given in Table 7.1.

Nowadays, computational methods such as computational fluid dynamics (CFD) are used to directly estimate the heat transfer coefficient for complex fluid flows for which correlations do not exist. Interested readers can refer to books given under

**Table. 7.1** Typical values of $h$ are given below

| Type of flow | $h(W/m^2K)$ |
|---|---|
| *Free convection* | |
| Gases | 2–25 |
| Liquids | 50–1000 |
| *Forced convection* | |
| Gases | 2–25 |
| Liquids | 50–1000 |
| Boiling | 2500–100,000 |

reference at the end of the chapter for an introduction to CFD. We shall restrict ourselves to the application of convective heat transfer as a boundary condition in the finite element formulation of conductive heat transfer problems.

### 7.2.3 Radiation

Radiation is the mode of heat transfer by means of electromagnetic waves or photons. Radiation is primarily through the change in the electron configuration of the substance at elevated temperatures. Hence, all substances above zero Kelvin temperature can emit radiation. Radiation is the only mode of heat transfer, which does not require a medium for energy transport. For example, the heating of the Earth by the Sun is an example of the transfer of energy by radiation through vacuum as electromagnetic waves. The heating of a room with an open-hearth fireplace is another example.

Radiation can be broadly classified into surface radiation, where only the surface of the body participates in radiative heat transfer, and volumetric radiation, where the entire substance is radiatively active. Gaseous radiation is an example of volumetric radiation in which gaseous molecules absorb, emit, and even scatter radiation. A discussion on gaseous radiation is beyond the scope of this book and we shall limit ourselves to surface radiation.

The extent to which a surface at a temperature $T_s$ can emit radiation is governed by Stefan–Boltzmann law

$$E_b = \sigma T_s^4, \tag{7.10}$$

where

$E_b$ is the emissive power in $W/m^2$;

$T_s$ is the surface temperature in $K$; and

$\sigma$ is the Stefan–Boltzmann constant = $5.67 \times 10^{-08}$ W/m$^2$ K.

Such a body, which follows Stefan–Boltzmann law, is a perfect radiator or in heat transfer parlance, is called a blackbody. Hence, $E_b$ is the emissive power of a blackbody. Real surfaces emit less than a blackbody. The ratio of the emissive power of a surface to the emissive power of a blackbody at the same temperature is known as emissivity

$$\epsilon = \frac{E}{E_b}. \tag{7.11}$$

For a real surface, the emitted radiation becomes

$$E = \epsilon\sigma T_s^4. \tag{7.12}$$

If the surroundings of the body are at $T_f$ (in Kelvin $K$), the net heat transfer between the surface and surroundings is written as

$$Q_{\text{rad}} = \left(\varepsilon\sigma T_s^4 - \alpha G\right)A, \tag{7.13}$$

where

$G$ is the irradiation on the body in $W/m^2 = \sigma T_f^4$;

$\propto$, is the absorptivity of the body for the irradiation $G$.

As an engineering approximation, real surfaces are assumed to be gray and diffuse, which results in equality of emissivity and absorptivity. The radiative heat transfer between a real surface and surroundings now becomes

$$Q_{\text{rad}} = \epsilon\sigma A\left(T_s^4 - T_f^4\right). \tag{7.14}$$

Radiative heat transfer can be expressed in a manner similar to convective heat transfer as

$$Q_{\text{rad}} = h_{\text{rad}}\text{A}\left(T_s - T_f\right), \tag{7.15}$$

where

$h_{\text{rad}}$, is the radiative heat transfer coefficient $= \epsilon\sigma\left(T_s + T_f\right)\left(T_s^2 + T_f^2\right)$.

Thus, radiation can also be incorporated as a boundary condition similar to convection in the finite element formulation.

## 7.3 Governing Equations

We shall now derive the governing equations of heat transfer within a solid body or a fluid medium with negligible velocities, such that conduction is the only mode of heat transfer within the medium. Let the medium have a volume $\Omega$ and surface area $\Gamma$. Let our region of interest (also known as system) be an infinitesimal element within the medium with volume $d\vartheta$ and surface area $d\Gamma$. We shall resort to the fundamental law of conservation of energy for derivation. This law is also known as the First Law of Thermodynamics:

$$Q_{|\text{in}} + W_{|\text{in}} = \Delta E_{|\text{sys}}, \tag{7.16}$$

where

$Q_{|\text{in}}$ , is the net heat transfer to the system in $J$;

$W_{|\text{in}}$, is the net work transfer to the system in $J$; and

$\Delta E_{|\text{sys}}$, is the change in energy of the system in $J$.

For a stationary system like a solid, the change in energy $\Delta E_{|\text{sys}}$ corresponds to the change in internal energy $\Delta U_{|\text{sys}}$, which is related to temperature change of the system as

$$\Delta U_{|\text{sys}} = m_{|\text{sys}} C_v \Delta T_{|\text{sys}}, \tag{7.17}$$

where

$m_{|\text{sys}}$ , is the mass of system in *kg*;

$C_v$, is specific heat for a constant volume interaction in *J/kgK*.

Since solids and, to a large extent, liquids are incompressible, specific heat is independent of the type of interaction, and hence we will denote the specific heat as $C$. For a system with no work interaction, the conservation of energy can be expressed per unit change in time as

$$\dot{Q}_{|\text{in}} = \Delta \dot{U}_{|\text{sys}}, \tag{7.18}$$

where

$\dot{Q}_{|\text{in}}$ is the net rate of heat transfer to the system in $W$;

$\Delta \dot{U}_{|\text{sys}}$ is the rate of change in internal energy of the system in $W$.

For a system with constant mass and specific heat

$$\dot{Q}_{|\text{in}} = m_{|\text{sys}} C \Delta \dot{T}_{|\text{sys}}, \tag{7.19}$$

where

$\Delta \dot{T}_{|\text{sys}}$ is the time rate of change of temperature in *K/s*.

Heat can enter the system in two ways:

- Heat generated within the system. For example, through chemical/nuclear reactions happening within the system. This can be accounted for if we know the heat generation per unit volume $\ddot{q}$.

- Heat transferred into the system through surface area $d\Gamma$ via conduction

  Hence

$$\dot{Q}_{\text{lin}} = Q_{\text{condlin}} + Q_{\text{gen}}. \tag{7.20}$$

Divergence of the heat flux vector $\vec{q}$ gives the net heat emulating from a unit volume through its surface area. Therefore, we can express $Q_{\text{condlin}}$ for the system in terms of $\vec{q}$ as

$$Q_{\text{condlin}} = -\vec{\nabla}\,.\,\vec{q}d\vartheta. \tag{7.21}$$

Knowing that $Q_{\text{gen}} = \ddot{q}\text{d}\vartheta$ and $m_{|\text{sys}} = \rho\text{d}\vartheta$, where $\rho$ is the density, we can write

$$-\vec{\nabla}\,.\,\vec{q}\text{d}\vartheta + \ddot{q}\text{d}\vartheta = \rho\text{C}\text{d}\vartheta\,\Delta\dot{T}_{|\text{sys}} \tag{7.22}$$

or

$$\rho\text{C}\Delta\dot{T}_{|\text{sys}} = -\vec{\nabla}\,.\,\vec{q} + \ddot{q}. \tag{7.23}$$

Substituting for $\vec{\text{q}}$ from Fourier's law for an isotropic medium

$$\rho\text{C}\frac{\text{d}T}{\partial t} = \vec{\nabla}\,.\left(k\vec{\nabla}T\right) + \ddot{q} \tag{7.24}$$

In terms of Einstein's summation convention

$$\rho\text{C}\frac{\partial T}{\partial t} = -\frac{\partial q_i}{\partial x_i} + \ddot{q} = \frac{\partial}{\partial x_i}\left(k\frac{\partial T}{\partial x_i}\right) + \ddot{q}. \tag{7.25}$$

This is the governing equation for conduction in a three-dimensional isotropic medium with heat generation inside. Note that thermal conductivity $k$ can be a function of temperature $T$.

Since this equation is valid for any infinitesimal element $\text{d}\vartheta$, it is valid for the entire domain $\Omega$. The left-hand side of the equation shows the time dependence of $T$, whereas the right-hand side shows the spatial dependence of $T$. Solving this equation, we will obtain the field temperature $T = T(x_i, t)$ over the domain $\Omega$ at any instant of time $t$.

## 7.4 Initial and Boundary Conditions

According to the classification of PDEs, the governing equation is parabolic in time and elliptic in space. To obtain a unique solution for the governing equation, we need to specify initial conditions in time for the entire domain $\Omega$ and boundary conditions on the boundary $\Gamma$.

### 7.4.1 Initial Conditions

Initial conditions are conditions in which the dependent variable T is specified for the entire domain $\Omega$ at a particular time instant $t_o$. Initial conditions can be taken as constant throughout the entire domain or they can be specified as a distribution, $T^0 = T(x_i, t_o)$.

### 7.4.2 Boundary Conditions

Boundary conditions are conditions specified on the boundary $\Gamma$ of the domain $\Omega$. Different types of boundary conditions can be specified:

### 7.4.3 Dirichlet Boundary Condition

In this type, the dependent variable $T$ is specified on the boundary as

$$T_{|\Gamma} = C_\Gamma(t) \tag{7.26}$$

Note that the known quantity $C_\Gamma$ can be constant along the boundary or it can have spatial as well as temporal variation.

### 7.4.4 Neumann Boundary Condition

In this type, the gradient of the dependent variable $\frac{\partial T}{\partial x_i}$ is specified on the boundary as

$$\frac{\partial T}{\partial x_i}|_\Gamma = C_\Gamma(t). \tag{7.27}$$

The flux boundary condition also falls under this type where heat flux at the boundary is specified as

$$q|_\Gamma = -k\frac{\partial T}{\partial x_i}|_\Gamma = C_\Gamma(t) \tag{7.28}$$

Again $C|_\Gamma$ can be constant along the boundary or it can have spatial as well as temporal variation. A special case where $\frac{\partial T}{\partial x_i}|_\Gamma = 0$ is called insulated boundary condition, because this corresponds to zero heat flux at the boundary.

### 7.4.5 Robin Boundary Condition

Boundary condition of this type is a linear combination of Dirichlet and Neumann type of boundary conditions, where

$$\alpha T|_\Gamma + \beta \frac{\partial T}{\partial x_i}|_\Gamma = C_\Gamma(t) \tag{7.29}$$

This boundary condition is also aptly called a mixed type boundary condition. Recall that convection and surface radiation are specified at the boundary $\Gamma$ of the domain $\Omega$. Both these fall under the mixed type boundary condition. By convective boundary condition

$$-k\frac{\partial T}{\partial x_i}|_\Gamma = h\left(T - T_f\right)|_\Gamma \tag{7.30}$$

or

$$k\frac{\partial T}{\partial x_i}|_\Gamma + hT|_\Gamma = hT_f|_\Gamma. \tag{7.31}$$

Similarly, the radiative boundary condition is specified as

$$k\frac{\partial T}{\partial x_i}|_\Gamma + h_R T|_\Gamma = h_R T_f|_\Gamma, \tag{7.32}$$

where

$h_R$, is the radiative heat transfer coefficient $= \varepsilon\sigma\left(T + T_f\right)\left(T_s^2 + T_f^2\right)$.

To summarize, in order to find the temperature field $T = T(x_i, t)$ with a domain $\Omega$, one has to solve the governing equation with initial conditions specified at time $T_0$ and boundary conditions specified along the boundary $\Gamma$. Note that the governing equation, derived from the fundamental principle of conservation of energy, remains the same for all conduction problems in isotropic medium, whereas the boundary and initial conditions are dependent on the specific problem to be solved. Therefore, to arrive at a realistic temperature field $T = T(x_i, t)$, one has to appropriately identify and specify the boundary and initial conditions. This comes with experience in modeling real-life thermal problems and validating with experimental measurements.

The solution of the governing equation subject to the given boundary conditions, $T = T(x_i, t)$ is a field variable, which is continuous throughout the domain of space and time. Such a solution, if it can be obtained mathematically, is called a closed form or analytical solution. Since analytical solutions are difficult to obtain in many real-life problems, we pursue an approximate solution $\hat{T}$ by considering a finite number of points in the domain so that $\hat{T} = \hat{T}(x_0, x_1, x_2, \ldots, x_N)$ can be considered as an approximate solution. Instead of having a single governing equation, we will have a system of equations, which needs to be simultaneously solved to obtain $\hat{T}$. This process of obtaining a system of equations is called *discretization*. We shall explore the process of discretization by means of FEM in heat transfer.

## 7.5 One-Dimensional Steady State Conduction

From the general governing equation, we can obtain the governing equation for one-dimensional steady state conduction in Cartesian coordinates by dropping the transient term and restricting the number of spatial dimensions to one

$$-\frac{\mathrm{d}q}{\mathrm{d}x} + \ddot{q} = \frac{d}{\mathrm{d}x}\left(k\frac{\mathrm{d}T}{\mathrm{d}x}\right) + \ddot{q} = 0. \tag{7.33}$$

This is applicable to a medium in which the heat transfer is predominant in the $X$ direction and heat transfer in other directions can be neglected. For example, consider a rod of length $L$ oriented along the $X$-axis. The rod is insulated along its curved surface so that heat transfer takes place only along the $X$-direction.

Since there are no time-dependent terms in the equation, the equation is elliptic in nature (any of a class of partial differential equations describing phenomena that do not change from moment to moment, as when a flow of heat or fluid takes place within a medium with no accumulations) and the solution is independent of initial conditions. The boundary conditions need to be specified at $x = 0$ and $x = L$ (the two ends of the rod), respectively. Note that the boundary conditions are no longer time dependent. To obtain a unique solution, at least one of the boundary conditions should be of Dirichlet or Robin type. The solution will be the temperature distribution $T = T(x)$ along the length $L$.

### *7.5.1 Equivalence with Elasticity*

One can appreciate the equivalence of the equation of one-dimensional steady state conduction with the one-dimensional equilibrium equation in elasticity for an axially loaded rod, which is given by

$$\frac{\mathrm{d}\sigma}{\mathrm{d}x} + \ddot{f_b} = \frac{d}{\mathrm{d}x}\left(E\frac{\mathrm{d}u}{\mathrm{d}x}\right) + \ddot{f_b} = 0, \tag{7.34}$$

where

$\sigma$ is the stress in $N/m^2$;

$\ddot{f}_b$ is the body force per unit volume along $x$ direction in $N/m^3$;

$E$ is the modulus of elasticity from Hooke's law in $N/m^2$; and

$u$ is the displacement in $m$.

Note the equivalence of Fourier's law of conduction and Hooke's law of elasticity in one dimension

$$q = -k\frac{\mathrm{d}T}{\mathrm{d}x} \quad \sigma = E\frac{\mathrm{d}u}{\mathrm{d}x}. \tag{7.35}$$

One can also find equivalence in the boundary condition types. Dirichlet condition of known temperature in conduction is equivalent to known displacement in elasticity. Neumann's condition of known heat flux in conduction is equivalent to known traction or pressure in elasticity.

### 7.5.2 *Method of Weighted Residuals for One-Dimensional Steady State Conduction*

Instead of exact solution $T$ in the governing Eq. 7.33, we seek an approximate solution $\hat{T}$, such that

$$\frac{d}{\mathrm{d}x}\left(k\frac{d\hat{T}}{\mathrm{d}x}\right) + \ddot{q} = R, \tag{7.36}$$

where $R$ is called the residual or error due to the approximation.

By method of weighted residuals, we multiply Eq. 7.36 by an arbitrary function $v$, which is also known as the test function, integrate over the interval $(0, L)$ and equate it to zero

$$\int_0^L vR\mathrm{d}x = \int_0^L v\left(\frac{\mathrm{d}}{\mathrm{d}x}\left(k\frac{\mathrm{d}\hat{T}}{\mathrm{d}x}\right) + \ddot{q}\right)\mathrm{d}x = 0 \tag{7.37}$$

Integrating by parts, we get

$$vk\frac{\mathrm{d}\hat{T}}{\mathrm{d}x}\bigg|_0^L - \int_0^L \frac{\mathrm{d}v}{\mathrm{d}x}k\frac{\mathrm{d}\hat{T}}{\mathrm{d}x}\mathrm{d}x + \int_0^L \ddot{q}v\mathrm{d}x = 0$$

or

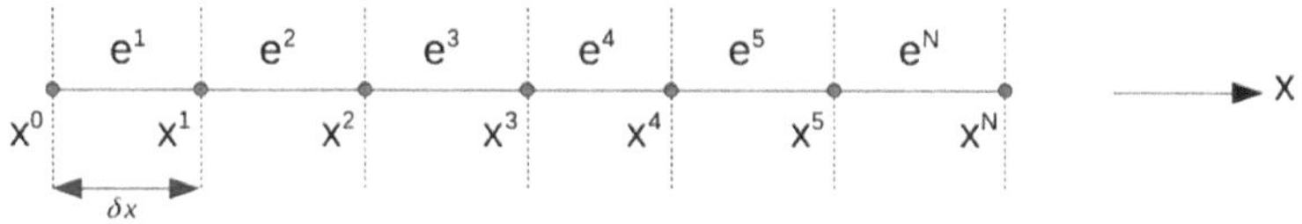


**Fig. 7.1** Nodes and Elements in a one-dimensional domain

$$-vq\Big|_0^L - \int_0^L \frac{\mathrm{d}v}{\mathrm{d}x}k\frac{\mathrm{d}\hat{T}}{\mathrm{d}x} + \int_0^L \ddot{q}\, v\mathrm{d}x = 0. \tag{7.38}$$

The governing equation being a differential equation of second order, the solution $T$ must be a continuous function up to second order so that second-order derivatives of $T$ can be defined. We say that $T$ is a $C^2$ continuous function. By integrating, we see that second-order terms are no longer there in the equation and $T$ can be a $C^1$ continuous function. Since the restriction of $T$ to be a $C^2$ continuous function is relaxed, the equation is also called the *weak formulation* of the governing equation. Note that the first term in the weak form exists only at boundary points 0 and $L$. Hence, boundary conditions are imposed through this term.

We will divide the continuous domain $(0, L)$ into $N$ non-overlapping sub-domains as shown in Fig. 7.1. The non-overlapping sub-domains are known as *elements*. Let $(x_0, x_1, x_2, \ldots, x_N)$ be discrete points in the domain where an approximate solution $\hat{T}$ will be obtained. These discrete points are known as *nodes*. In this case, there are two nodes per element. Hence, an approximate solution $\hat{T}$ in the entire domain can be considered as a linear combination of nodal values

$$\hat{T} = \sum_{j=0}^{N} \emptyset_j(x)T_j, \tag{7.39}$$

where

$\emptyset_j$ is called the *trial function* which is used to approximate the variation between nodal values.

Substituting $T$ with $\hat{T}$ given in equation in the weak form, we get

$$-vq\Big|_0^L - \sum_{j=0}^{N}\int_0^L \frac{\mathrm{d}v}{\mathrm{d}x}k\frac{\mathrm{d}\emptyset_j}{\mathrm{d}x}T_j\mathrm{d}x + \int_0^L \ddot{q}\, v\mathrm{d}x = 0. \tag{7.40}$$

By the Galerkin method, the choice of the test function $v$ is the same as the trial function $\emptyset$

$$\sum_{j=0}^{N}\int_0^L k\frac{\mathrm{d}\emptyset_i}{\mathrm{d}x}\frac{\mathrm{d}\emptyset_j}{\mathrm{d}x}\mathrm{d}xT_j = -\emptyset_i q\Big|_0^L + \int_0^L \ddot{q}\emptyset_i\mathrm{d}x. \tag{7.41}$$

Since there are $N + 1$ trial functions, there will be $N + 1$ test functions, and accordingly, we will get $N + 1$ number of equations. This system of $N + 1$ equations can be expressed in matrix form as

$$[K_{ij}]\{T\} = \{Q\} + \left\{\ddot{Q}\right\}, \tag{7.42}$$

where

$[K_{ij}] = \int_0^L k\frac{d\emptyset_i}{dx}\frac{d\emptyset_j}{dx}dx$, is the conductance matrix which is equivalent to the stiffness matrix in elasticity;

$\{Q\}$ is the load vector corresponding to the boundary nodes $x_0$ and $x_N$; and.

$\left\{\ddot{Q}\right\}$ is heat source vector.

Since the heat flux or Neumann boundary condition is contained in the load vector $[Q]$ and is already accounted for in the formulation, this is also known as a *natural boundary condition.* In contrast, the Dirichlet boundary condition is also known as an *essential boundary condition* since this has to be explicitly accounted for in the matrix. One way is to set all $K_{ij}$ s in the i[th] row other than $K_{ii}$ to zero and the right-hand side to the product of known temperature at node $I$, $T_i$, and $K_{ii}$.

$[K_{ij}]$ is an $(N + 1) \times (N + 1)$ square matrix, whereas $[Q]+\left[\ddot{Q}\right]$ is an $(N + 1) \times 1$ column matrix. This can be solved by using any of the matrix solving algorithms like Gauss–Seidel or Gauss–Jordan algorithm discussed in Chapter 2. The convergence of the solution depends upon the diagonal dominance of the conductance matrix, which is ensured by the formulation and the number of off-diagonal terms in each row, i.e. the bandwidth of the matrix. The trial functions are chosen to limit the matrix bandwidth size as small as possible. This is done by adopting piecewise or elemental functions $N_i$, such that

$$N_i(x_j) = 0 \text{ if}(x_i \text{ lies outside element } i)$$

$$N_i(x_j) \neq 0, \text{ otherwise.} \tag{7.43}$$

$N_i$ is also known as *shape function* since this defines the variation of the unknown variable $T$ within an element. Usually, the shape function is of polynomial type with linear, quadratic or cubic variation within the element though other types of functions can also be used.

### 7.5.3 Examples

***Example 7.1*** A bar of 10 mm × 10 mm uniform cross section and length 1 m with constant thermal conductivity of 10 W/mK is dissipating heat of 1W. The end temperatures of the bar are maintained at 300 K. Find the steady state temperature distribution in the bar if the heat transfer within the bar is one-dimensional.

Assuming that the heat dissipation is uniform within the bar, we get the heat generation rate per unit volume $\ddot{q}$

$$\ddot{q} = \frac{1}{\left(10 \times 10^{-3}\right)\left(10 \times 10^{-3}\right)(1)} = 10^4\ \mathrm{W/m^3}$$

For the one-dimensional domain $0 \leq x \leq L$, we can write by Galerkin method

$$\sum_{j=0}^{N} \int_0^L k \frac{\mathrm{d}N_i}{\mathrm{d}x}\frac{\mathrm{d}N_j}{\mathrm{d}x}\mathrm{d}x T_j = -N_i q \Big|_0^L + \int_0^L \ddot{q} N_i \mathrm{d}x.$$

Instead of integrating for the entire domain, we integrate for a typical element $l$

$$\int_{x_0^{(l)}}^{x_1^{(l)}} k \frac{\mathrm{d}N_i}{\mathrm{d}x}\frac{\mathrm{d}N_j}{\mathrm{d}x}\mathrm{d}x T_j = -N_i q \Big|_{x_0^{(l)}}^{x_1^{(l)}} + \int_{x_0^{(l)}}^{x_1^{(l)}} \ddot{q} N_i \mathrm{d}x,$$

where

$x_1^{(l)}$ and , $x_0^{(l)}$ are end points corresponding to element $l$.

To make the integration applicable to all elements in the domain, we go for global to local coordinate transformation. Transforming the coordinate system from global in $x$ to local in $\xi$, such that

$$\xi = \frac{x - x_0^{(l)}}{x_1^{(l)} - x_0^{(l)}} = \frac{x - x_0^{(l)}}{h^{(l)}} \quad \xi_x = \frac{1}{h^{(l)}} \quad \mathrm{d}\xi = \frac{\mathrm{d}x}{h^{(l)}},$$

where

$h^{(l)}$ is the size of element $l$.

Now choosing the variation within the element as linear, we go for first order Lagrange polynomial

$$N_0(\xi) = l_0^1(\xi) = \frac{\xi - \xi_1^{(l)}}{\xi_1^{(l)} - \xi_0^{(l)}} = 1 - \xi; \quad N_1(\xi) = l_1^1(\xi) = \frac{\xi - \xi_0^{(l)}}{\xi_1^{(l)} - \xi_0^{(l)}} = \xi.$$

Now the variation of temperature within an element will be

$$\hat{T} = N_0 T_0 + N_1 T_1.$$

The weak formulation in local coordinates for an element with $0 \leq \xi \leq 1$ becomes

$$\frac{k}{h^{(l)}}\sum_{j=0}^{1}\int_0^1 \frac{\mathrm{d}N_i}{\mathrm{d}\xi}\frac{\mathrm{d}N_j}{\mathrm{d}\xi}\mathrm{d}\xi\, T_j = -N_i q\Big|_0^1 + h^{(l)}\ddot{q}\int_0^1 N_i \mathrm{d}\xi,$$

where

$N_i$, can be $N_0$ or $N_1$.

This can be expressed in matrix form as

$$\begin{bmatrix} K_{00} & K_{01} \\ K_{10} & K_{11} \end{bmatrix}\begin{bmatrix} T_0^{(l)} \\ T_1^{(l)} \end{bmatrix} = \begin{bmatrix} q|_0^{(l)} \\ -q|_1^{(l)} \end{bmatrix} + \ddot{q}\begin{bmatrix} S_0 \\ S_1 \end{bmatrix},$$

where

$$K_{ij} = \frac{k}{h^{(l)}}\int_0^1 \frac{\mathrm{d}N_i}{\mathrm{d}\xi}\frac{\mathrm{d}N_j}{\mathrm{d}\xi}\mathrm{d}\xi,\ S_i = h^{(l)}\int_0^1 N_i \mathrm{d}\xi \text{ and } N_0(1) = N_1(0) = 0.$$

Evaluating and substituting for constant $K_{ij}$ and $S_i$, we get

$$\frac{k}{h^{(l)}}\begin{bmatrix} 1 & -1 \\ -1 & 1 \end{bmatrix}\begin{bmatrix} T_0^{(l)} \\ T_1^{(l)} \end{bmatrix} = \begin{bmatrix} q|_0^{(l)} \\ -q|_1^{(l)} \end{bmatrix} + \frac{h^{(l)}\ddot{q}}{2}\begin{bmatrix} 1 \\ 1 \end{bmatrix}.$$

This is called elemental or local matrix form which is valid for any element $l$ within the domain. To get the global matrix form for the entire domain, we have to assemble all the local matrices together knowing that the relation between the local node index and global index is

$$x_0^{(l)} = x_{l-1},\ x_1^{(l)} = x_l,\ T_0^{(l)} = T_{l-1},\ T_1^{(l)} = T_l\ 1 \le l \le N.$$

To ensure that there is a unique temperature and flux for the node shared between two elements, we need

$$T_1^{(l)} = T_0^{(l+1)}.$$

If there are $N$ uniform elements in the domain with $h^{(1)} = \cdots = h^{(N)} = h = \frac{L}{N}$, the global matrix form becomes

$$\frac{k}{h}\begin{bmatrix} 1 & -1 & 0 & 0 & \cdots \\ -1 & 2 & -1 & 0 & \cdots \\ \vdots & \vdots & \ddots & \cdots & \cdots \\ \cdots & 0 & -1 & 2 & -1 \\ \cdots & 0 & 0 & -1 & 1 \end{bmatrix}\begin{bmatrix} T_0 \\ T_1 \\ \vdots \\ T_{N-1} \\ T_N \end{bmatrix} = \begin{bmatrix} q|_0 \\ 0 \\ \vdots \\ 0 \\ -q|_L \end{bmatrix} + \frac{h\ddot{q}}{2}\begin{bmatrix} 1 \\ 2 \\ \vdots \\ 2 \\ 1 \end{bmatrix}.$$

Note that all internal flux terms cancel each other during assembly due to *compatibility*. Since the boundary conditions are specified as known temperatures (Dirichlet type), the first and last rows are modified and the global matrix can be written as

$$\begin{bmatrix} 1 & 0 & 0 & 0 & \cdots \\ -1 & 2 & -1 & 0 & \cdots \\ \vdots & \vdots & \ddots & \cdots & \cdots \\ \cdots & 0 & -1 & 2 & -1 \\ \cdots & 0 & 0 & 0 & 1 \end{bmatrix} \begin{bmatrix} T_0 \\ T_1 \\ \vdots \\ T_{N-1} \\ T_N \end{bmatrix} = \begin{bmatrix} 300.0 \\ 0.0 \\ \vdots \\ 0.0 \\ 300.0 \end{bmatrix} + \frac{h^2 \ddot{q}}{2k} \begin{bmatrix} 0.0 \\ 2 \\ \vdots \\ 2 \\ 0.0 \end{bmatrix}.$$

where

$N$ is the total number of elements $> 1$.

This is in the form $[A][X] = [B]$, with [A] a tri-diagonal matrix, which can be solved using the standard *Thomas algorithm* procedure. The exact solution is given by

$$T(x) = T_0 + \frac{\ddot{q}L^2}{2k}\left[\frac{x}{L} - \left(\frac{x}{L}\right)^2\right].$$

The numerical solution for a different number of elements is shown in Fig. 7.2. Increasing the number of elements brings us closer to the exact solution $T(x)$.

In order to quantify the difference between the approximate solution $\hat{T}$ and the exact solution $T(x)$, we define the error norm for the entire domain as

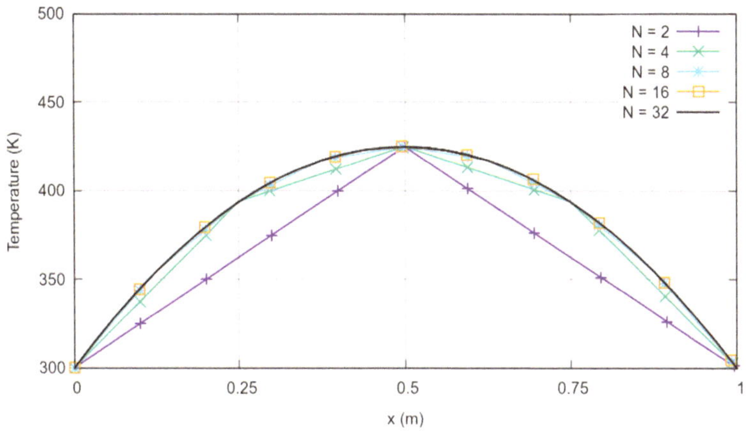


**Fig. 7.2** Temperature distribution inside the domain

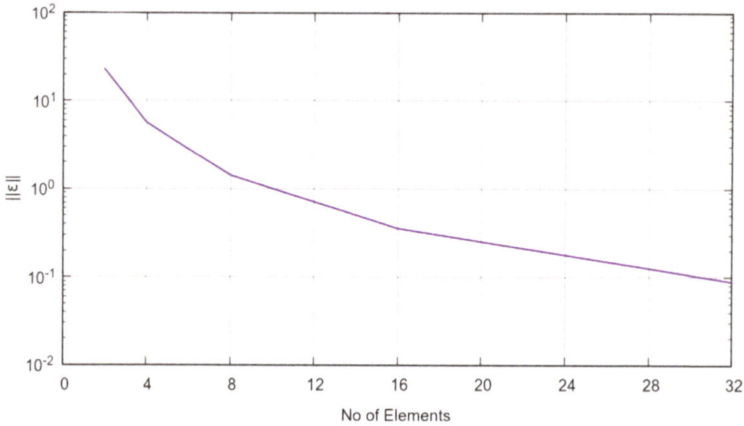


**Fig. 7.3** Error norm versus number of elements

$$||\varepsilon|| = \sqrt{\sum_{l=0}^{N} \int_{x_0^{(l)}}^{x_1^{(l)}} \left(T - \hat{T}\right)^2 \mathrm{d}x}.$$

A graph of the variation of the error norm with a number of elements is shown in Fig. 7.3. We see that there is a reduction in $||\varepsilon||$ as the number of elements increases and the approximate solution approaches the exact solution. This is called *convergence.*

In the above example, we could quantify the error because the exact solution was known. This is less often as exact solutions to most real-life problems are not known a priori. One can, however, still do a *grid independence* study by checking the difference in numerical solutions between two successive grid refinements. If the difference is within an acceptable limit, we say the solution has converged and has approached the exact solution.

***Example 7.2*** Do Example 7.1 with quadratic elements and compare this against the linear elements used earlier.

For quadratic elements, we choose Lagrange polynomial of second order. There will be three nodes per element.

Transforming the coordinate system from global in $x$ to local in $\xi$, such that

$$\xi = \frac{x - x_0^{(l)}}{x_2^{(l)} - x_0^{(l)}} = \frac{x - x_0^{(l)}}{h^{(l)}} \quad \xi_x = \frac{1}{h^{(l)}} \quad \mathrm{d}\xi = \frac{\mathrm{d}x}{h^{(l)}},$$

where

$x_2^{(l)}$ *and* , $x_0^{(l)}$ are endpoints corresponding to element $l$,$x_1^{(l)}$ is the mid-point, and $h^{(l)}$ is the size of element $l$.

Now the variation of temperature within the element will be

$$\hat{T} = N_0 T_0 + N_1 T_1 + N_2 T_2.$$

With Lagrange polynomial of second order, the shape functions are:

$$N_0 = l_0^2(\xi) = \left(\frac{\xi - \xi_1^{(l)}}{\xi_0^{(l)} - \xi_1^{(l)}}\right)\left(\frac{\xi - \xi_2^{(l)}}{\xi_0^{(l)} - \xi_1^{(l)}}\right) = (1 - 2\xi)(1 - \xi);$$

$$N_1 = l_1^2(\xi) = \left(\frac{\xi - \xi_0^{(l)}}{\xi_1^{(l)} - \xi_0^{(l)}}\right)\left(\frac{\xi - \xi_2^{(l)}}{\xi_1^{(l)} - \xi_2^{(l)}}\right) = 4\xi(1 - \xi);$$

$$N_2 = l_2^2(\xi) = \left(\frac{\xi - \xi_0^{(l)}}{\xi_2^{(l)} - \xi_0^{(l)}}\right)\left(\frac{\xi - \xi_1^{(l)}}{\xi_2^{(l)} - \xi_1^{(l)}}\right) = \xi(2\xi - 1).$$

The weak formulation in local coordinates for an element becomes

$$\frac{k}{h^{(l)}} \sum_{j=0}^{N} \int_0^1 \frac{\mathrm{d}N_i}{\mathrm{d}\xi}\frac{\mathrm{d}N_j}{\mathrm{d}\xi}\mathrm{d}\xi\, T_j = -N_i q \Big|_0^1 + \int_0^1 N_i \mathrm{d}\xi,$$

where

$N_i$, can be $N_0$, $N_1$ or $N_2$.

This can be expressed in matrix form as

$$\begin{bmatrix} K_{00} & K_{01} & K_{02} \\ K_{10} & K_{11} & K_{12} \\ K_{20} & K_{21} & K_{22} \end{bmatrix}\begin{bmatrix} T_0^{(l)} \\ T_1^{(l)} \\ T_2^{(l)} \end{bmatrix} = \begin{bmatrix} q|_0^{(l)} \\ 0 \\ -q|_2^{(l)} \end{bmatrix} + \ddot{q}\begin{bmatrix} S_0 \\ S_1 \\ S_2 \end{bmatrix}, \tag{7.44}$$

where

$$K_{ij} = \int_0^1 \frac{k}{h^{(l)}} \frac{\mathrm{d}N_i}{\mathrm{d}\xi}\frac{\mathrm{d}N_j}{\mathrm{d}\xi}\mathrm{d}\xi \quad S_i = \int_0^1 h^{(l)} N_i \mathrm{d}\xi.$$

Evaluating and substituting for constant $K_{ij}$ and $S_i$, we get

$$\frac{k}{3h^{(l)}}\begin{bmatrix} 7 & -8 & 1 \\ -8 & 16 & -8 \\ 1 & -8 & 7 \end{bmatrix}\begin{bmatrix} T_0^{(l)} \\ T_1^{(l)} \\ T_2^{(l)} \end{bmatrix} = \begin{bmatrix} q|_0^{(l)} \\ 0 \\ -q|_2^{(l)} \end{bmatrix} + \frac{h^{(l)}\ddot{q}}{6}\begin{bmatrix} 1 \\ 4 \\ 1 \end{bmatrix}.$$

This is the elemental matrix form for quadratic elements. To get the global matrix form for the entire domain, we assemble all the elemental matrices together using the relation between the local node index and the global index

$$x_0^{(l)} = x_{2l-2}, \quad x_1^{(l)} = x_{2l-1}, \quad x_2^{(l)} = x_{2l}$$

$$T_0^{(l)} = T_{2l-2}, \quad T_1^{(l)} = T_{2l-1}, \text{ and } T_2^{(l)} = T_{2l}; \quad 1 \le l \le N.$$

To ensure that there is a unique temperature for the node shared between two elements, we need

$$T_2^{(l)} = T_0^{(l+1)}.$$

If there are $N$ uniform elements in the domain with $h^{(i)} = \cdots = h^{(N)} = \frac{L}{N}$, the global matrix form becomes

$$\frac{k}{3h}\begin{bmatrix} 7 & -8 & 1 & 0 & 0 & \cdots & \cdots \\ -8 & 16 & -8 & 0 & 0 & \cdots & \cdots \\ 1 & -8 & 14 & -8 & 1 & \cdots & \cdots \\ \vdots & \vdots & \vdots & \ddots & \cdots & \cdots & \cdots \\ \cdots & \cdots & 1 & -8 & 14 & -8 & 1 \\ \cdots & \cdots & 0 & 0 & -8 & 16 & -8 \\ \cdots & \cdots & 0 & 0 & 1 & -8 & 7 \end{bmatrix}\begin{bmatrix} T_0 \\ T_1 \\ T_2 \\ \vdots \\ T_{2N-2} \\ T_{2N-1} \\ T_{2N} \end{bmatrix} = \begin{bmatrix} q|_0 \\ 0 \\ 0 \\ \vdots \\ 0 \\ 0 \\ -q|_L \end{bmatrix} + \frac{h\ddot{q}}{6}\begin{bmatrix} 1 \\ 4 \\ 2 \\ \vdots \\ 2 \\ 4 \\ 1 \end{bmatrix}.$$

Again all internal flux terms cancel each other during assembly due to *compatibility*. Since the boundary conditions are specified as known temperatures (Dirichlet type), the global matrix becomes

$$\begin{bmatrix} 7 & 0 & 0 & 0 & 0 & \cdots & \cdots \\ -8 & 16 & -8 & 0 & 0 & \cdots & \cdots \\ 1 & -8 & 14 & -8 & 1 & \cdots & \cdots \\ \vdots & \vdots & \vdots & \ddots & \cdots & \cdots & \cdots \\ \cdots & \cdots & 1 & -8 & 14 & -8 & 1 \\ \cdots & \cdots & 0 & 0 & -8 & 16 & -8 \\ \cdots & \cdots & 0 & 0 & 0 & 0 & 7 \end{bmatrix}\begin{bmatrix} T_0 \\ T_1 \\ T_2 \\ \vdots \\ T_{2N-2} \\ T_{2N-1} \\ T_{2N} \end{bmatrix} = \begin{bmatrix} 2100.0 \\ 0 \\ 0 \\ \vdots \\ 0 \\ 0 \\ 2100.0 \end{bmatrix} + \frac{L^2\ddot{q}}{2N^2k}\begin{bmatrix} 0 \\ 4 \\ 2 \\ \vdots \\ 2 \\ 4 \\ 0 \end{bmatrix},$$

where

$N$ is the total number of elements $\ge 1$.

Note that the assembled matrix has a higher bandwidth of 5 compared to the bandwidth of 3 for the linear element formulation. This can be solved by any of the iterative methods like the Gauss–Seidel method to obtain the solution. Figure 7.4 shows the solution obtained for a single quadratic element with 3 nodes. The error norm for this solution is less than $10^{-13}$. This is because the exact solution itself is a quadratic distribution.

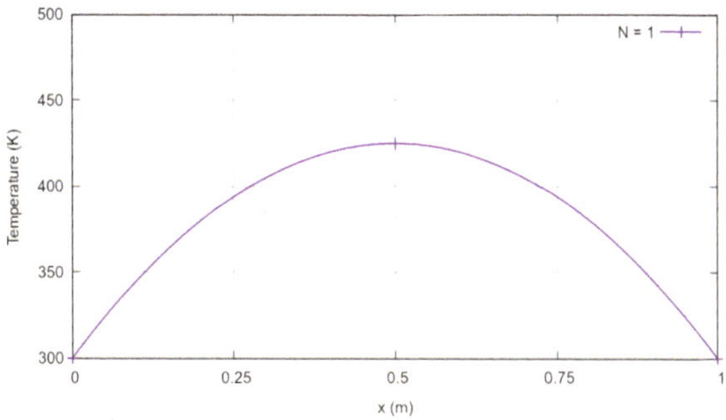


**Fig. 7.4** Temperature distribution inside the domain for a quadratic element

## 7.6 One-Dimensional Transient Conduction

We obtain the governing equation of transient one-dimensional conduction by restricting the governing equation to one spatial dimension

$$\rho C \frac{\partial T}{\partial t} = \frac{\partial}{\partial x}\left(k \frac{\partial T}{\partial x}\right) + \ddot{q}\,. \tag{7.45}$$

Note that this is a partial differential equation which when solved will give the solution, $T = T(x_i, t)$. Since the equation has a second-order differential w.r.t $x$ and a first-order differential w.r.t. to $t$, the equation is elliptic w.r.t $x$ and parabolic w.r.t to $t$. This means to obtain a solution, we need to specify one condition w.r.t $t$, $T_i(x)$, also known as *initial condition* and as in the case of steady state conduction, boundary conditions w.r.t $x$. For a one-dimensional rod of length $L$, the boundary conditions specified at $x = 0$ and $x = L$ can now be a function of $t$. The restriction of having at least one Dirichlet/Robin type boundary condition does not exist for the transient case as the solution is obtained by marching with respect to $t$ from the initial condition $T_i(x)$.

### *7.6.1 Method of Weighted Residuals for One-Dimensional Transient Conduction*

We multiply the transient equation by an arbitrary test function $v$ and integrate over the interval $(0, L)$

$$\int_0^L \rho C \frac{\partial T}{\partial t} v \mathrm{d}x = \int_0^L \frac{\mathrm{d}}{\mathrm{d}x}\left(k \frac{\mathrm{d}T}{\mathrm{d}x}\right) v \mathrm{d}x + \int_0^L \ddot{q} v \mathrm{d}x. \tag{7.46}$$

Integrating the second-order term by parts and bringing all terms to left-hand side, we get

$$\int_0^L \rho C \frac{\partial T}{\partial t} v \mathrm{d}x - vk \frac{\mathrm{d}T}{\mathrm{d}x} \bigg|_0^L + \int_0^L \frac{\mathrm{d}v}{\mathrm{d}x} k \frac{\mathrm{d}T}{\mathrm{d}x} \mathrm{d}x - \int_0^L \ddot{q} v \mathrm{d}x = 0 \tag{7.47}$$

or

$$\int_0^L \rho C \frac{\partial T}{\partial t} v \mathrm{d}x + vq \bigg|_0^L + \int_0^L \frac{\mathrm{d}v}{\mathrm{d}x} k \frac{\mathrm{d}T}{\mathrm{d}x} \mathrm{d}x - \int_0^L \ddot{q} v \mathrm{d}x = 0. \tag{7.48}$$

Taking the natural boundary condition terms and source terms to the right side, we get

$$\int_0^L \rho C \frac{\partial T}{\partial t} v \mathrm{d}x + \int_0^L \frac{\mathrm{d}v}{\mathrm{d}x} k \frac{\mathrm{d}T}{\mathrm{d}x} \mathrm{d}x = -vq \bigg|_0^L + \int_0^L \ddot{q} v \mathrm{d}x. \tag{7.49}$$

This is the weak formulation of one-dimensional transient conduction. We again seek an approximate solution $\hat{T}$ by dividing the continuous domain (0, $L$) into $N$ non-overlapping elements, such that

$$\hat{T} = \sum_{j=0}^{N} \emptyset_j(x) T_j(t). \tag{7.50}$$

Note that the trial function $\emptyset_j$ is only a function of $x$ and $T_j(t)$ is dependent on $t$. Substituting $T$ with $\hat{T}$ given in equation in the weak form, we get

$$\sum_{j=0}^{N} \int_0^L \rho C \emptyset_j \frac{\mathrm{d}T_j}{\mathrm{d}t} v \mathrm{d}x + \sum_{j=0}^{N} \int_0^L k \frac{\mathrm{d}v}{\mathrm{d}x} \frac{\mathrm{d}\emptyset_j}{\mathrm{d}x} T_j \mathrm{d}x = -vq \bigg|_0^L + \int_0^L \ddot{q} v \mathrm{d}x. \tag{7.51}$$

By the Galerkin technique, the choice of $v$ is $\phi$. The equation becomes

$$\sum_{j=0}^{N} \int_0^L \rho C \emptyset_j \frac{dT_j}{\mathrm{d}t} \emptyset_i \mathrm{d}x + \sum_{j=0}^{N} \int_0^L k \frac{d\emptyset_i}{\mathrm{d}x} \frac{d\emptyset_j}{\mathrm{d}x} T_j \mathrm{d}x = -\emptyset_i q \bigg|_0^L + \int_0^L \ddot{q} \emptyset_i \mathrm{d}x. \tag{7.52}$$

Since there are $N + 1$ trial functions, there will be $N + 1$ test functions, and accordingly, we will get $N + 1$ equations. This system of $N + 1$ equations can be expressed in matrix form as

$$[K_{ij}]\left\{\frac{\mathrm{d}T}{\mathrm{d}t}\right\} + [K_{ij}]\{T\} = \{Q\} + \left\{\ddot{Q}\right\}, \tag{7.53}$$

where

$\left[M_{ij}\right] = \int_0^L \rho C \emptyset_i \emptyset_j \mathrm{d}x$, is the mass matrix;

$\left[K_{ij}\right] = \int_0^L k \frac{d\emptyset_i}{\mathrm{d}x}\frac{d\emptyset_j}{\mathrm{d}x}\hat{T}_j \mathrm{d}x$, is the conductance matrix;

$\{Q\}$, is the load vector corresponding to the boundary nodes $x_0$ and $x_N$; and

$\left\{\ddot{Q}\right\}$, is a heat source vector.

This equation is in semi discrete form as we have discretized only with respect to x. To understand time integration, for the time being, let us imagine we have an ordinary differential equation that can be expressed as

$$\frac{\mathrm{d}T}{\mathrm{d}t} = f(T, t). \tag{7.54}$$

whose solution is of the form $T(t)$. If we know the initial condition $T_i = T(0)$, then the solution at any later time $\Delta t$ can be found out by integrating the above

$$T(0 + \Delta t) = T_\mathrm{i} + \int_0^{\Delta t} \frac{\mathrm{d}T}{\mathrm{d}t}\mathrm{d}t = T_\mathrm{i} + \int_0^{\Delta t} f(T, t)\mathrm{d}t. \tag{7.55}$$

There are various ways to approximate the integrand on the right-hand side

$$\int_0^{\Delta t} f(T, t)\mathrm{d}t \approx \Delta t f(T, 0)$$

$$\int_0^{\Delta t} f(T, t)\mathrm{d}t \approx \Delta t f(T, \Delta t)$$

$$\int_0^{\Delta t} f(T, t)\mathrm{d}t \approx \frac{\Delta t}{2}(f(T, 0) + f(T, \Delta t)). \tag{7.56}$$

Substituting into the equation, we get the solution at $T(\Delta t)$

$$T(\Delta t) = T_i + \Delta t f(T, 0)$$

$$T(\Delta t) = T_i + \Delta t f(T, \Delta t)$$

$$T(\Delta t) = T_\mathrm{i} + \frac{\Delta t}{2}(f(T, 0) + f(T, \Delta t)). \tag{7.57}$$

Since this holds for between any two consecutive time steps say $n$ and $n+1$ with $T(n)$ being known, we can write

$$T^{n+1} = T^n + \Delta t f^n = T^n + \Delta t \left.\frac{\mathrm{d}T}{\mathrm{d}t}\right|^n$$

$$T^{n+1} = T^n + \Delta t f^{n+1} = T^n + \Delta t \left.\frac{\mathrm{d}T}{\mathrm{d}t}\right|^{n+1}$$

$$T^{n+1} = T^n + \frac{\Delta t}{2}\left(\left.\frac{\mathrm{d}T}{\mathrm{d}t}\right|^n + \left.\frac{\mathrm{d}T}{\mathrm{d}t}\right|^{n+1}\right). \tag{7.58}$$

The equations can be combined and written as

$$T^{n+1} = T^n + \Delta t\left(\theta \left.\frac{\mathrm{d}T}{\mathrm{d}t}\right|^n + (1-\theta)\left.\frac{\mathrm{d}T}{\mathrm{d}t}\right|^{n+1}\right) \quad \text{with } 0 \le \theta \le 1. \tag{7.59}$$

If $\theta = 1$, the formulation or scheme is called *explicit scheme* as there is no dependency on time derivative at $n + 1^{th}$ time. Conversely, if $\theta = 0$, the scheme is *implicit* as there is no dependency on time derivative at the $n^{th}$ time. Explicit FEM is used to calculate the state of a given system at a different time from the current time. In contrast, an implicit analysis finds a solution by solving an equation that includes both the current and later states of the given system. A special case when $\theta = 0.5$ is called the *Crank Nicholson scheme.*

Now we are ready to perform time integration for the semi discrete equation. We multiply the above equation by the mass matrix $[M_{ij}]$ and substitute for the time gradient term from the semi discrete equation

$$\begin{aligned}[M_{ij}]\{T^{n+1}\} = [M_{ij}]\{T^n\} + \Delta t[\ominus(-[K_{ij}]\{T^n\} + \{Q^n\} + \{\ddot{Q}^n\}) \\ + (1-\ominus)(-[K_{ij}]\{T^{n+1}\} + \{Q^{n+1}\} + \{\ddot{Q}^{n+1}\})]\end{aligned} \tag{7.60}$$

or

$$\begin{aligned}[M_{ij} + (1-\ominus)\Delta t K_{ij}]\{T^{n+1}\} &= [M_{ij} - \Delta t \ominus K_{ij}]\{T^n\} \\ &+ \Delta t\begin{bmatrix}\ominus(\{Q^n\} + \{\ddot{Q}^n\}) \\ +(1-\ominus)(\{Q^{n+1}\} + \{\ddot{Q}^{n+1}\})\end{bmatrix}.\end{aligned} \tag{7.61}$$

The right-hand side terms are all known and this can be solved to obtain by any matrix method to obtain $[T]^{n+1}$. The implicit and explicit schemes have a time accuracy of $O(\Delta t)$, whereas Crank Nicholson scheme has a time accuracy of $O\left(\Delta t^2\right)$. In terms of stability, the implicit scheme is unconditionally stable, whereas the explicit and Crank Nicholson schemes are conditionally stable, which puts a restriction on $\Delta t$. The stability criterion being

$$\left[M_{ij} - \Delta t \ominus K_{ij}\right] \geq 0. \tag{7.62}$$

In finite element method, we have to solve a matrix even in the explicit formulation due to the presence of a mass matrix $\left[\mathrm{M_{ij}}\right]$ unlike other methods like Finite Difference or Finite Volume Methods which have a diagonal mass matrix and the explicit formulation can be solved by direct substitution.

### 7.6.2 Examples

***Example 7.3*** In a residence, a brick wall of 25 cm thickness is exposed to solar radiation of 500 W/m$^2$ from 07:00 h onwards. Find the temperature of the outer and inner sides of the wall from 07:00 to 10:00 h if the temperature of the wall at 07:00 h is uniform at 30 °C, given the thermal properties of brick is as follows:

| Thermal conductivity | Density | Specific heat | Solar absorptivity | Emissivity |
|---|---|---|---|---|
| $k$ | $\rho$ | $C\rho$ | $\alpha_s$ | $\epsilon$ |
| 0.72 W/mK | 1920 kg/m$^3$ | 835 J/kgK | 0.67 | 0.93 |

Study the following cases:

1. Assume there is no convection and radiation from the front side.
2. Assume that there is only free convection to an ambient of 30 °C from the front.
3. Assume that there are both free convection and radiation to an ambient of 30 °C. from the front

We shall consider $x = 0$ m as the front side and $x = 0.25$ m as the back side of the brick wall. Given that $q|_0 = \propto_s q_s = 0.67 \times 500 = 335$ w/m$^2$. Continuing from Example 7.1, we shall assume a linear variation of temperature within an element with $N_0 = \xi$, $N_1 = 1 - \xi$ and thermal properties to be constant. Considering an implicit formulation with $\theta = 0$, the elemental form for transient conduction becomes

$$\rho C h^l \begin{bmatrix} M_{00} & M_{01} \\ M_{10} & M_{11} \end{bmatrix} \begin{bmatrix} T_0^{(l)n+1} \\ T_1^{(l)n+1} \end{bmatrix} + \Delta t \begin{bmatrix} K_{00} & K_{01} \\ K_{10} & K_{11} \end{bmatrix} \begin{bmatrix} T_0^{(l)n+1} \\ T_1^{(l)n+1} \end{bmatrix}$$

$$= \rho C h^l \begin{bmatrix} M_{00} \; M_{01} \\ M_{10} \; M_{11} \end{bmatrix} \begin{bmatrix} T_0^{(l)n} \\ T_1^{(l)n} \end{bmatrix} + \Delta t \begin{bmatrix} q\Big|_0^{(l)} \\ -q\Big|_1^{(l)} \end{bmatrix},$$

where $\left[M_{ij}\right] = \int\limits_0^L N_i N_j \mathrm{d}\xi$. Evaluating $M_{ij}$ and $K_{ij}$ and defining $A = \frac{k\Delta t}{\rho C (h^l)^2}$ and $B = \frac{6\Delta t}{\rho C h^l}$, we get the element matrix form as

$$\begin{bmatrix} 2+6A \;\; 1-6A \\ 1-6A \;\; 2+6A \end{bmatrix} \begin{bmatrix} T_0^{(i)n+1} \\ T_1^{(i)n+1} \end{bmatrix} = \begin{bmatrix} 2 \; 1 \\ 1 \; 2 \end{bmatrix} \begin{bmatrix} T_0^{(i)n} \\ T_1^{(i)n} \end{bmatrix} + B \begin{bmatrix} q\Big|_0^{(i)} \\ -q\Big|_1^{(i)} \end{bmatrix}.$$

If there are $N$ uniform elements in the domain with $h^{(l)} = \cdots = h^{(N)} = \frac{L}{N}$, the global matrix form becomes

$$\begin{bmatrix} 2+6A & -1 & 0 & 0 & \cdots \\ 1-6A & 4+12A & 1-6A & 0 & \cdots \\ \vdots & \vdots & \ddots & \cdots & \cdots \\ \cdots & 0 & 1-6A & 4+12A & 1-6A \\ \cdots & 0 & 0 & 1-6A & 2+6A \end{bmatrix} \begin{bmatrix} T_0^{n+1} \\ T_1^{n+1} \\ \vdots \\ T_{N-1}^{n+1} \\ T_N^{n+1} \end{bmatrix} = \begin{bmatrix} 2 & 1 & 0 & 0 & \cdots \\ 1 & 4 & 1 & 0 & \cdots \\ \vdots & \vdots & \ddots & \cdots & \cdots \\ \cdots\, 0 & 1 & 4 & 1 \\ \cdots\, 0 & 0 & 1 & 2 \end{bmatrix} \begin{bmatrix} T_0^n \\ T_1^n \\ \vdots \\ T_{N-1}^n \\ T_N^n \end{bmatrix} + B \begin{bmatrix} q|_0^0 \\ 0 \\ \vdots \\ 0 \\ -q|_1^N \end{bmatrix},$$

where $q|_0 = \alpha_s \times 500 = 335\,\mathrm{W/m^2}$ and since nothing has been said about the inner side conditions, we can assume it is insulated i.e.$q|_L = 0$. Since the implicit formulation is unconditionally stable, there's no restriction on the choice of $\Delta t$ for a given $h$ or vice versa. Choosing a $\Delta t$ of 10 s, we perform a time marching solution from $t = 0$ to $t = 10800$ s by solving the above matrix for each time step. The results obtained are for different number of elements and are shown in Fig. 7.5. We see that grid convergence is established when number of elements = 32.

Before we move on to other cases, let us formulate the same case in explicit form with $\theta = 1$ to get an idea of the issue of stability. The element form for the explicit formulation is

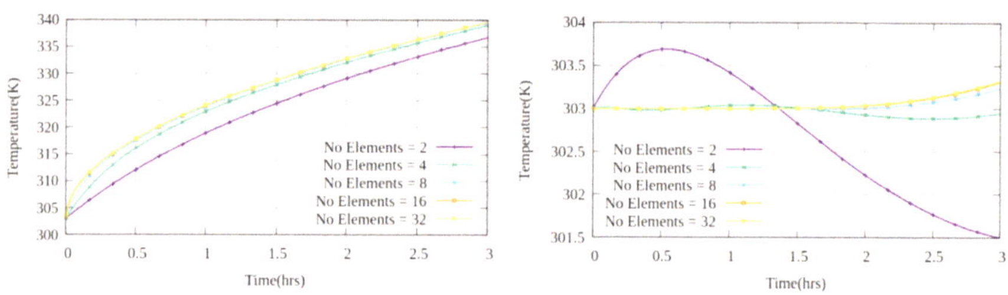


**Fig. 7.5** Temperature history of the inner and outer side of brick wall

$$\rho Ch^i \begin{bmatrix} M_{00} & M_{01} \\ M_{10} & M_{11} \end{bmatrix} \begin{bmatrix} T_0^{(i)n+1} \\ T_1^{(i)n+1} \end{bmatrix} = \rho Ch^i \begin{bmatrix} M_{00} & M_{01} \\ M_{10} & M_{11} \end{bmatrix} \begin{bmatrix} T_0^{(i)n} \\ T_1^{(i)n} \end{bmatrix} - \frac{k\Delta t}{h^{(i)}} \begin{bmatrix} K_{00} & K_{01} \\ K_{10} & K_{11} \end{bmatrix} \begin{bmatrix} T_0^{(i)n} \\ T_1^{(i)n} \end{bmatrix} + \Delta t \begin{bmatrix} q\big|_o^{(i)} \\ -q\big|_1^{(i)} \end{bmatrix}$$

or

$$\begin{bmatrix} 2 & 1 \\ 1 & 2 \end{bmatrix} \begin{bmatrix} T_0^{(i)n+1} \\ T_1^{(i)n+1} \end{bmatrix} = \begin{bmatrix} 2-6A & 1+6A \\ 1+6A & 2-6A \end{bmatrix} \begin{bmatrix} T_0^{(i)n} \\ T_1^{(i)n} \end{bmatrix} + B \begin{bmatrix} q\big|_o^{(i)} \\ -q\big|_1^{(i)} \end{bmatrix}.$$

By the stability criteria, $2 - 6A \geq 0$ or $\frac{k\Delta t}{\rho Cph^2} \leq \frac{1}{3}$. For the above case, with the number of elements $= 32$, this would restrict the time step to $\Delta t \leq 45$ s.

Let us now consider the second case where there is free convection from the outer wall in addition to the solar radiation. Knowing that the convective heat transfer coefficient for free convection $h \approx 5W/m^2$, the outer wall heat flux becomes

$$q|_0 = \propto_s q_s + h(T_\infty - T|_0).$$

With $T_\infty$ being the ambient temperature given as 30 °C.

The global matrix for the implicit formulation for this case is

$$\begin{bmatrix} 2+6A+hB & -1 & 0 & 0 & \cdots \\ 1-6A & 4+12A & 1-6A & 0 & \cdots \\ \vdots & \vdots & \ddots & \cdots & \cdots \\ \cdots & 0 & 1-6A & 4+12A & 1-6A \\ \cdots & 0 & 0 & 1-6A & 2+6A \end{bmatrix} \begin{bmatrix} T_0^{n+1} \\ T_1^{n+1} \\ \vdots \\ T_{N-1}^{n+1} \\ T_N^{n+1} \end{bmatrix} = \begin{bmatrix} 2 & 1 & 0 & 0 & \cdots \\ 1 & 4 & 1 & 0 & \cdots \\ \vdots & \vdots & \ddots & \cdots & \cdots \\ \cdots & 0 & 1 & 4 & 1 \\ \cdots & 0 & 0 & 1 & 2 \end{bmatrix} \begin{bmatrix} T_0^n \\ T_1^n \\ \vdots \\ T_{N-1}^n \\ T_N^n \end{bmatrix} + B \begin{bmatrix} \propto_s q_s + hT_\infty^{n+1} \\ 0 \\ \vdots \\ 0 \\ 0 \end{bmatrix}.$$

For the case when convective and radiative heat transfer from the outer wall is present, we additionally account for the radiative heat transfer in a similar manner to the convective heat transfer

$$q|_0 = \propto_s q_s + h(T_\infty - T|_0) + h_R(T_\infty - T|_0),$$

where

$$h_R = \epsilon\sigma\left(T_\infty^2 + T|_0^2\right)(T_\infty - T|o).$$

We see that the radiative heat transfer is a nonlinear function of $T$. One way to address non-linearity is to approximate $h_R$ and iteratively correct it each time step. The approximation can be taken from the previous known temperatures

$$h_R = \epsilon\sigma\left(T_\infty^{n2} + T|_0^{n2}\right)\left(T_\infty^n + T|_0^n\right).$$

The global matrix for the implicit formulation for this case is

$$\begin{bmatrix} 2+6A+hB+h_RB & -1 & 0 & 0 & \cdots \\ 1-6A & 4+12A & 1-6A & 0 & \cdots \\ \vdots & \vdots & \ddots & \cdots & \cdots \\ \cdots & 0 & 1-6A & 4+12A & 1-6A \\ \cdots & 0 & 0 & 1-6A & 2+6A \end{bmatrix} \begin{bmatrix} T_0^{n+1} \\ T_1^{n+1} \\ \vdots \\ T_{N-1}^{n+1} \\ T_N^{n+1} \end{bmatrix}$$

$$= \begin{bmatrix} 2 & 1 & 0 & 0 & \cdots \\ 1 & 4 & 1 & 0 & \cdots \\ \vdots & \vdots & \ddots & \cdots & \cdots \\ \cdots 0 & 1 & 4 & 1 \\ \cdots 0 & 0 & 1 & 2 \end{bmatrix} \begin{bmatrix} T_0^n \\ T_1^n \\ \vdots \\ T_{N-1}^n \\ T_N^n \end{bmatrix}$$

$$+ B \begin{bmatrix} \propto_s q_s + hT_\infty^{n+1} + h_R T_\infty^{n+1} \\ 0 \\ \vdots \\ 0 \\ 0 \end{bmatrix}.$$

Figure 7.6 compares the outer wall temperature for the cases discussed.

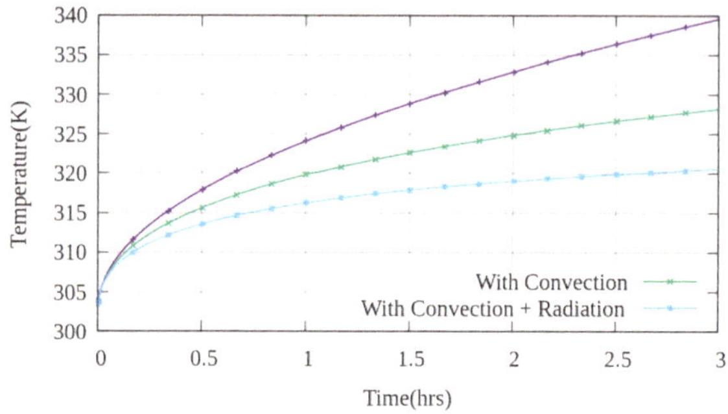


**Fig. 7.6** Temperature history of outer wall for three cases

## 7.7 Higher dimensional heat transfer problems

### 7.7.1 *Method of Weighted Residuals for Steady State Conduction*

Consider again the governing equation for steady state conduction in an isotropic medium in vector form

$$\vec{\nabla}.\left(k\vec{\nabla}T\right) + \ddot{q} = 0.$$

We multiply the above equation by the test function $v$, and integrate over the domain $\Omega$

$$\int \vec{\nabla}.\left(k\vec{\nabla}T\right)v\mathrm{d}\Omega + \int \ddot{q}v\mathrm{d}\Omega = 0. \tag{7.63}$$

Integrating the first term by parts, we get

$$\oint vk\vec{\nabla}T.\mathrm{d}\vec{\Gamma} - \int \vec{\nabla}v.\vec{\nabla}T\mathrm{d}\Omega + \int \ddot{q}v\mathrm{d}\Omega = 0. \tag{7.64}$$

Substituting for $\vec{q} = -k\vec{\nabla}T$ and rearranging, we get

$$\int k\vec{\nabla}v.\vec{\nabla}T\mathrm{d}\Omega = -\oint v\vec{q}|_{\Gamma}.\mathrm{d}\vec{\Gamma} + \int \ddot{q}v\mathrm{d}\Omega. \tag{7.65}$$

where $\vec{q}|_{\Gamma}$, is the heat conducted at the boundary $\boldsymbol{\Gamma}$ of the domain $\boldsymbol{\Omega}$. Again, boundary conditions are imposed through the first term on the right-hand side. If there are N + 1 nodes in an element, we seek an approximate solution $\hat{T}$ as a linear combination of nodal values as $\widehat{T} = \sum_{j=0}^{N} \phi_j \hat{T}_j$.

Substituting $T$ with $\hat{T}$ given in equation in the weak form, we get

$$\sum_{j=0}^{N} \int k\vec{\nabla}v.\vec{\nabla}\phi\hat{T}_j\mathrm{d}\Omega = -\oint v\vec{q}|_{\Gamma}.\mathrm{d}\vec{\Gamma} + \int \ddot{q}v\mathrm{d}\Omega. \tag{7.66}$$

By the Galerkin method, the choice of the test function $v$ is the same as the trial function $\phi$

$$\sum_{j=0}^{N} \int k\vec{\nabla}\phi_i.\vec{\nabla}\phi_j\hat{T}_j\mathrm{d}\Omega = -\oint \phi_i\vec{q}|_{\Gamma}.\mathrm{d}\vec{\Gamma} + \int \ddot{q}\phi_i\mathrm{d}\Omega. \tag{7.67}$$

This system of equations can be expressed in matrix form as

$$[K_{ij}]\{T\} = \{Q\} + \{\ddot{Q}\},$$

where

$[K_{ij}] = \int k\vec{\nabla}\phi_i.\vec{\nabla}\phi_j d\Omega$, is the conductance matrix

$\{Q\}$ is load vector and $\{\ddot{Q}\}$ is heat source vector.

### 7.7.2 Examples

***Example 7.4*** Cooling fins are extended surfaces used for enhancing heat transfer to the environment. They are used in cooling high heat dissipating processors, IC engines, pumps, compressors, etc. Fig. 7.7 shows a typical extended surface from a wall, which is at 100 °C. Find out the effectiveness of the fin if effectiveness is defined as

$$\varepsilon = \frac{Q_{\text{with fin}}}{Q_{\text{without fin}}}.$$

The fin is made of aluminum alloy with thermal conductivity of 120 W/mK. Assume that there is free convection between the fin and ambient at 30 °C and the order of free convection is 5 W/m$^2$ K. Since the problem is symmetric about $X$-axis, a symmetry boundary condition can be used and only half the domain needs to be considered. The domain of the problem along with boundary conditions is shown in Fig. 7.8.

**Fig. 7.7** Heat transfer through fin

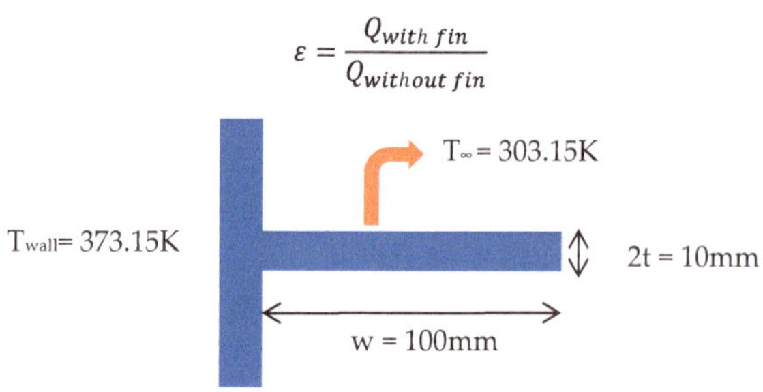


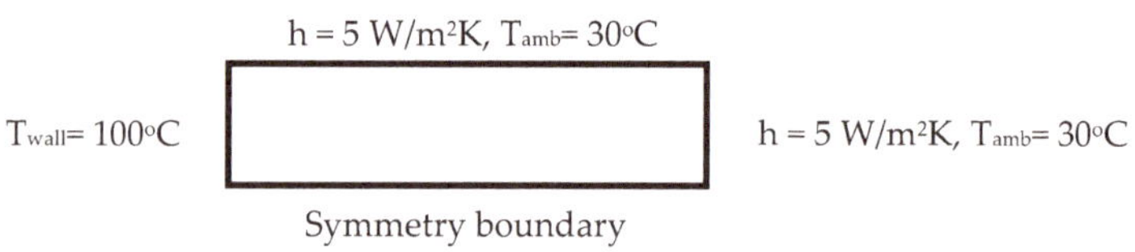


**Fig. 7.8** Fin domain with boundary conditions

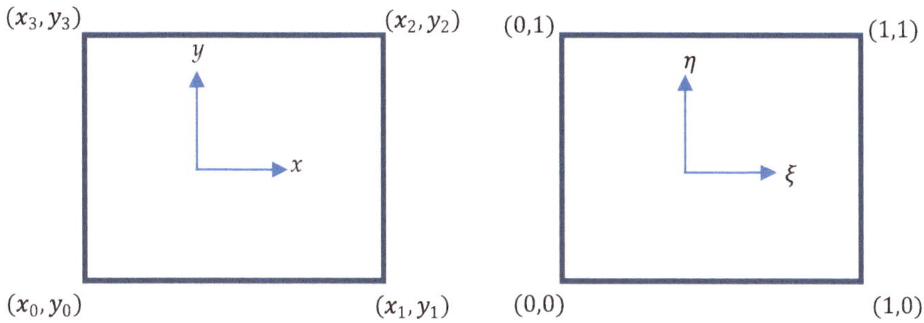


**Fig. 7.9** Transformation from global to local coordinates

We will consider a single quadrilateral element as shown in Fig. 7.9.

Defining the transformation from global coordinates to local coordinates as

$$\begin{bmatrix} \mathrm{d}x \\ \mathrm{d}y \end{bmatrix} = \begin{bmatrix} x_\xi & x_\eta \\ y_\xi & y_\eta \end{bmatrix} \begin{bmatrix} \mathrm{d}\xi \\ \mathrm{d}\eta \end{bmatrix} = [J] \begin{bmatrix} \mathrm{d}\xi \\ \mathrm{d}\eta \end{bmatrix}.$$

For a non-rotating coordinate transformation, the off-diagonal terms in the Jacobian matrix are zero.

$$\begin{bmatrix} \mathrm{d}x \\ \mathrm{d}y \end{bmatrix} = \begin{bmatrix} x_\xi & 0 \\ 0 & y_\eta \end{bmatrix} \begin{bmatrix} \mathrm{d}\xi \\ \mathrm{d}\eta \end{bmatrix}.$$

The inverse relation is

$$\begin{aligned} \begin{bmatrix} \mathrm{d}\xi \\ \mathrm{d}\eta \end{bmatrix} &= \begin{bmatrix} x_\xi & 0 \\ 0 & y_\eta \end{bmatrix}^{-1} \begin{bmatrix} \mathrm{d}x \\ \mathrm{d}y \end{bmatrix} \\ &= \frac{1}{|J|} \begin{bmatrix} y_\eta & 0 \\ 0 & x_\xi \end{bmatrix} \begin{bmatrix} \mathrm{d}x \\ \mathrm{d}y \end{bmatrix} \\ &= \begin{bmatrix} \xi_x & 0 \\ 0 & \eta_y \end{bmatrix} \begin{bmatrix} \mathrm{d}x \\ \mathrm{d}y \end{bmatrix}. \end{aligned}$$

We assume linear variation within the element as

$$T = a\xi + b\eta + c\xi\eta + d.$$

For the four nodal temperatures, the above equation can be expressed in matrix form as

$$\begin{bmatrix} \xi_0 & \eta_0 & \xi_0\eta_0 & 1 \\ \xi_1 & \eta_1 & \xi_1\eta_1 & 1 \\ \xi_2 & \eta_2 & \xi_2\eta_2 & 1 \\ \xi_3 & \eta_3 & \xi_3\eta_3 & 1 \end{bmatrix} \begin{bmatrix} a \\ b \\ c \\ d \end{bmatrix} = \begin{bmatrix} T_0 \\ T_1 \\ T_2 \\ T_3 \end{bmatrix}.$$

If we take the variation in $(\xi, \eta)$ as shown in Fig. 7.9, the matrix becomes

$$\begin{bmatrix} 0 & 0 & 0 & 1 \\ 1 & 0 & 0 & 1 \\ 1 & 1 & 1 & 1 \\ 0 & 1 & 0 & 1 \end{bmatrix} \begin{bmatrix} a \\ b \\ c \\ d \end{bmatrix} = \begin{bmatrix} T_0 \\ T_1 \\ T_2 \\ T_3 \end{bmatrix}.$$

The coefficients can be obtained by taking the inverse

$$\begin{bmatrix} a \\ b \\ c \\ d \end{bmatrix} = \begin{bmatrix} 0 & 0 & 0 & 1 \\ 1 & 0 & 0 & 1 \\ 1 & 1 & 1 & 1 \\ 0 & 1 & 0 & 1 \end{bmatrix}^{-1} \begin{bmatrix} T_0 \\ T_1 \\ T_2 \\ T_3 \end{bmatrix}.$$

Temperature at any point in the element becomes

$$T = \begin{bmatrix} \xi & \eta & \xi\eta & 1 \end{bmatrix} \begin{bmatrix} 0 & 0 & 0 & 1 \\ 1 & 0 & 0 & 1 \\ 1 & 1 & 1 & 1 \\ 0 & 1 & 0 & 1 \end{bmatrix}^{-1} \begin{bmatrix} T_0 \\ T_1 \\ T_2 \\ T_3 \end{bmatrix}$$

$$= \begin{bmatrix} 1 - \xi - \eta + \xi\eta & \xi(1-\eta) & \xi\eta & \eta(1-\xi) \end{bmatrix} \begin{bmatrix} T_0 \\ T_1 \\ T_2 \\ T_3 \end{bmatrix}.$$

Therefore, the shape functions are $N_0 = 1 - \xi - \eta + \xi\eta$, $N_1 = \xi(1-\eta)$, $N_2 = \xi\eta$ and $N_3 = \eta(1-\xi)$.

The derivative of the shape function can now be defined in terms of the local coordinates

$$\begin{bmatrix} \frac{\partial N_i}{\partial x} \\ \frac{\partial N_i}{\partial y} \end{bmatrix} = \begin{bmatrix} \xi_x & 0 \\ 0 & \eta_y \end{bmatrix} \begin{bmatrix} \frac{\partial N_i}{\partial \xi} \\ \frac{\partial N_i}{\partial \eta} \end{bmatrix}$$

$$= \frac{1}{|J|} \begin{bmatrix} y_\eta & 0 \\ 0 & x_\xi \end{bmatrix} \begin{bmatrix} \frac{\partial N_i}{\partial \xi} \\ \frac{\partial N_i}{\partial \eta} \end{bmatrix}.$$

The Jacobian coefficients are defined as $x_\xi = x_1 - x_0 = w$ and $y_\eta = y_3 - y_0 = t$. The conductance matrix terms become

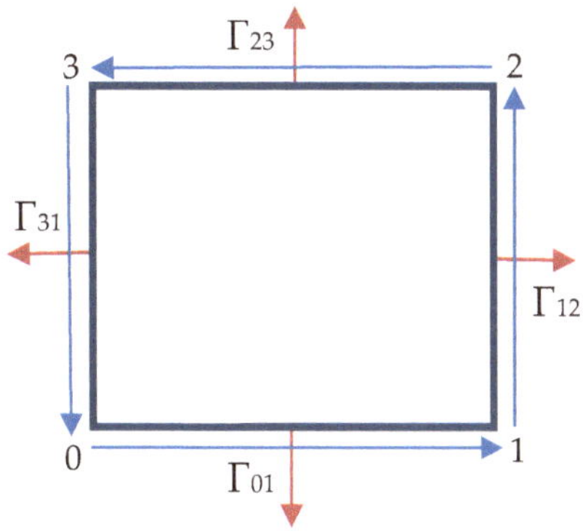


**Fig. 7.10** Integration along the boundary edges of element

$$
\begin{aligned}
[K_{ij}] &= \int k\vec{\nabla}N_i.\vec{\nabla}N_j \mathrm{d}\Omega \\
&= \iint k\left(\frac{\partial N_i}{\partial x}\frac{\partial N_j}{\partial x} + \frac{\partial N_i}{\partial y}\frac{\partial N_j}{\partial y}\right)\mathrm{d}x\mathrm{d}y \\
&= \iint k\left(\frac{y_\eta}{x_\xi}\frac{\partial N_i}{\partial \xi}\frac{\partial N_j}{\partial \xi} + \frac{x_\xi}{y_\eta}\frac{\partial N_i}{\partial \eta}\frac{\partial N_j}{\partial \eta}\right)\mathrm{d}\xi\mathrm{d}\eta.
\end{aligned}
$$

For an isotropic material, the conductance matrix is

$$
[K] = \frac{k}{6}\begin{bmatrix} 2b+2b^{-1} & -2b+b^{-1} & -b-b^{-1} & b-2b^{-1} \\ -2b+b^{-1} & 2b+2b^{-1} & b-2b^{-1} & -b-b^{-1} \\ -b-b^{-1} & b-2b^{-1} & 2b+2b^{-1} & -2b+b^{-1} \\ b-2b^{-1} & -b-b^{-1} & -2b+b^{-1} & 2b+2b^{-1} \end{bmatrix},
$$

$b = \frac{y_\eta}{x_\xi} = \frac{t}{w}$ being the aspect ratio of the element.

The load vector $[Q]$ is defined by integrating along the boundary as shown in Fig. 7.10. The directions of boundary edges are also shown.

If $q$ is the heat flux normal to the edge in the outward direction, the load vector can be defined as

$$
[Q] = -([Q_{01}] + [Q_{12}] + [Q_{23}] + [Q_{30}])
$$

$$[Q] = -\left(\begin{array}{c}\begin{bmatrix}\int N_0 q_{01} d\Gamma_{01}\\ \int N_1 q_{01} d\Gamma_{01}\\ \int N_2 q_{01} d\Gamma_{01}\\ \int N_3 q_{01} d\Gamma_{01}\end{bmatrix} + \begin{bmatrix}\int N_0 q_{12} d\Gamma_{12}\\ \int N_1 q_{12} d\Gamma_{12}\\ \int N_2 q_{12} d\Gamma_{12}\\ \int N_3 q_{12} d\Gamma_{12}\end{bmatrix} + \\ \begin{bmatrix}\int N_0 q_{23} d\Gamma_{23}\\ \int N_1 q_{23} d\Gamma_{23}\\ \int N_2 q_{23} d\Gamma_{23}\\ \int N_3 q_{23} d\Gamma_{23}\end{bmatrix} + \begin{bmatrix}\int N_0 q_{30} d\Gamma_{30}\\ \int N_1 q_{30} d\Gamma_{30}\\ \int N_2 q_{30} d\Gamma_{30}\\ \int N_3 q_{30} d\Gamma_{30}\end{bmatrix}\end{array}\right).$$

The elemental matrix for a quadrilateral element can now be written as

$$\frac{k}{6}\begin{bmatrix} 2b+2b^{-1} & -2b+b^{-1} & -b-b^{-1} & b-2b^{-1}\\ -2b+b^{-1} & 2b+2b^{-1} & b-2b^{-1} & -b-b^{-1}\\ -b-b^{-1} & b-2b^{-1} & 2b+2b^{-1} & -2b+b^{-1}\\ b-2b^{-1} & -b-b^{-1} & -2b+b^{-1} & 2b+2b^{-1}\end{bmatrix}\begin{bmatrix}T_0\\T_1\\T_2\\T_3\end{bmatrix}$$

$$= -\left(\begin{array}{c}\begin{bmatrix}\int N_0 q_{01} d\Gamma_{01}\\ \int N_1 q_{01} d\Gamma_{01}\\ \int N_2 q_{01} d\Gamma_{01}\\ \int N_3 q_{01} d\Gamma_{01}\end{bmatrix} + \begin{bmatrix}\int N_0 q_{12} d\Gamma_{12}\\ \int N_1 q_{12} d\Gamma_{12}\\ \int N_2 q_{12} d\Gamma_{12}\\ \int N_3 q_{12} d\Gamma_{12}\end{bmatrix} + \\ \begin{bmatrix}\int N_0 q_{23} d\Gamma_{23}\\ \int N_1 q_{23} d\Gamma_{23}\\ \int N_2 q_{23} d\Gamma_{23}\\ \int N_3 q_{23} d\Gamma_{23}\end{bmatrix} + \begin{bmatrix}\int N_0 q_{30} d\Gamma_{30}\\ \int N_1 q_{30} d\Gamma_{30}\\ \int N_2 q_{30} d\Gamma_{30}\\ \int N_3 q_{30} d\Gamma_{30}\end{bmatrix}\end{array}\right).$$

Substituting for values of shape functions N along edges $\Gamma_{01}, \Gamma_{12}, \Gamma_{23}, \Gamma_{23}$, respectively, the element matrix becomes

$$\frac{k}{6}\begin{bmatrix} 2b+2b^{-1} & -2b+b^{-1} & -b-b^{-1} & b-2b^{-1}\\ -2b+b^{-1} & 2b+2b^{-1} & b-2b^{-1} & -b-b^{-1}\\ -b-b^{-1} & b-2b^{-1} & 2b+2b^{-1} & -2b+b^{-1}\\ b-2b^{-1} & -b-b^{-1} & -2b+b^{-1} & 2b+2b^{-1}\end{bmatrix}\begin{bmatrix}T_0\\T_1\\T_2\\T_3\end{bmatrix}$$

$$= -\left(\begin{bmatrix} w\int(1-\xi)q_{01}\mathrm{d}\xi\\ w\int \xi q_{01}\mathrm{d}\xi\\ 0\\ 0\end{bmatrix} + \begin{bmatrix}0\\ t\int(1-\eta)q_{12}\mathrm{d}\eta\\ t\int \eta q_{12}\mathrm{d}\eta\\ 0\end{bmatrix}\right.$$

$$\left. + \begin{bmatrix}0\\0\\ w\int \xi q_{23}\mathrm{d}\xi\\ w\int(1-\xi)q_{23}\mathrm{d}\xi\end{bmatrix} + \begin{bmatrix} t\int(1-\eta)q_{30}\mathrm{d}\eta\\ 0\\ 0\\ t\int \eta q_{30}\mathrm{d}\eta\end{bmatrix}\right).$$

Assuming that there is only one element in the domain. Temperatures $T_0$, $T_3$ are known, convection is present along the edges $\Gamma_{12}$, $\Gamma_{23}$, and $q_{01}$ is zero because of symmetry. Accounting for these in the above matrix

$$\begin{bmatrix} 2b+2b^{-1} & -2b+b^{-1} & -b-b^{-1} & b-2b^{-1} \\ -2b+b^{-1} & 2b+2b^{-1} & b-2b^{-1} & -b-b^{-1} \\ -b-b^{-1} & b-2b^{-1} & 2b+2b^{-1} & -2b+b^{-1} \\ b-2b^{-1} & -b-b^{-1} & -2b+b^{-1} & 2b+2b^{-1} \end{bmatrix} \begin{bmatrix} T_0 \\ T_1 \\ T_2 \\ T_3 \end{bmatrix}$$
$$= -\frac{6}{k}\left( \begin{bmatrix} 0 \\ t\int_1^2 (1-\eta)h(T-T_\infty)\mathrm{d}\eta \\ t\int_1^2 \eta h(T-T_\infty)\mathrm{d}\eta \\ 0 \end{bmatrix} + \begin{bmatrix} 0 \\ 0 \\ w\int_3^2 \xi h(T-T_\infty)\mathrm{d}\xi \\ w\int_3^2 (1-\xi)h(T-T_\infty)\mathrm{d}\xi \end{bmatrix} \right).$$

We will evaluate the first term on the right side. Along edge $\Gamma_{12}$, $N_0 = 0$, $N_1 = 1-\eta$, $N_2 = \eta$, and $N_3 = 0$. Therefore, $T = (1-\eta)T_1 + \eta T_2$.

$$\begin{aligned} t\int_1^2 (1-\eta)h_{12}(T-T_\infty)\mathrm{d}\eta &= t\int_1^2 (1-\eta)h_{12}((1-\eta)T_1 + \eta T_2 - T_\infty)\mathrm{d}\eta \\ &= h_{12}tT_1 \int_1^2 (1-\eta)^2\mathrm{d}\eta \\ &\quad + h_{12}tT_2 \int_1^2 \eta(1-\eta)\mathrm{d}\eta - h_{12}tT_\infty \int_1^2 (1-\eta)\mathrm{d}\eta \\ &= \frac{h_{12}tT_1}{3} + \frac{h_{12}tT_2}{6} - \frac{h_{12}tT_\infty}{2} \end{aligned}$$

$$\begin{aligned} t\int_1^2 \eta h_{12}(T-T_\infty)\mathrm{d}\eta &= t\int_1^2 \eta h_{12}((1-\eta)T_1 + \eta T_2 - T_\infty)\mathrm{d}\eta \\ &= h_{12}tT_1 \int_1^2 \eta(1-\eta)\mathrm{d}\eta + h_{12}tT_2 \int_1^2 \eta^2\mathrm{d}\eta - h_{12}tT_\infty \int_1^2 \eta\mathrm{d}\eta \\ &= \frac{h_{12}tT_1}{6} + \frac{h_{12}tT_2}{3} - \frac{h_{12}tT_\infty}{2}. \end{aligned}$$

A similar methodology can be used for other terms. The right-hand side now becomes

$$-\frac{6}{k}\left( \begin{bmatrix} 0 \\ \frac{th_{12}T_2}{6} + \frac{th_{12}T_1}{3} - \frac{th_{12}T_\infty}{2} \\ \frac{th_{12}T_2}{3} + \frac{th_{12}T_1}{6} - \frac{th_{12}T_\infty}{2} \\ 0 \end{bmatrix} + \begin{bmatrix} 0 \\ 0 \\ \frac{wh_{23}T_2}{3} + \frac{wh_{23}T_3}{6} - \frac{wh_{23}T_\infty}{2} \\ \frac{wh_{23}T_2}{6} + \frac{wh_{23}T_3}{3} - \frac{wh_{23}T_\infty}{2} \end{bmatrix} \right).$$

Taking the temperature coefficients to the left-hand side and rearranging, we get

$$\begin{bmatrix} 2b+2b^{-1} & -2b+2b^{-1} & -b-b^{-1} & b-2b^{-1} \\ -2b+b^{-1} & 2b+2b^{-1}+\frac{th_{12}}{k} & b-2b^{-1}+\frac{2th_{12}}{k} & -b-b^{-1} \\ -b-b^{-1} & b-2b^{-1}+\frac{th_{12}}{k} & 2b+2b^{-1}+\frac{2th_{12}}{k}+\frac{2wh_{23}}{k} & -2b+b^{-1}+\frac{wh_{23}}{k} \\ b-2b^{-1} & -b-b^{-1} & -2b+b^{-1}+\frac{wh_{23}}{k} & 2b+2b^{-1}+\frac{2wh_{23}}{k} \end{bmatrix}$$

$$\begin{bmatrix} T_0 \\ T_1 \\ T_2 \\ T_3 \end{bmatrix} = \frac{3T_\infty}{k}\begin{bmatrix} 0 \\ th_{12} \\ wh_{23}+th_{12} \\ wh_{23} \end{bmatrix}.$$

With temperatures, $T_0$ and $T_3$ being known and h being uniform, the matrix can be rewritten as

$$\begin{bmatrix} 2b+2b^{-1} & 0 & 0 & 0 \\ -2b+b^{-1} & 2b+2b^{-1}+\frac{th}{k} & b-2b^{-1}+\frac{2th}{k} & -b-b^{-1} \\ -b-b^{-1} & b-2b^{-1}+\frac{th}{k} & 2b+2b^{-1}+\frac{2th}{k}+\frac{2wh}{k} & -2b+b^{-1}+\frac{wh}{k} \\ 0 & 0 & 0 & 2b+2b^{-1}+\frac{2wh}{k} \end{bmatrix}$$

$$\begin{bmatrix} T_0 \\ T_1 \\ T_2 \\ T_3 \end{bmatrix} = T_{wall}\begin{bmatrix} 2b+2b^{-1} \\ 0 \\ 0 \\ 2b+2b^{-1}+\frac{2wh}{k} \end{bmatrix} + \frac{3T_\infty}{k}\begin{bmatrix} 0 \\ th \\ wh+th \\ 0 \end{bmatrix}$$

$$\begin{bmatrix} 20.2 & 0 & 0 & 0 \\ 19.9 & 40.100208 & -39.949583 & -20.05 \\ -20.05 & -39.949792 & 40.10875 & 19.904167 \\ 0 & 0 & 0 & 20.208264 \end{bmatrix}\begin{bmatrix} T_0 \\ T_1 \\ T_2 \\ T_3 \end{bmatrix} = \begin{bmatrix} 7537.63 \\ 0.189469 \\ 3.978844 \\ 7540.7138 \end{bmatrix}.$$

The solution is $\begin{bmatrix} T_0 \\ T_1 \\ T_2 \\ T_3 \end{bmatrix} = \begin{bmatrix} 373.15\text{K} \\ 369.99\text{K} \\ 369.98\text{K} \\ 373.15\text{K} \end{bmatrix}$. The effectiveness of the fin can be calculated by computing the total heat convected from the fin

$$Q_{\text{with fin}} = 2\left(\int q_{12}\mathrm{d}\Gamma_{12} + \int q_{23}\mathrm{d}\Gamma_{23}\right) = 2\left(\int h(T-T_\infty)\mathrm{d}\Gamma_{12} + \int h(T-T_\infty)\mathrm{d}\Gamma_{23}\right)$$
$$= 2\left(ht\left(\frac{T_1+T_2}{2} - T_\infty\right) + hw\left(\frac{T_2+T_3}{2} - T_\infty\right)\right) = 71.757\,\text{W/m}$$

$= 71.757$ W/m

$$Q_{\text{without fin}} = h2t(373.15 - 303.15) = 3.5\,\text{W/m}.$$

Therefore, the effectiveness of the fin,$\varepsilon = \frac{71.757}{3.5} = 20.502$.

## 7.8 Exercises

7.8.1 Formulate the Example 7.1 problem with cubic elements using the methodology followed in Examples 7.1 and 7.2.

7.8.2 A composite slab made of 10 mm aluminum alloy AA2014 and 10 mm Stainless SS304L is exposed to thermal environments as shown in Fig. 7.11. Determine the steady state temperature distribution within the slab. Use standard properties of AA2014 and SS304L.

7.8.3 A 2 inch diameter stainless steel pipe of 1 mm thickness carrying hot water at 350 K is insulated with a porous insulator of thickness 10 mm having a thermal conductivity of 0.1 W/mK. Determine the temperature distribution within this composite system if the thermal environments are as shown in Fig. 7.12 (*Hint: Use the one-dimensional steady state axisymmetric conduction equation without heat generation to formulate the problem*).

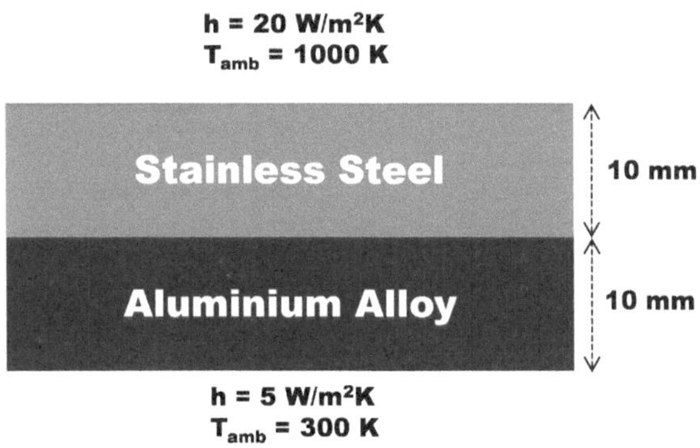


**Fig. 7.11** Composite slab

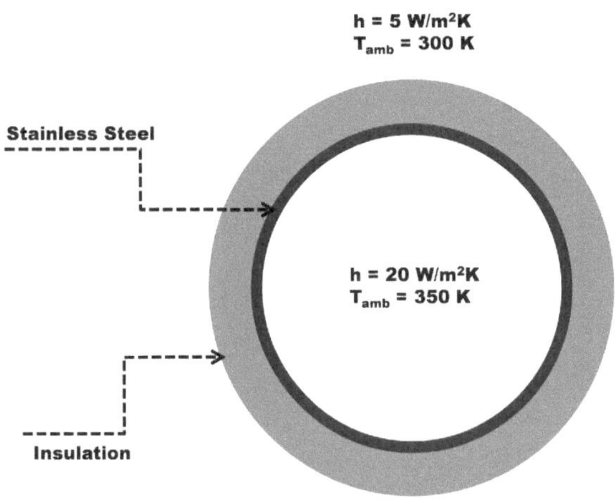


**Fig. 7.12** Pipe with insulation

7.8.4 Do a convergence study for the fin problem given in Example 7.4 by increasing the number of elements in the *X* direction. A 2-element assembly along with node numbering is shown in Fig. 7.13.

Find the effectiveness of the fin for 2-element and 4-element cases.

7.8.5 Do the fin problem given in Example 7.4 by using a quadratic 8-node element as shown in Fig. 7.14.

Find the effectiveness of the fin and compare it with the previous problem.

7.8.6 Do the fin problem given in Example 7.4 by using 2 linear triangular elements as shown in Fig. 7.15.

Find the effectiveness of the fin and compare it with the previous problem.

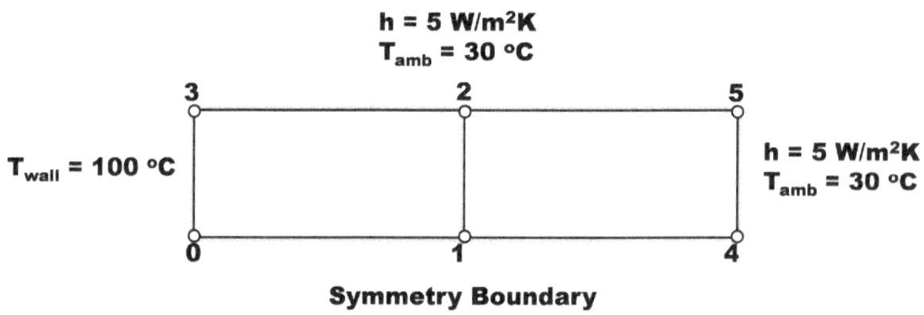


**Fig. 7.13** Fin with 2 4-node quadrilateral elements

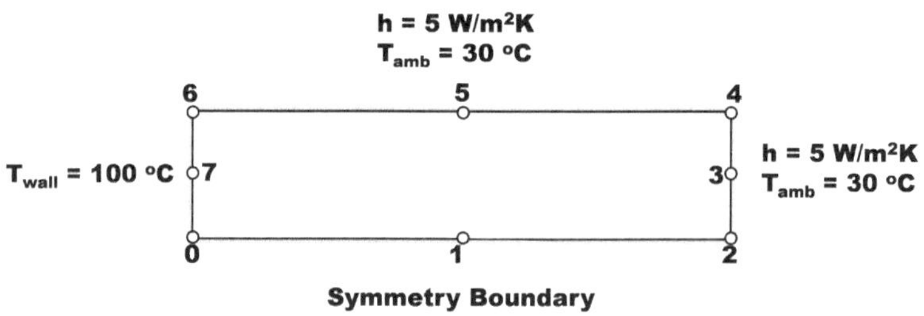


**Fig. 7.14** Fin with 1 8-node quadrilateral element

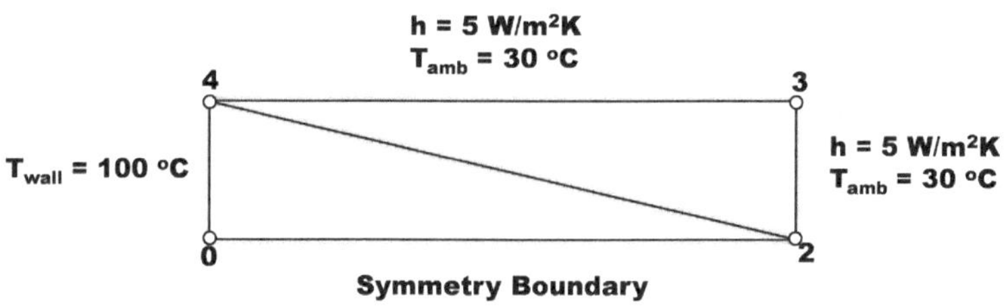


**Fig. 7.15** Fin with 2 3-node triangular elements

7.8.7 Formulate and solve the two-dimensional steady state problem using a single 4-noded quadrilateral element as shown in Fig. 7.16.

7.8.8 Formulate and solve problem 7 using an assembly of 4 linear quadrilateral elements as shown in Fig. 7.17.

7.8.9 Formulate and solve problem 7 using an assembly of 4 linear triangular elements as shown in Fig. 7.18.

7.8.10 A unit cube made of Al Alloy AA2014 is at $T = 500$ K at time $t = 0$ s. Determine the temperature distribution of the cube at $t = 1000$ s if it is having free convection ($h = 5$ W/m$^2$ K) on all sides to an ambient of 300 K. Do the same problem if the cube is made up of stainless steel SS304L. Formulate the three-dimensional transient conduction problem using an 8-node brick element.

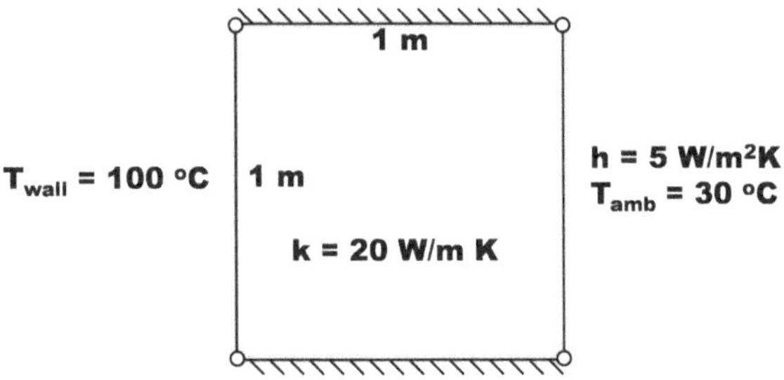


**Fig. 7.16** Two-dimensional problem with single quadrilateral element

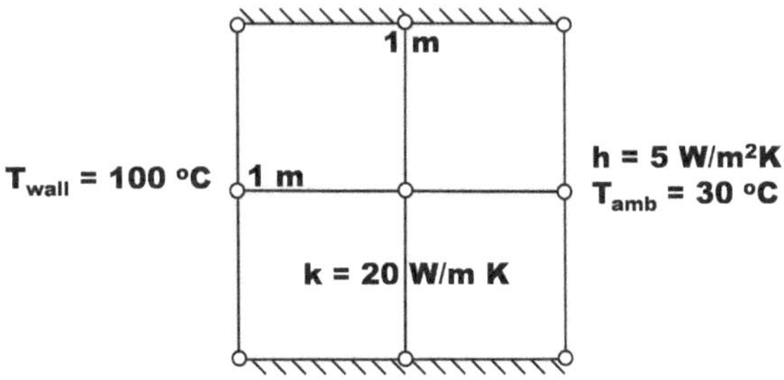


**Fig. 7.17** Two-dimensional problem with four quadrilateral elements

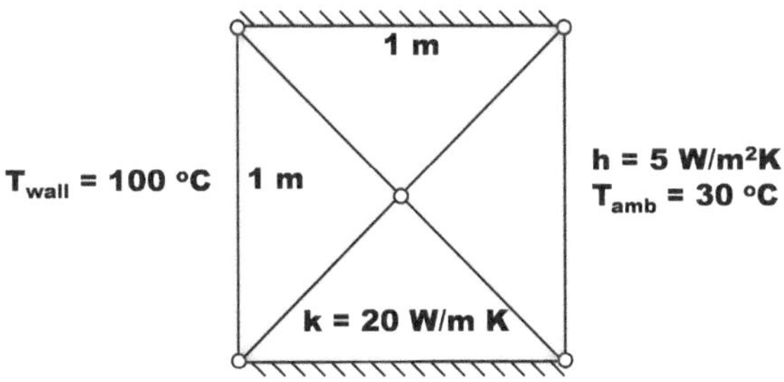


**Fig. 7.18** Two-dimensional problem with four triangular elements

## References

Anderson JD (1995) Computational fluid dynamics: the basics with applications. McGraw Hill
Incropera FP, Dewitt DP (2007) Fundamentals of heat and mass transfer. Wiley

# Chapter 8
# Dynamic Analysis

**K. Jayakumar**

*Everything in life is vibration.*
*—Albert Einstein*

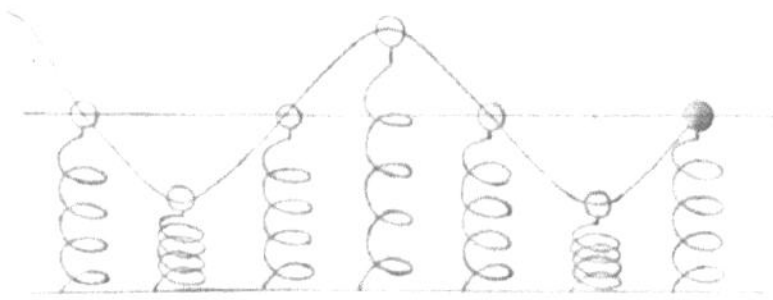

## 8.1 Introduction

The term dynamic in general means any phenomenon that evolves over time. Mathematically this dependency is defined as a function of time. Drawing parallels to the above definition, structural dynamics is the study of structures that undergo time varying displacements, velocities and acceleration at every point of their continuous extent in space. This causes motions on the medium which we perceive as vibrations of the structure.

For conservative systems, vibration can be defined as the interplay between potential energy and kinetic energy. This definition is very general as it is true from the atomic to cosmic scale. Kinetic energy is a physical quantity associated with mass and depending on associated field of study, the potential energy assumes special forms. In atoms, it is the electrostatic and/or magneto static potential, between planets it is gravitational potential and in structural dynamics it is the elastic potential that interacts or interplays with the kinetic energy to cause vibrations. This fact is utilized to infer that a particle possesses mass by observing its vibrations or oscillations.

K. Jayakumar (✉)
FSDD/SDAG/STR, Veli, VSSC, Thiruvananthapuram, Kerala 695022, India
e-mail: k_jayakumar@vssc.gov.in
 
S. U. Nair and S. Somanath (eds.), *Introduction to Finite Element Analysis*,
https://doi.org/10.1007/978-981-19-7989-7_8

Studies on interaction between motion and its causes (forces) are carried out in dynamics as the phenomena evolve over time. It forms one of the branches of classical mechanics in physics with the other being statics. This is an important subject of study in engineering fields such as aerospace, civil, mechanical and marine engineering. This is due to its heavy applications in designing various systems such as bridges, dams, machineries, ships, cranes, etc. Newton's law forms the basis of classical mechanics.

A fair understanding of response due to the external environment is necessary to realize a satisfactory design of systems working in these settings. This becomes all the more important for sensitive designs with narrow margins, such as aerospace and medical applications. The oscillatory phenomena observed in nature, for instance, electromagnetism, propagation of sound in air, elastic response in solids, etc., can, fortunately, be studied using standard wave equations. It quantitatively describes the oscillatory phenomenon in spatial and temporal domains.

## 8.2 Primer to Dynamics

Newton's laws were formulated for single particles based on the notion of absolute time and relative space. Relative space means that there is no fixed frame of reference for studying a physical phenomenon. But, to study motion of particles, a frame of reference is required when the vector method (Newton's approach) is used. This method becomes cumbersome when many particles are involved and the dynamic equilibrium of every particle and their interactions are accounted for. From an engineering perspective, drawing the free-body diagram of idealized systems becomes a herculean task. The following sub-sections discuss notions of analytical dynamics, which is a prerequisite for understanding Lagrange's equations and Hamilton's principle.

## 8.3 Work and Energy

In Fig. 8.1, incremental work done $\overline{dW}$ by the force $\boldsymbol{F}$ for moving a particle from $\boldsymbol{r}$ to $\boldsymbol{r} + d\boldsymbol{r}$ is

$$\overline{dW} = \boldsymbol{F} \cdot d\boldsymbol{r}. \tag{8.1}$$

Note that Eq. (8.1) corresponds to a scalar (dot) product.

But $\boldsymbol{F} = m\ddot{\boldsymbol{r}}$, therefore,

$$\overline{dW} = m\ddot{\boldsymbol{r}} \cdot d\boldsymbol{r} = m\dot{\boldsymbol{r}} \cdot d\dot{\boldsymbol{r}} = d\left(\frac{1}{2}m\dot{\boldsymbol{r}} \cdot \dot{\boldsymbol{r}}\right) = dT,$$

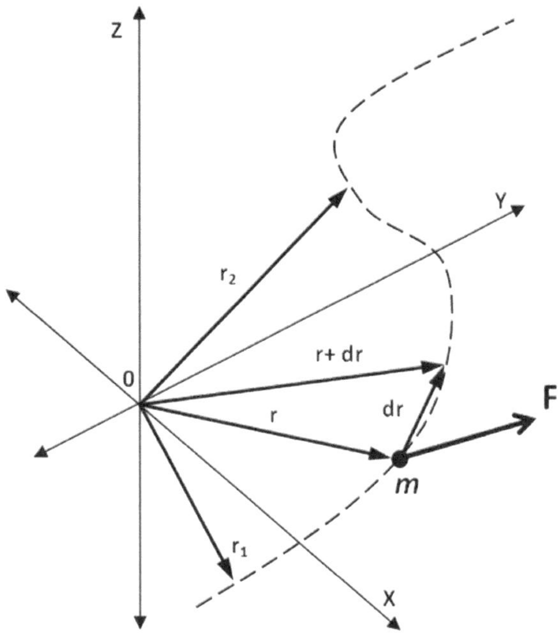


**Fig. 8.1** Motion of a particle in space with mass 'm' acted upon by force F

where T is kinetic energy of the particle. Equation (8.1) states that the increment in work associated with displacement of mass from $\boldsymbol{r}$ to $\boldsymbol{r} + \mathrm{d}\boldsymbol{r}$ is change in kinetic energy, which is a scalar quantity.

Formulating physical problems using scalar fields greatly simplifies the mathematical expressions while solving for field variables.

## 8.4 Conservative Force Field

If the force field is dependent upon position alone, i.e. $\boldsymbol{F} = \boldsymbol{F}(\boldsymbol{r})$, then it can be expressed as negative gradient of some scalar function $U$, Eq. (8.2), which is the potential energy

$$\boldsymbol{F} = -\nabla U. \tag{8.2}$$

The force field described above is known as a conservative force field. Further, the following properties given by Stoke's theorem, Eqs. (8.3) and (8.4), are applicable to a conservative force field

$$\oint F \cdot dr = 0 \tag{8.3}$$

$$-\nabla \times F = 0. \quad (8.4)$$

## 8.5 Single Degree of Freedom System

These systems are classified as the following:

- First-order system
- Second-order system

### 8.5.1 First-Order System

A schematic representation of first-order system is shown in Fig. 8.2, where the applied force $F(t)$ is in equilibrium with forces induced in spring ($k$) and damper ($c$). This is briefly discussed for the sake of completeness.

The equation of motion of this first-order system is given as

$$c\dot{x} + kx = F(t). \quad (8.5)$$

Equation (8.5) contains a first derivative. and hence contains only one arbitrary constant in the complementary function of solution. Therefore, the name first-order system. The constant coefficients c and k are known as system parameters. From elementary differential equation theory, the solution of the homogenous form of Eq. (8.5) is

$$x(t) = \begin{cases} x_0 e^{\frac{-t}{\tau}} & t > 0 \\ & , \\ 0 & t < 0 \end{cases} \quad (8.6)$$

where $A = x_0$ is initial specified displacement and $\tau = \frac{c}{k}$ is known as time constant and gives a measure of the speed of return of the system to equilibrium. Figure 8.3

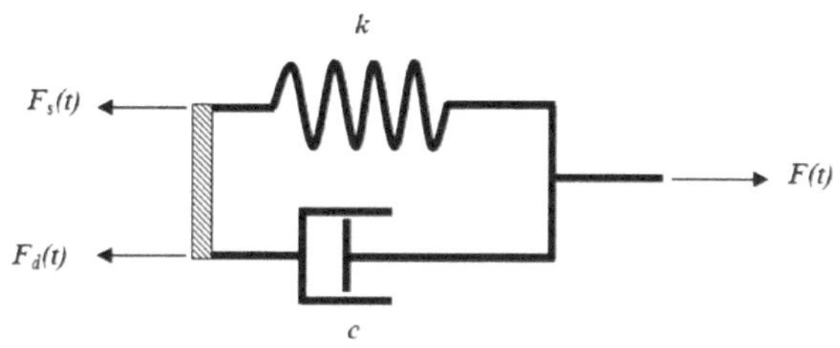


**Fig. 8.2** A first-order system with spring and damper (no mass)

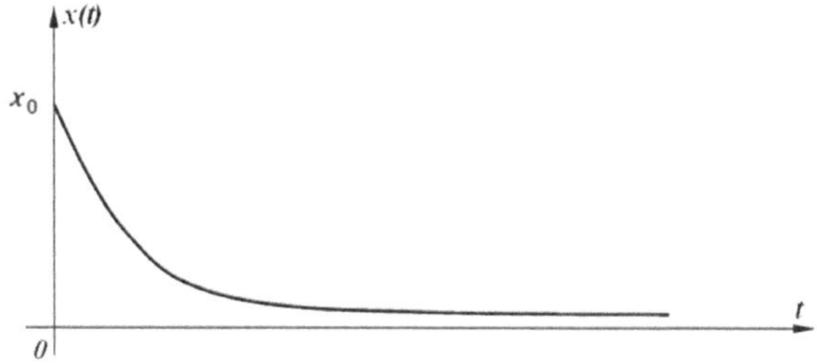


**Fig. 8.3** Solution of Eq. (8.5) when $F(t) = 0$

shows the plot of homogeneous part of the solution (complimentary part) to Eq. (8.5), i.e. solution of the equation when $F(t) = 0$.

For steady-state response or the particular solution of Eq. (8.5), it is common to consider a harmonic excitation of frequency $\omega$. Recall that total solution of an ordinary differential equation is the sum of homogenous and particular solution. Steady state response has same frequency as that of excitation frequency, because the system is required to vibrate with time varying force. Therefore, the forcing function can be conveniently expressed in complex form as given below

$$F(t) = kAe^{i\omega t}. \tag{8.7}$$

Solution to Eq. (8.5) has the same form of forcing function as

$$x(t) = X(i\omega)e^{i\omega t}. \tag{8.8}$$

Substituting Eqs. (8.7) and (8.8) in (8.5)

$$X(i\omega) = \frac{A}{1 + i\omega\tau}. \tag{8.9}$$

A non-dimensional ratio given by Eq. (8.10) $H(\mathrm{i}\omega)$ is introduced, which is defined as

$$H(i\omega) = \frac{X(i\omega)}{A}. \tag{8.10}$$

Here, $H(\mathrm{i}\omega)$ is known as frequency response function of the system, $X(i\omega)$ and $A$ are the amplitude of response and forcing function, respectively.

Equation (8.11) gives time response definition of the system as

$$x(t) = AH(i\omega)e^{i\omega t}. \tag{8.11}$$

But, $H(i\omega) = |H(i\omega)|e^{-i\phi}$, where $|H(\mathrm{i}\omega)|$ is the magnitude and $\Phi$ is the phase angle of $H(\mathrm{i}\omega)$. Hence, Eq. (8.11) can be rewritten as (8.12)

$$x(t) = A|H(i\omega)|e^{i(\omega t-\phi)}. \tag{8.12}$$

In forced vibration studies, more insight into system behavior can be gained by examining the system response to varying driving frequencies than studying the response as a function of time.

From complex algebra, the modulus of frequency response function is given by Eq. (8.13)

$$|H(i\omega)| = \left[Re^2 H(i\omega) + Im^2 H(i\omega)\right]^{1/2} = \frac{1}{\left[1 + (\omega\tau)^2\right]^{1/2}}. \quad (8.13)$$

Figure 8.4a shows the plot of $|H(i\omega)|$ for varying $\omega\tau$. For smaller driving frequencies, $|H(i\omega)|$ approaches unity and it tends to zero for higher frequencies. This system permits low-frequency harmonics undistorted and attenuates high-frequency harmonics. Hence, a first-order system is also known as a low-pass filter, i.e. it allows only forces associated with lower frequencies.

The phase angle of $H(i\omega)$ is given as

$$\phi = \tan^{-1}\left[\frac{-ImH(i\omega)}{ReH(i\omega)}\right] = \tan^{-1}(\omega\tau). \quad (8.14)$$

Figure 8.4b shows the variation of $\phi$ with respect to $\omega\tau$. At higher frequencies, the velocity is out of phase with the displacement by $\pi/2$

**Fig. 8.4** Frequency-response plot of a first-order system

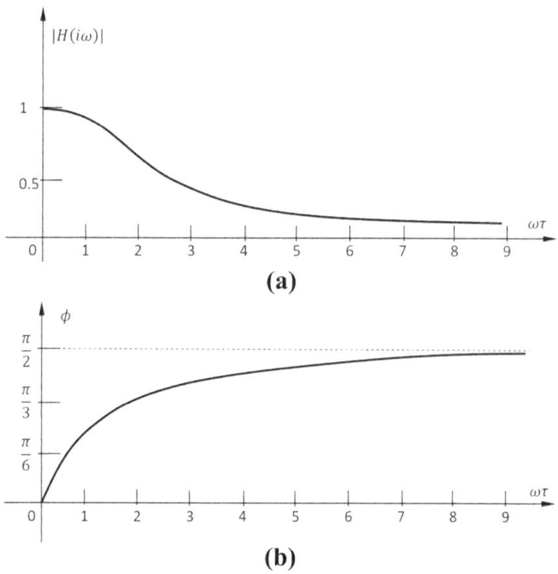

### 8.5.2 Second-Order System

The studies on first-order system can be naturally extended to second-order systems. Most of the applied solid mechanics problems are of second-order type. The differential equation given by (8.15), of motion for a second-order system, is

$$m\ddot{x} + c\dot{x} + kx = F(t), \tag{8.15}$$

where $m$ is mass of the system, other coefficients have been mentioned earlier. Single-degree-of-freedom system represented by Eq. (8.15) is shown in Fig. 8.5.

Homogeneous solution for Eq. (8.15) gives damped free vibration solution of the system. For convenience, Eq. (8.15) can be expressed in non-dimensional form by factoring out the mass

$$\ddot{x} + 2\xi\omega_n\dot{x} + \omega_n^2 x = 0, \tag{8.16}$$

where $\omega_n = \sqrt{\frac{k}{m}}$ is the natural frequency and

$$\xi = \frac{c}{2m\omega_n} \tag{8.17}$$

is the viscous damping factor or damping ratio. Following standard methods, and focusing on oscillatory behavior, the homogeneous solution can be expressed as

$$x(t) = \left[A_1 exp\left(i\sqrt{1-\xi^2}\omega_n t\right)\right] + \left[A_2 exp\left(-i\sqrt{1-\xi^2}\omega_n t\right)\right]e^{-\xi\omega_n t}. \tag{8.18}$$

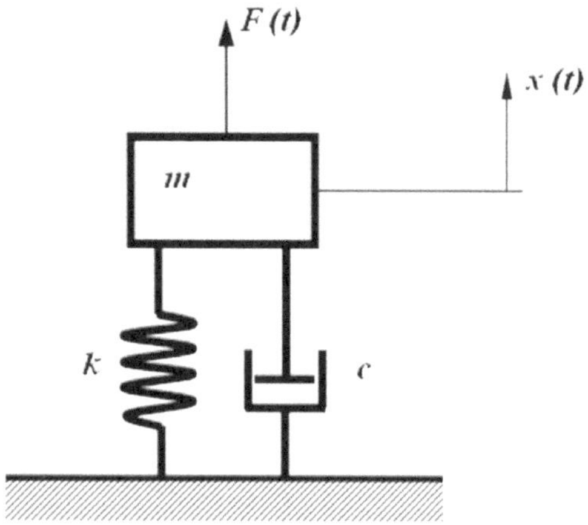


**Fig. 8.5** A single-degree-of-freedom system of second order, with mass-spring-damper

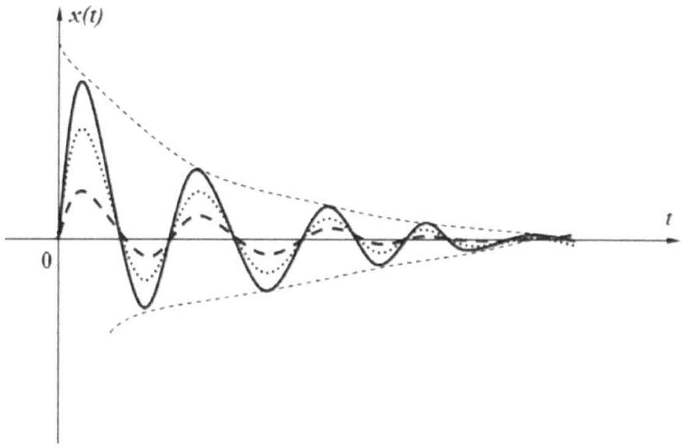


**Fig. 8.6** Damped free vibration response curves for different damping factors

Figure 8.6 shows the response curves obtained from Eq. (8.18) for various damping factors. As time, $t \rightarrow \infty$, $x(t) \rightarrow 0$, when the system is imparted initial displacement and/or initial velocity. In real systems, this is observed phenomenon as damping is always present in some form. On the specification of initial displacement and/or initial velocity, the constants $A_1$ and $A_2$ can be determined.

The steady state response of Eq. (8.15) is sought by assuming a harmonic forcing function with amplitude $\omega_n^2 A$

$$F(t) = \omega_n^2 A e^{i\omega t}. \tag{8.19}$$

Inserting Eq. (8.19) into (8.15) and knowing that response frequency is same as forcing frequency, the expression for steady state response is assumed as

$$x(t) = X(i\omega)e^{i\omega t}. \tag{8.20}$$

Substituting (8.19) and (8.20) into (8.15) yields

$$Z(i\omega)X(i\omega)e^{i\omega t} = \omega_n^2 A e^{i\omega t}. \tag{8.21}$$

The impedance function $Z(i\omega)$ in Eq. (8.21) is defined as $Z(i\omega) = \omega_n^2 - \omega^2 + i2\xi\omega\omega_n$.

Solving $X(i\omega)$, of Eq. (8.21) gives

$$X(i\omega) = \frac{\omega_n^2 A}{\omega_n^2 - \omega^2 + i2\omega\omega_n} = \frac{A}{1 - (\omega/\omega_n)^2 + i2\xi(\omega/\omega_n)}. \tag{8.22}$$

Therefore, the frequency response function from Eq. (8.22) is

$$H(i\omega) = \frac{X(i\omega)}{A} = \frac{1}{1 - (\omega/\omega_n)^2 + i2\xi(\omega/\omega_n)}.$$

Similar to first-order systems, the harmonic response in time domain is expressed in Eq. (8.23) as

$$x(t) = AH(i\omega)e^{i\omega t} = A|H(i\omega)|e^{i(\omega t-\phi)}, \tag{8.23}$$

where, $|H(i\omega)| = \frac{1}{\left[\left\{1-(\omega/\omega_n)^2\right\}^2+(2\xi\omega/\omega_n)^2\right]^{1/2}}$ is the magnitude of frequency response function and known as magnification factor and $\phi$ is the phase angle given by Eq. (8.24)

$$\phi = \tan^{-1}\left[\frac{2\xi\omega/\omega_n}{1-(\omega/\omega_n)^2}\right]. \tag{8.24}$$

The plot of frequency response and phase plot diagram of a single-degree-of-freedom system is given in Fig. 8.7.

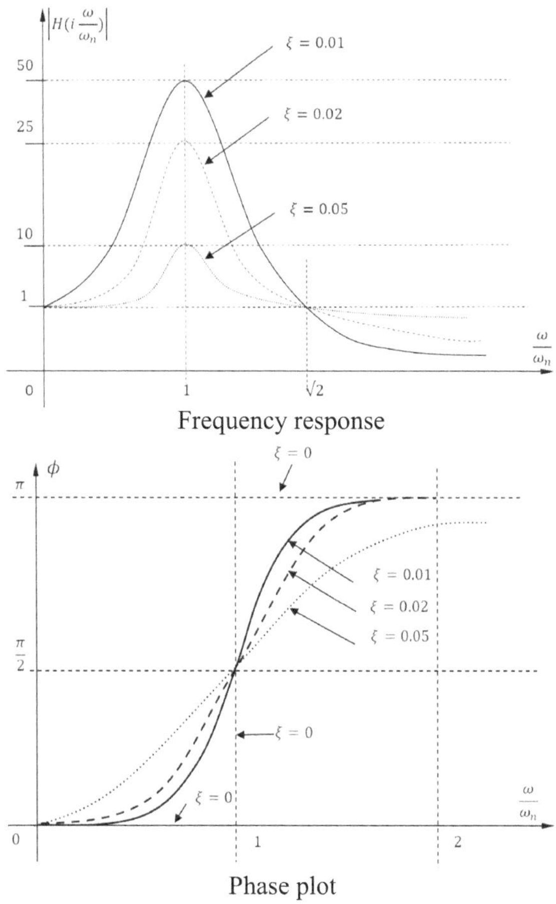


**Fig. 8.7** Frequency response and phase plot diagram of single-degree-of-freedom system. Frequency response, Phase plot

To get complete solution of the system, homogeneous solution (transients) and particular integral (steady state) are summed. For continuous systems, such as beams, plates, shells, and solids, the response functions are well documented in standard literature mostly for simple geometries and constraints. Naturally, response expressions of continuous systems are obtained from solving partial differential equations, i.e. the models, representing these simple systems. Depending on the situation, one seeks required type of solution from these models, viz. impulse/shock, transient and/or steady state response. Closed form or analytical solutions give physical insights into the system behavior where solutions clearly enunciate interdependence of system parameters. Furthermore, they serve as benchmarks to validate experimental and numerical models.

## 8.6 Equation of Motion for *N* Particles

Newton's second law for a single particle with mass m and acceleration ü is

$$f = m\ddot{u} \tag{8.25}$$

where $f$ is the force experienced by mass, $m$.

In most real-life cases, mass of the particle does not vary with time. Recall that Eq. (8.25) relates force experienced by mass to the rate of change in momentum. This can be expressed as

$$f = m\frac{d\dot{u}}{dt} = \frac{d(m\dot{u})}{dt}. \tag{8.26}$$

Generalizing Newton's laws of motion for $N$ particles

$$f_j - \frac{d\left(m_j\dot{u}_j\right)}{dt} = 0 \quad j = 1, \ldots, N \tag{8.27}$$

$$L_j - \frac{d\left(J_j\right)}{dt} = 0 \quad j = 1, \ldots, N, \tag{8.28}$$

where

$f_j$ is the resultant force acting on particles with mass $m_j$,

$L_j$ is the resultant moment causing angular momentum $L_j$ Eqs. (8.27) and (8.28) are by virtue of conservation of momentum.

If the motion is not independent, then these equations have to be modified with Eq. (8.29), which is a statement of constraints

$$g(\dot{u}_1, \dot{u}_2, \ldots, \dot{u}_N) = 0. \tag{8.29}$$

This results in a coupled set of equations that needs to be solved simultaneously (Simultaneous equations are two or more algebraic equations that share variables, e.g. $x$ and $y$. They are called simultaneous equations because the equations are solved together). If the number of particles is large, then number of equations will be large, making the solution process difficult and time-consuming. The reason for ending up with a coupled set of equations is the choice of reference frame, which is usually arbitrary, chosen from compelling suggestive cues of the problem. To formulate the equations, a linearly independent set of vectors or functions are sufficient, without any bearing to physical 'features' of the problem or reference frames. This is the essence of analytical dynamics.

## 8.7 Lagrange's Equations

The content of this section is advanced for undergraduate students. This forms the core of analytical dynamics. While writing this section, the author has kept subject depth to level required by undergraduate students without compromising on essential concepts. The students are requested to refer to standard texts given in references for further details on this topic. For graduate and postgraduate students, this section will be a primer for advanced applications if they have studied variational calculus.

**Principle of virtual work**

*If a system, which is in equilibrium under an action of a set of forces, is subjected to a virtual displacement then the total work done by the forces will be zero.* It implies that for bodies in equilibrium, undergoing a small arbitrary displacements, total work done by the system is zero.

The virtual displacement is a physically possible one, i.e. any displacement which is compatible with system constraints. Variation in true displacement is caused by variation of actual loading, whereas virtual displacement is one that can take place without violation of constraints. The advantage of this method is that virtual work (energy) contributions are scalar quantities, which can be added algebraically.

Virtual displacements overcome vector addition of forces, but virtual work itself is calculated from scalar product of two vectors (force and displacements). Hamilton's method overcomes this disadvantage.

Hamilton's principle states that first variation of the Lagrangian integrated between any arbitrary times of a mechanical system is zero. In a conservative system (where there is no dissipation of energy and the law of conservation of energy is valid), of all possible paths between two points along which a dynamical system may move from one point to another within a given time interval, actual path followed by the system is the one which minimizes line integral of Lagrangian. This is expressed mathematically as

$$\delta \int_{t_1}^{t_2} \mathbb{L}\mathrm{dt} = 0, \tag{8.30}$$

where in Eq. (8.30), the Lagrangian $\mathbb{L} = \mathrm{T} - \mathrm{U}$; T is kinetic energy and U is the potential energy.

First variation of the work is $\delta \mathrm{W} = \delta \mathrm{W}_{\mathrm{con}} + \delta \mathrm{W}_{\mathrm{ncon}}$. The subscripts mean conservative and non-conservative work components.

But $\delta \mathrm{W}_{\mathrm{con}} = -\delta \mathrm{U}$, i.e. variation of work is a negative variation of potential energy for a conservative system. In other words, total energy of the system remains constant. As an example, a single-degree-of-freedom shown in Fig. 8.5 is a non-conservative system if the damping $c$ is present. This results in dissipation or addition of energy of oscillating system by damper. After a certain amount of time, the system comes to rest if $c > 0$ or attains a stable state and will oscillate with increasing amplitudes for subsequent time intervals when $c < 0$. This signifies addition of energy to the system as in case of divergence and flutter observed in aircraft wings caused by interaction between aerodynamic and elastic forces. These instabilities eventually cause structural failure of wings. Absence of damper, $c = 0$ results in a conservative system, wherein the system will continue to vibrate for an infinite amount of time if left undisturbed. If non-conservative forces are included, Eq. (8.30) is modified as

$$\int_{t_1}^{t_2} \{\delta(\mathrm{T} - \mathrm{U}) + \delta W_{ncon}\} dt = 0. \tag{8.31}$$

Variational operator in Eq. (8.31) cannot be applied to the whole of integral as there is no potential function associated with non-conservative forces. Hence, to apply Eq. (8.31), the variation of non-conservative forces has to be determined using the virtual work principle.

The application of Hamilton's principle leads to an equation of motion for any system. Hamilton's principle can be applied to discrete multi-degrees-of- freedom as well as continuous systems. Application of Hamilton's principle to complex system simplifies the mathematics, as the involved quantities are scalar (energies), which need simple algebraic manipulations. Vector quantities are required in accounting work done by non-conservative forces.

For complicated problems involving time and boundaries, the equations of motion and boundary conditions are easily identified. In Lagrange's method, equations of motion are determinable but for a solution, physics of the problem needs to be understood in-depth to specify boundary conditions and initial conditions. This is possible only for limited cases of problems. In Hamilton's method, not only the equation of motion, but also the associated boundary and initial conditions for complex cases can be obtained, where it is not possible to determine these constraints *a priori*.

If $q_i$'s are $n$ independent variables, the kinetic energy T, potential energy U, and dissipation function D assume the following form:

$$\mathrm{T} = \mathrm{T}(\dot{q}_1, \dot{q}_2, \ldots, \dot{q}_n); \tag{8.32}$$

$$\mathrm{D} = \mathrm{D}(\dot{q}_1, \dot{q}_2, \ldots, \dot{q}_n); \tag{8.33}$$

$$\mathrm{U} = \mathrm{U}(q_1, q_2, \ldots, q_n). \tag{8.34}$$

Equation (8.35) gives the work done by the non-conservative forces, which is

$$\delta W_{\mathrm{ncon}} = \sum_{j=1}^{n} \left( \mathrm{Q}_j - \frac{\partial \mathrm{D}}{\partial \dot{q}_j} \right) \delta \dot{q}_j, \tag{8.35}$$

where $\mathrm{Q}_j$ are the generalized forces.

Equation (8.36) gives Lagrange's equations for the system

$$\frac{d}{dt}\left(\frac{\partial \mathrm{T}}{\partial \dot{q}_j}\right) + \frac{\partial \mathrm{D}}{\partial \dot{q}_j} + \frac{\partial \mathrm{U}}{\partial q_j} = \mathrm{Q}_j; \quad j = 1, 2, \ldots, n. \tag{8.36}$$

Using matrix notation, the equation of motion for a multi-degree-of-freedom system can be systematically derived. The kinetic energy, potential energy, and dissipation function are, therefore

$$\mathrm{T} = \frac{1}{2}\{\dot{q}\}^{\mathrm{T}}[\mathrm{M}]\{\dot{q}\}; \tag{8.37}$$

$$\mathrm{D} = \frac{1}{2}\{\dot{q}\}^{\mathrm{T}}[\mathrm{C}]\{\dot{q}\}; \tag{8.38}$$

$$\mathrm{U} = \frac{1}{2}\{q\}^{\mathrm{T}}[\mathrm{K}]\{q\}. \tag{8.39}$$

Using Lagrange's Eqs. (8.36) on (8.37), (8.38), and (8.39), each term becomes

$$\frac{d}{dt}\left(\frac{\partial \mathrm{T}}{\partial \dot{q}_j}\right) = [\mathrm{M}]\{\ddot{q}\} \tag{8.40}$$

$$\frac{\partial \mathrm{D}}{\partial \dot{q}_j} = [\mathrm{C}]\{\dot{q}\} \tag{8.41}$$

$$\frac{\partial \mathrm{U}}{\partial q_j} = [\mathrm{K}]\{q\}. \tag{8.42}$$

Therefore, the equation of motion of a multi-degree-of-freedom system is

$$[M]\{\ddot{q}\} + [C]\{\dot{q}\} + [K]\{q\} = \{Q\}. \tag{8.43}$$

To determine matrix coefficients in the equation of motion (8.43), it is just enough to obtain energy expressions in matrix form for a discrete multi-degree-of-freedom system.

Generalized quantities [M], [C], and [K] given in Eqs. (8.40)–(8.42) refer to those physical parameters when they are expressed in generalized coordinates. If a system's equations are expressed in generalized coordinates, it results in an independent set of equations, which can be solved independently and is desired mathematical representation of the system. When arbitrary coordinates are chosen, which usually is the situation, to begin with in most of the formulation instances; it results in a coupled system of equations requiring mathematically cumbersome simultaneous solution of the equations. This is the primary reason, when a system's behavior is cast in arbitrary coordinates, eigenfunctions or vectors are sought. These mathematical constructs aid in expressing the equations represented by arbitrary coordinates into new basis vectors that are linearly independent. To emphasize, eigenfunctions or vectors are linearly independent. Therefore, when the coupled equations are suitably transformed in terms of eigenfunctions or vectors, it results in a decoupled system of equations that can be independently solved.

## 8.8 Motion of a Multi-degree-of-Freedom System Under Constraints

If energy functions are expressed in terms of displacements which are not independent and are easier to express, the linear constraint equations are

$$\begin{aligned} &g_j(q_1, q_2, \ldots, q_n) = 0 \\ &j = 1, 2, \ldots, m \end{aligned} \tag{8.44}$$

where n is number of degrees-of-freedom and m is the number of constraints.

In matrix form, the constraints are

$$[G]\{q\} = \{0\}. \tag{8.45}$$

Partitioning G in Eq. (8.45), such that $G_2$ is a square matrix of size $m \times m$

$$\begin{bmatrix} G_1 & G_2 \end{bmatrix} \begin{Bmatrix} q_1 \\ q_2 \end{Bmatrix} = \{0\}. \tag{8.46}$$

Equation (8.46) can be expressed as

$$\{q_2\} = -[G_2]^{-1}[G_1]\{q_1\}. \tag{8.47}$$

Equation (8.47) expresses the relationship between dependent and independent displacements.

Combining the two sets of displacements gives

$$\{q\} = \begin{Bmatrix} q_1 \\ q_2 \end{Bmatrix} = \begin{bmatrix} I \\ G_2^{-1}G_1 \end{bmatrix}\{q_1\} = [T_G]\{q_1\}. \tag{8.48}$$

Substituting Eq. (8.48) into energy equations for displacement

$$T = \frac{1}{2}\{\dot{q}_1\}^T[\overline{M}]\{\dot{q}_1\}; \tag{8.49}$$

$$D = \frac{1}{2}\{\dot{q}_1\}^T[\overline{C}]\{\dot{q}_1\}; \tag{8.50}$$

$$U = \frac{1}{2}\{\dot{q}_1\}^T[\overline{K}]\{\dot{q}_1\}; \tag{8.51}$$

where in Eqs. (8.49), (8.50), and (8.51), $[\overline{M}] = [T_G]^T[M][T_G]$, $[\overline{C}] = [T_G]^T[C][T_G]$, and $[\overline{K}] = [T_G]^T[K][T_G]$.

The virtual work done by applied forces is also transformed as:

$$\delta W = \sum_{j=1}^{n} Q_j \delta q_j = \{\delta q_1\}^T\{Q\}$$

$$= \{\delta q_1\}^T[T_G]^T\{Q\} = \{\delta q_1\}^T\{\overline{Q}\}. \tag{8.52}$$

The energy and virtual work functions are expressed in terms of independent displacements and are substituted into Lagrange's equations to give equation of motion under constraints as

$$[\overline{M}]\{\ddot{q}_1\} + [\overline{C}]\{\dot{q}_1\} + [\overline{K}]\{q_1\} = \{\overline{Q}\}. \tag{8.53}$$

## 8.9 Solution to Multi-degrees-of-Freedom System

It is necessary to solve the coupled Eq. (8.53) of motion to determine responses. In structural dynamics, the solution to Eq. (8.53) corresponds to determining displacement, velocity and acceleration vectors. Solution to equation is sought either in time or frequency domains which are equivalent, i.e. information about state of the system is preserved. Essential theory of vibration is given in this section to appreciate its applications for many engineering problems. The reader is advised to read many standard publications such as those given in references.

Equation of motion given in Eq. (8.53) under constraints can be considered as a generalization of a single-degree-of-freedom system to a multi-degree-of-freedom system. The equations are coupled and characterized by presence of off-diagonal elements of system matrices. The mass matrix obtained from first principles is called consistent as it is formulated in the same way as stiffness matrix. The lumped mass matrix representation consists of mass components corresponding to translation degrees-of-freedom and rotational mass is assumed zero. Certain industry-standard proprietary software uses coupled mass, which is an average of consistent and lumped mass formulations. The computation of stiffness and mass matrices is explained in later part of this section.

### *8.9.1 Expressions for System Parameters: Mass and Stiffness*

The primary objective of finite element method is to evaluate system parameters of the problem concerned. In solid mechanics and structural engineering applications, the system parameters are mass, damping, and stiffness. For heat transfer problems, the system parameters are conductance and capacitance. Similar quantities exist for electromagnetic problems.

Mechanical systems undergoing vibrations are represented by their system properties- mass, stiffness, and damping. Derivation of these system matrices can be found in the finite element formulation-related chapter of this book or in many published literature. Expressions for these parameters are reproduced here for the sake of completeness and role of finite element ends after determination of these parameters. Solutions to these equations are determined using linear algebra techniques with the aid of digital computers.

Stiffness $K_{ij}^e$ of an element $e$ is defined as

$$[K]_{ij}^e = \int_{V^e} B^{e^T} D^e B^e dV^e, \tag{8.54}$$

where $[B]$ is strain-displacement matrix, $D$ the material matrix, and $V$ is volume of the element.

Mass matrix [m] of an element $e$ is defined as

$$[\mathrm{m}]^{\mathrm{e}} = \int_{\mathrm{V^e}} \rho[\mathrm{N}]^{\mathrm{T}}[\mathrm{N}]\mathrm{dV^e}, \tag{8.55}$$

where $\rho$ is density of the material and [N] is interpolation function.

Values of stiffness and mass given by the above expressions (8.54 and 8.55) are used to solve dynamic Eq. (8.53).

In published literature, damping parameter has many mathematical expressions but the fact remains that it is never near the values obtained in tests or experiments. Therefore, the reader can refer to these derivations if interested. This chapter will explain the way damping is handled in realistic design situations. Here, the damping values considered are estimated rather than using derived values. Although it may not be mathematically on a sound foundation, the heuristic does reasonably predict test results for all practical purposes. Such examples abound in applied sciences where certain mathematical models do not completely agree with observations for reasons not that obvious.

## 8.10 Free Vibration Analysis

Free vibration analysis is used for dynamic characterization of a structure conveyed by its eigenvalue and eigenvector (functions in case of the continuum). The system of Eq. (8.55) corresponding to a constrained structural system is in general coupled. This implies, all unknown variables appear in every equation requiring a simultaneous solution, which for systems with more than three unknowns is quite difficult to solve. It is needless to state the order of difficulty for systems with a large number of unknowns. Mathematical coupling of equations appears due to arbitrary choice of coordinates, in other words, arbitrary basis functions. It is not straightforward to obtain *a priori* a set of basis functions that offers the prospect of decoupled equations, enabling independent solution of system equations, which helps in uncoupling of equations of motion.

The eigenvalues and eigenvectors of a vibrating mechanical system correspond to its natural frequencies and mode shapes, respectively; also their inherent characteristics. This information is used to study its response to dynamic excitations, which is the objective of response analysis.

From basic linear algebra, we know that when elements of a square coefficient matrix are non-zeros, it implies that equations are coupled in their unknowns. Hence, it is required to determine basis vectors using the coefficient matrices of the Eq. (8.53), which aid in decoupling the equations of motion.

Toward this end, $\left[\overline{C}\right]$ is omitted from Eq. (8.53), hence un-damped equation of motion for free vibration is

$$[M]\{\ddot{\mathrm{q}}_1\} + [K]\{q_1\} = \{0\}. \tag{8.56}$$

Eigenvalue problem corresponding to Eq. (8.56) is obtained by considering the oscillations to be harmonic

$$\left[\mathrm{K} - \omega_i^2\mathrm{M}\right]\{\phi\} = \{0\}, \tag{8.57}$$

where $\omega_i$ and $\{\phi\}$ are the eigenvalues and corresponding eigenvectors, or natural frequencies and mode shapes. Conditions for Eq. (8.57) to have non-trivial solution is

$$\left|\mathrm{K} - \omega_i^2 \mathrm{M}\right| = 0, \tag{8.58}$$

i.e. the determinant value should be zero. M is always positive definite,K is positive definite or positive semi-definite, and the eigenvalues are all real and either positive or zero. If the eigenvalues are distinct, then corresponding to each one, there is a non-trivial solution to Eq. (8.56). These solutions are known as eigenvectors. Eigenvectors are usually normalized so that the maximum value is unity corresponding to the largest component. Further, for linear structural systems in most situations, the system matrices are square and symmetric. The eigenvectors for square symmetric matrices exhibit orthogonal properties. Mathematically, the orthogonal properties of eigenvectors are given by (8.59)

$$\phi_i^T \phi_j = \delta_{ij}, \tag{8.59}$$

where Kronecker delta $\delta_{ij} = \begin{cases} 1 & i = j \\ 0 & i \neq j \end{cases}$.

If $[\phi]$ is a matrix whose columns are eigenvectors of the system, then

$$[\phi]^T [\phi] = I. \tag{8.60}$$

Equation (8.60) uses the property given by Eq. (8.59).

The expressions given above correspond to mathematical characterization of the system and are used for response analyses.

## 8.11 Time Domain Solution/Transient Analysis

When a structure is subjected to sudden external stimulus/stimuli, it exhibits a transient response. The external excitation is in general non-periodic. Strictly speaking, the term 'transient' should be applied to situation when forces are applied for a short interval of time. This corresponds to time of application of force, which is of the order comparable with the fundamental period of the system that results in excitation of inertial forces. Subsequent motion of the structure is free vibration, which will decay due to the damping present. The term 'transient' can be applied to a situation when applied load is an arbitrarily varying function of time.

Transient response analysis is important in aerospace structural engineering applications where structures are excited by gust loads, handling and transportation forces, stage separation shocks, collision, etc. Load augmentation on sensitive structures needs to be understood when transient loads are acting on them.

The transient response solution of solids subjected to time varying loads can be determined using either modal or direct integration method for continuous or discrete systems:

- Modal transient method: Free vibration analysis for an un-damped free oscillation is carried out to determine eigenvalues and eigenvectors. This method utilizes mode shapes of the structure, reduces solution degrees-of-freedom, and can significantly impact the run time. Using this, the physical representation of system equations is expressed in terms of modal coordinates and solved to obtain response as superposition of responses from different modes of the structure. This approach replaces the physical degrees-of-freedom with a reduced number of modal degrees-of-freedom. Fewer degrees-of-freedom means a faster solution. This can be a big time saver for transient models with a large number of time steps.
- Direct transient method: In direct transient response analysis, the structural response is computed by solving a set of coupled equations using direct numerical integration. Direct method has the advantage of not solving eigenproblem prior to response analysis. This method involves computation time as system matrices need to be inverted for every time step which calculates the response of a system to a load, over time. The load applied to the system can vary over time or simply be an initial condition that is allowed to evolve over time. This method may be more efficient for models where high-frequency excitation requires the extraction of a large number of modes.

### 8.11.1 Modal Transient Method

The unknown $\boldsymbol{q}_1$ of Eq. (8.56) are eliminated using another variable $\eta$ using the transformation

$$\{q_1\} = [\phi]\{\eta\}. \tag{8.61}$$

On substituting Eq. (8.61) into Eq. (8.56) and considering forcing function gives

$$[M][\phi]\{\ddot{\eta}\} + [K][\phi]\{\eta\} = \{\overline{\mathrm{Q}}\}. \tag{8.62}$$

Pre-multiplying Eq. (8.62) with $[\phi]^T$

$$[\phi]^T[M][\phi]\{\ddot{\eta}\} + [\phi]^T[K][\phi]\{\eta\} = [\phi]^T\{\overline{\mathrm{Q}}\}. \tag{8.63}$$

On comparing Eq. (8.56), it can be easily verified that the resulting Eq. (8.63) will be decoupled. In other words, a $N$ coupled system of equations will result in $N$ uncoupled/independent set of equations.

The $r$th modal equation is obtained by normalizing with generalized mass (in $r$th mode)

$$\ddot{\eta}_r + \omega_r^2 \eta_r = Q_r(t). \tag{8.64}$$

It is at this stage that damping is introduced. From the test data, measured value of damping ratio $\gamma$ is added to left-hand side of the Eq. (8.64) and expressed to resemble that of single-degree-of-freedom system with damping.

For a linear structural system, the $r$th modal equation in generalized coordinate $\eta$ is

$$\ddot{\eta}_r + 2\gamma_r \omega_r \dot{\eta}_r + \omega_r^2 \eta_r = Q_r(t). \tag{8.65}$$

The term generalized coordinate means, it is the basis vector when used, resulting in an independent set of equations (uncoupled). This basis vector can be linearly combined to represent any state of configuration when the system undergoes vibrations.

Duhamel's integral, equation (8.66), (a way of calculating the response of linear systems and structures to arbitrary time varying external perturbation) is used for solving equation (8.65)

$$\eta_r(t) = \int_0^t Q_r(\tau) h_r(t - \tau) d\tau, \tag{8.66}$$

where the impulse response function $h_r$ is given as

$$h_r(t) = \frac{1}{\omega_{\mathrm{d}r}} e^{-\gamma_r \omega_r t} \sin(\omega_{\mathrm{d}r} t) \tag{8.67}$$

and

$$\omega_{\mathrm{d}r} = \omega_r \left(1 - \gamma_r^2\right)^{\frac{1}{2}}. \tag{8.68}$$

On solving Eq. (8.68) for $\eta$, the transient or steady state solution is expressed in physical coordinates using Eq. (8.61).

### 8.11.2 Direct Transient Method

The load in Eq. (8.68) is seldom an analytic function in real-life situations. It is available mostly as tabular data. Numerical methods are resorted to in such situations. Most popular among them is Newmark-$\beta$method. Following expressions give state vectors of the equation motion for a multi-degrees-of-freedom using direct transient method (refer 8.14.2).

$$
\begin{aligned}
[a_1 M + a_2 C + K]q_{j+1} = & f_{j+1} + [a_1 M + a_2 C]q_j \\
& + [a_3 M - a_4 C]\dot{q}_j + [a_5 M - a_6 C]\ddot{q}_j
\end{aligned}
\tag{8.69}
$$

$$
\ddot{q}_{j+1} = a_1\left\{q_{j+1} - q_j\right\} - a_3\dot{q}_j - a_5\ddot{q}_j \tag{8.70}
$$

$$
\dot{q}_{j+1} = a_2\left\{u_{j+1} - q_j\right\} + a_4\dot{q}_j + a_6\ddot{q}_j, \tag{8.71}
$$

where

$$
\boldsymbol{C} = \left(\frac{G}{\omega^*}\right)\boldsymbol{K}
$$

$$
G = 2\zeta
$$

$$
a_1 = \frac{1}{\beta(\Delta t)^2};\ a_2 = \frac{\gamma}{\beta\Delta t};
$$

$$
a_3 = \frac{1}{\beta\Delta t};\ a_4 = \left(1 - \frac{\gamma}{\beta}\right);
$$

$$
a_5 = \left\{\left(\frac{1}{2\beta}\right) - 1\right\};\ a_6 = \left\{1 - \left(\frac{\gamma}{2\beta}\right)\right\}\Delta t.
$$

These expressions, Eqs. (8.69)–(8.71), can be coded in any computer language for determining solutions using digital computers. Though advantage of the direct method is that, eigenvalue problem need not be solved to carry out response analysis, it is not recommended when the problem size or degrees-of-freedom is large.

Modal transient approach will yield exactly same result as direct approach when all modal degrees-of-freedom (DOF) are included in the transformation. However, the strength of this approach is due to a solution that is very close to exact, which can usually be obtained with significantly fewer modal DOF than there are physical degrees-of-freedom. With fewer DOF, the solution can proceed much faster; hence this method will be efficient for large models and for models that require many time steps.

## 8.12 Frequency-Response Analysis

No mechanical system vibrates unless it is imparted some initial displacement and/or velocity (initial conditions). If these stimuli act initially for an infinitesimal time, the subsequent response is called a free response, whereas it is known as a forced response if it continues to act throughout the duration of interest. In the mathematical solution used to determine response of such a system, these two responses are usually obtained separately as particular solution and complementary or homogenous solution. Sum of these two responses is referred to as total response of the system.

Time domain analytic models are represented by differential equations with time as independent variable. Similarly, frequency is the independent variable in frequency-domain analysis, where it is a set of input–output and transfer functions. Time domain approach is adopted when response predictions are required for a time varying force. Frequency domain methods are most efficient when forces are functions of frequencies as is the case for random vibrations and periodic loadings.

When forced response is studied in frequency domain, the mathematical manipulations are relatively simple. Response of a mechanical system to an external stimulus expressed in the frequency domain is known as its frequency response. In other words, frequency response is the response of dynamic system to harmonic excitation. The responses are determined by varying amplitude and frequency of excitation. Thus, for analytical purposes, system response is studied for harmonic excitations. Although real systems are seldom excited by purely harmonic forces with single frequency components, analysis of the effects of such excitations on vibrating systems provides foundation for analyses of more complicated forms of excitation encountered in real systems, such as periodic excitations of various types containing multiple frequency components of random excitations and so forth.

Any periodic excitation can be decomposed into harmonic components using Fourier series and the effect of each such harmonics can be studied independently and added to obtain total response. Frequency domain considerations are applicable even when excitations are not periodic. For arbitrary excitations, one has to resort to integral transforms such as Laplace and Fourier in frequency domain and convolution integral (Duhamel) in time domain. In fact, a time signal can be transformed into its frequency spectrum through the Fourier transform. Frequency components of a given time signal can be obtained using Fourier transform analytically or numerically. Without loss of generality, it can be stated that linear dynamical systems have equivalent time and frequency-domain representation and analysis. Hence, frequency response analysis is widely used as vibration signals are periodic in nature.

When subjected to dynamic forces, total response of a structure is sum of the responses of its modes of vibration. From a structural designer's point of view, only a relatively small number of structural modes; typically those with low frequencies are of concern. The deformation associated with higher order modes is not enough to produce significant stress.

Equation (8.62) can be written by including the generalized damping as

$$\overline{\mathrm{M}}\ddot{\eta} + \overline{\mathrm{C}}\dot{\eta} + \overline{\mathrm{K}}\eta = \overline{\mathrm{Q}}, \tag{8.72}$$

where,
$\overline{\mathrm{M}} = \Phi^{\mathrm{T}}\mathrm{M}\Phi\overline{\mathrm{C}} = \Phi^{\mathrm{T}}\mathrm{C}\Phi\overline{\mathrm{K}} = \Phi^{\mathrm{T}}\mathrm{K}\Phi$ and $\overline{\mathrm{Q}} = \Phi^{\mathrm{T}}\mathrm{Q}$.

Both $\overline{M}$ and $\overline{K}$ are diagonal matrices. If columns of Φ are mass normalized, then $\overline{M} = I$ and $\overline{K} = \wedge I$,

Where $\wedge = \begin{bmatrix} \omega_1^2 & \cdots & 0 \\ \vdots & \ddots & \vdots \\ 0 & \cdots & \omega_n^2 \end{bmatrix}$.

If nodal forces are harmonic, with the same frequency $\omega$, Eq. (8.72) reduces to

$$\ddot{\eta} + \overline{C}\dot{\eta} + \wedge\eta = Q. \tag{8.73}$$

The steady state solution is obtained by assuming that response is harmonic with frequency $\omega$. The derivation is same as that of single-degree-of-freedom, where $e^{i\omega t}$ is factored to obtain following relations:

$$\left[\wedge - \omega^2 I + i\omega\overline{C}\right]\{\eta\} = \{Q\} \tag{8.74}$$

$$\{\eta\} = \left[\wedge - \omega^2 I + i\omega\overline{C}\right]^{-1}\{Q\}. \tag{8.75}$$

Substituting Eq. (8.75) into Eq. (8.61)

$$\{q_1\} = [\Phi]\left[\wedge - \omega^2 I + i\omega\overline{C}\right]^{-1}[\Phi]^T\left\{\overline{Q}\right\}. \tag{8.76}$$

In familiar form, the relation between displacement vector $\{u(i\omega)\}$, transfer function matrix $[H(i\omega)]$ and forcing vector $\{F(i\omega)\}$ is

$$\{u(i\omega)\} = [H(i\omega)]\{F(i\omega)\}, \tag{8.77}$$

where $H(i\omega) = \Phi\left[\wedge - \omega^2 I + i\omega\overline{C}\right]^{-1}\Phi^T$.

$H_{jk}(i\omega)$ is the transfer function which represents response in DOF $j$ due to a harmonic force of unit magnitude at $k$ DOF. The transfer function or frequency response function assumes the following form for modal damping:

$$H_{jk}(i\omega) = \sum_{r=1}^{n} \frac{\phi_{jr}\phi_{kr}}{\left(\omega_r^2 - \omega^2 + i2\gamma_r\omega_r\omega\right)}, \tag{8.78}$$

$\gamma_r$ is the modal damping ratio.

The term $\phi_{jr}\phi_{kr}$ is referred to as modal constant or modal gain.

## 8.13 Random Response Analysis

The accurate analysis, design, and assessment of mechanical and structural systems, subjected to realistic dynamic environments, must consider potential for random loads and randomness in structural and material properties. The engineering field that deals with these issues is known as random vibrations. Modern theory of random vibrations is product of generations of works in the fields of deterministic structural vibrations, probabilistic analysis of mechanical system response and random signal analysis. When the response of a mechanical system is forced by a random vibration excitation, the specific response to a future event cannot be computed because the dynamic environments to be realized in the future cannot be precisely predicted.

In real-life situations, problems associated with vibrations are seldom deterministic. The systems and excitations are probabilistic. Therefore, statistical descriptions of parameters are needed. Example: buffeting in the vicinity of bulbous portions of launch vehicles during transonic flight regime, base excitation of components and structures during qualification and/or acceptance test, and response of aircraft structure during taxiing over runway.

Application of statistical principles explicitly implies that very large numbers of observations/events are involved. Rich insights into random dynamical problems can be acquired when it is investigated in frequency domain, besides the significant mathematical simplicity.

Fundamentals of modern theory of probability are essential to understand random vibration phenomenon. Modern theory of probability is based on axioms propounded by Andrey Nikolayevich Kolmogrov (1903–1987). Detailed treatment on this subject can be found in standard publications.

Random processes can be analyzed in time or frequency domains. Certain standard statistical relations and measures are listed below for time and frequency domains used to study random response phenomena.

Time domain relationships are:

1. The expectation operator $E[\cdot]$ is defined as $E[p(x)] = \int_{-\infty}^{\infty} p(x) f_X(x) dx$, where $f_X(x)$ is the probability density function of $X$.
2. The mean or expected value is

$$\mu_X(t) = E[x(t)] = \int_{-\infty}^{\infty} x f_X(t) dt.$$

3. The similarity between two quantities is measured using correlation function. The correlation of two quantities taken from the same record is called autocorrelation, given by

$$R_X(\tau) = \lim_{T\to\infty} \frac{1}{T} \int_{-T/2}^{T/2} x(t)x(t+\tau)dt.$$

The maximum value of the autocorrelation function is obtained at $\tau = 0$, $R_X(0) = \overline{x}^2$. Thus, the maximum value of the autocorrelation function is equal to the mean square value.

Cross correlation between two random quantities $x(t)$ and $y(t)$ is defined by the equation

$$R_{XY}(\tau) = E[x(t)y(t+\tau)] = \lim_{T\to\infty} \frac{1}{T} \int_{-\frac{T}{2}}^{\frac{T}{2}} x(t)y(t+\tau)dt.$$

4. Mean square value about the mean is defined as variance and is given by

$$\sigma_X^2(t) = E\left[\{x(t) - \mu_X(t)\}^2\right] = \int_{-\infty}^{\infty} \{x(t) - \mu_X(t)\}^2 f_X(t)dt.$$

5. Auto covariance $\sigma_{XX}(t)$—The covariance of random variables drawn from one stochastic process is called auto covariance. The auto-covariance function $\sigma_{XX}(t_1, t_2)$ is defined as 8.14.2.

$$\sigma_{XX}(t_1, t_2) = E[\{X(t_1) - \mu_X(t_1)\}\{X(t_2) - \mu_X(t_2)\}.$$

6. Cross covariance $\sigma_{XY}$(t)—The covariance of random variables drawn from two stochastic processes is called cross covariance 8.14.2.

$$\sigma_{XY}(t_1, t_2) = E[\{X(t_1) - \mu_X(t_1)\}\{Y(t_2) - \mu_Y(t_2)\}].$$

7. Coefficient of correlation $\rho_{XY}$—it gives the relation between covariance and standard deviation. Its value lies between $-1$ and $+1$

$$\rho_{XY} = \frac{\sigma_{XY}}{\sigma_X \sigma_Y}.$$

8. When covariance $\sigma_{XY} = 0$, the random variables X and Y are said to be uncorrelated.

Frequency domain relationships are:

9. Spectral densities function of X: $S_X(\&)$

It can be defined as the Fourier transform of autocorrelation function. The autocorrelation function provides information on random variables in time domain, whereas spectral density function gives similar information in the frequency domain.

$$S_X(\omega) = \int_{-\infty}^{\infty} R_X(\tau) e^{-i\omega\tau} d\tau.$$

10. Spectral densities function of X and Y:$S_{XY}(\&)$

The cross-spectral density function is a more generalized form of the spectral density function; allows cross relation between excitation in one location to another location on the structure.

$$S_{XY}(\omega) = \int_{-\infty}^{\infty} R_{XY}(\tau) e^{-i\omega\tau} d\tau.$$

11. The structural systems' response to random environments such as displacements, velocities, and accelerations are important from applications point of view. Information contained in time and frequency domains is equivalent. These are related to Wiener–Khinchine relations

$$S_X(\omega) = \frac{1}{2\pi} \int_{-\infty}^{\infty} R_X(\tau) e^{-i\omega\tau} d\tau$$

$$R_X(\tau) = \int_{-\infty}^{\infty} S_X(\omega) e^{i\omega\tau} d\omega.$$

12. Equations in serial number (11) forms Fourier transform pairs between spectral density $S_X(\&)$ and autocorrelation $R_X(t)$, respectively. In other words, spectral density can be defined as Fourier transform of autocorrelation function. The cross-spectral density function $S_{XY}(\&)$ is a more generalized form of spectral density function; allows cross relation between excitation in one location to another location on the structure [refer 8.14.8].

$$S_{XY}(\omega) = \int_{-\infty}^{\infty} R_{XY}(\tau) e^{-i\omega\tau} d\tau.$$

In reality, there will be input uncertainties or dispersions associated with the model, such as material properties, constraints, geometry, etc. These dispersions in the input values strongly influence structural responses. Therefore, it becomes necessary to consider the effect of uncertainty using a probabilistic approach.

Probabilistic analysis provides a rational basis for improvement of safety and quality of engineering systems as risk-informed design practices are adopted. Real-world dispersion in values of inputs is recognized and considered in the design process aiding comparisons of designs based on reliability. The sensitivity of cardinal input parameters on design can be assessed and focused process control can be carried out. The important 12 probabilistic measures given above are related to accounting of dispersions in the forcing function (environment), and further, it is assumed that system variables don't exhibit scatter. Most finite element software implement this aspect. The readers are encouraged to refer to publications on probabilistic treatment of scatter in properties of system. From a structural engineering perspective, the system parameters are related to geometry and mechanical material properties.

To conduct random vibration analysis in a finite element framework, the relations need to be expressed in matrix form as opposed to continuum representations found in standard literature. With due consideration to Central Limit Theorem, the random process encountered in most of the engineering problems can be assumed Gaussian. In this chapter, only stationary random processes are considered. This assumption is valid in most engineering situations as the statistical properties, such as mean and variance are invariant in the time interval considered for study.

For a multi-degree-of-freedom system, the random environment is characterized in frequency domain as spectral density quantities specified as $[S_{FF}(i\omega)]$. The spectral density of responses of interest is given as

$$\text{Displacement :}[S_{XX}(i\omega)] = [\Phi]\left[H^*(i\omega)\right][\Phi]^T[S_{FF}(i\omega)][\Phi][H(i\omega)][\Phi]^T \tag{8.79}$$

$$\text{Velocity :}\left[S_{\dot{X}\dot{X}}(i\omega)\right] = -\omega^2[S_{XX}(i\omega)]; \tag{8.80}$$

$$\text{Acceleration :}\left[S_{\ddot{X}\ddot{X}}(i\omega)\right] = -\omega^4[S_{XX}(i\omega)]. \tag{8.81}$$

Recall that, at the end of finite element modeling process, i.e. when a finite element mesh is available, the system characteristics such as mass, stiffness, and damping are computed. The transfer function $\mathrm{H}(i\omega)$ is a function of system characteristics. Hence, when a finite element mesh is available, it implicitly means that transfer function can be constructed using the approach mentioned in Sect. 8.12. This can be substituted in Eqs. (8.79), (8.80), and (8.81) to obtain power spectral density of the state of the system, viz. displacement, velocity, and acceleration.

## Bibliography

Elements of vibration analysis (1986). Leonard Meirovitch, McGraw Hill Inc

Introduction to finite element vibration analysis (1990) Maurice Petyt. Cambridge University Press, Cambridge

Mechanics of structures, variational and computational methods, 2nd edn (2003) Walter Wunderlich, Walter D Pilkey, CRC Press, NY

Theory of vibrations (1988) W T Thomson, Prentice Hall

Wirsching PH, Paez TL, Ortiz K (2006) Random vibrations: theory and practice. Dover Publications, New York

Lutes LD, Sarkani S (2004) Random vibrations: analysis of structural and mechanical systems. Elsevier Butterworth Heinemann, Massachusetts

Broussinos P, Kabe A (1990) Multi-mode random response analysis procedure, Technical report 90–53, The Aerospace Corporation, California

Giora M (2008) Structural dynamics and probabilistic analysis for engineers. Elsevier, UK

# Chapter 9
# Buckling of Column

**R. Marimuthu**

*There exists everywhere a medium in things, determined by equilibrium—Dmitri Mendeleev.*

## 9.1 Introduction

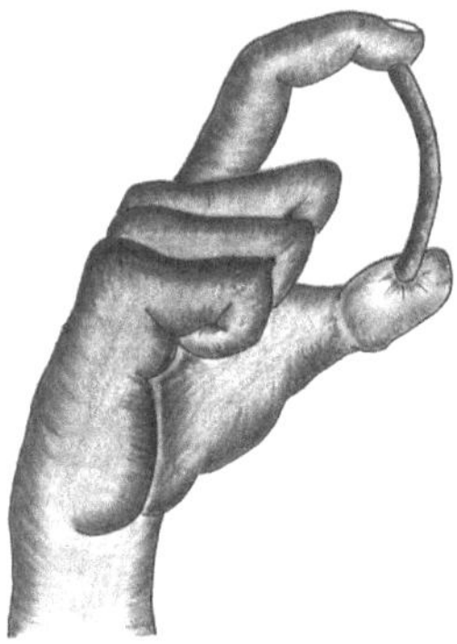

When a structure is loaded in compression incrementally, at a certain load level the configuration of the structure will start to alter considerably faster than before. Such behaviors are referred to by the general term instability or buckling. Most of the aerospace structures are thin-walled, and the designs are often driven by stability behavior. The theory of the behavior of columns was investigated by Leonhard Euler in the eighteenth century. He derived the Euler formula that gives the maximum axial load that a long, slender, ideal column can carry beyond which the column buckles. An ideal column, though difficult to see in the real world, is that which is perfectly straight, made of homogenous material, and free from all defects or imperfections. In

R. Marimuthu (✉)
DH, FSDD/SDAG/STR, Veli, VSSC, Thiruvananthapuram 695022, India

 
S. U. Nair and S. Somanath (eds.), *Introduction to Finite Element Analysis*,
https://doi.org/10.1007/978-981-19-7989-7_9

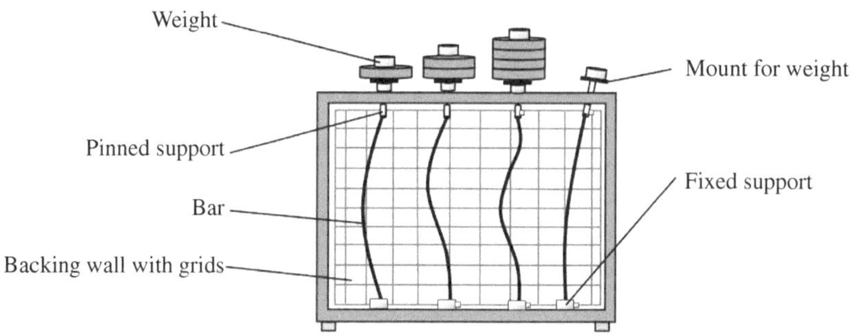


**Fig. 9.1** Experiment on first (fundamental) critical buckling load on thin stainless steel rod

this section, Euler column buckling is considered for the study. The critical buckling load factor for different boundary conditions is evaluated theoretically, and results are compared with Finite Element formulation and FEAST$^{SMT}$ results.

**Geometric stiffness or Stress stiffness or Initial stress matrix**: A structure is stiffened (or softened) under tensile or compressive loading. Stress stiffening is an important source of stiffness and must be considered when analyzing structures. For example, a carbonated soft drink can is stiffened by the pressure of the gas inside it, and that stress stiffness vanishes when the can is opened. In buckling analysis, the stress stiffness matrix (geometric stiffness matrix or initial stress matrix) is added to the regular stiffness matrix to get the total stiffness. The buckling problem can be considered as an eigenvalue problem involving stiffness matrix [K] and geometric stiffness matrix $[K_G]$ as shown below, where $p_i$ is the eigenvalue or critical buckling load and $\{q\}_i$ is the corresponding eigenvector or the mode shape of the ith mode under buckling:

$$[K]\{q\}_i - p_i[K_G]\{q\}_i = 0. \tag{9.1}$$

An experiment conducted on a stainless steel bar is shown in Fig. 9.1. The axial compressive load is gradually increased and when it reaches a particular load level, the bar buckles. Buckling load depends upon the end conditions and its effective length. The load that causes the rod to buckle is called critical load. The bar till just before buckling is straight. Figure 9.1 shows buckling loads and mode shapes for different end conditions.

## 9.2 Derivation of Shape Function for Euler Column

A two-node Euler column in a natural coordinate system with boundary conditions is shown in Fig. 9.2. The transverse displacement at any point within the element using a natural coordinate system is given as

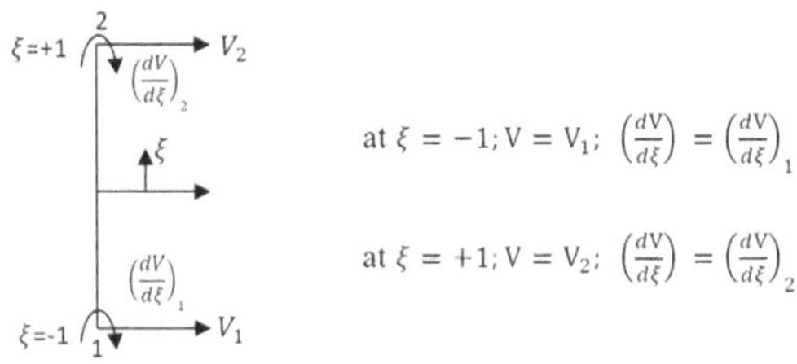


**Fig. 9.2** Two-node Euler column showing degrees of freedom at nodes

$$V(\xi) = a_0 + a_1\xi + a_2\xi^2 + a_3\xi^3. \tag{9.2}$$

Substituting the end conditions at nodes given in Fig. 9.2 into the above equation yields

$$\left.\begin{array}{l} V_1 = a_0 - a_1 + a_2 - a_3 \\ \left(\frac{dV}{d\xi}\right)_1 = a_1 - 2a_2 + 3a_3 \\ V_2 = a_0 + a_1 + a_2 + a_3 \\ \left(\frac{dV}{d\xi}\right)_2 = a_1 + 2a_2 + 3a_3 \end{array}\right\}. \tag{9.3}$$

In matrix form, the above equation can be written as

$$\begin{Bmatrix} V_1 \\ \left(\frac{dV}{d\xi}\right)_1 \\ V_2 \\ \left(\frac{dV}{d\xi}\right)_2 \end{Bmatrix} = \begin{bmatrix} 1 & -1 & 1 & -1 \\ 0 & 1 & -2 & 3 \\ 1 & 1 & 1 & 1 \\ 0 & 1 & 2 & 3 \end{bmatrix} \begin{Bmatrix} a_0 \\ a_1 \\ a_2 \\ a_3 \end{Bmatrix}. \tag{9.4}$$

On applying matrix operations, the equation reduces to

$$\begin{Bmatrix} a_0 \\ a_1 \\ a_2 \\ a_3 \end{Bmatrix} = \frac{1}{4}\begin{bmatrix} 2 & 1 & 2 & -1 \\ -3 & -1 & 3 & -1 \\ 0 & -1 & 0 & 1 \\ 1 & 1 & -1 & 1 \end{bmatrix} \begin{Bmatrix} V_1 \\ \left(\frac{dV}{d\xi}\right)_1 \\ V_2 \\ \left(\frac{dV}{d\xi}\right)_2 \end{Bmatrix}. \tag{9.5}$$

Hence

$$\left.\begin{array}{l} a_0 = \frac{1}{4}\left(2V_1 + \left(\frac{dV}{d\xi}\right)_1 + 2V_2 - \left(\frac{dV}{d\xi}\right)_2\right) \\ a_1 = \frac{1}{4}\left(-3V_1 - \left(\frac{dV}{d\xi}\right)_1 + 3V_2 - \left(\frac{dV}{d\xi}\right)_2\right) \\ a_2 = \frac{1}{4}\left(-\left(\frac{dV}{d\xi}\right)_1 + \left(\frac{dV}{d\xi}\right)_2\right) \\ a_3 = \frac{1}{4}\left(V_1 + \left(\frac{dV}{d\xi}\right)_1 - V_2 + \left(\frac{dV}{d\xi}\right)_2\right) \end{array}\right\}. \tag{9.6}$$

Substituting these constants from Eq. (9.6) in Eq. (9.2) and grouping the terms results in

$$V(\xi) = H_1(\xi)V_1 + H_2(\xi)\left(\frac{dV}{d\xi}\right)_1 + H_3(\xi)V_2 + H_4(\xi)\left(\frac{dV}{d\xi}\right)_2 \tag{9.7}$$

where

$$\left.\begin{array}{r} H_1(\xi) = \frac{1}{4}(2 - 3\xi + \xi^3) \\ H_2(\xi) = \frac{1}{4}(1 - \xi - \xi^2 + \xi^3) \\ H_3(\xi) = \frac{1}{4}(2 + 3\xi - \xi^3) \\ H_4(\xi) = \frac{1}{4}(-1 - \xi + \xi^2 + \xi^3) \end{array}\right\}. \tag{9.8}$$

These shape functions provided in Eq. (9.8) are known as Hermite polynomials and satisfy both displacement and its derivative continuity at nodes known as $C^1$ continuity. Rotation at the node is a derivative of transverse displacement with respect to global coordinate x; using chain rule, this can be obtained as given below

$$\theta = \frac{dV}{dx} = \frac{dV}{d\xi}\frac{d\xi}{dx}. \tag{9.9}$$

For two-node element, x coordinate at any point within the element can be expressed in terms of nodal coordinates $X_1$ and $X_2$ as

$$x = \frac{1-\xi}{2}X_1 + \frac{1+\xi}{2}X_2 \tag{9.10}$$

$$\frac{dx}{d\xi} = -\frac{X_1}{2} + \frac{X_2}{2} = \frac{X_2 - X_1}{2} = \frac{1}{2}. \tag{9.11}$$

Hence

$$\left(\frac{dV}{d\xi}\right) = \frac{1}{2}\frac{dV}{dx}. \tag{9.12}$$

Hence, the transverse displacement in terms of nodal degrees of freedom is

$$V(\xi) = N_1V_1 + N_2\theta_1 + N_3V_2 + N_4\theta_2 \tag{9.13}$$

where

$$\theta_1 = \left(\frac{dV}{dx}\right)_1 \text{ and } \theta_2 = \left(\frac{dV}{dx}\right)_2 \tag{9.14}$$

$$N_1 = H_1(\xi);\; N_2 = \frac{1}{2}H_2(\xi);\; N_3 = H_3(\xi);\; N_4 = \frac{1}{2}H_4(\xi). \tag{9.15}$$

Hence, the transverse displacement in terms of nodal degrees of freedom is

$$V = \{N\}^T\{q\}. \tag{9.16}$$

## 9.3 Governing Differential Equation and Finite Element Formulation Using Galerkin Method

Figure 9.3 shows a pin-ended column subjected to axial compressive load. Governing differential equation corresponding to Euler column buckling is given by

$$EI\frac{d^2V}{dx^2}+pV = 0 \tag{9.17}$$

where

E is Young's modulus of the material.
I is the moment of Inertia of the column.
V is the deflection in the direction of y.
p is the externally applied compressive load.

Equation (9.17) is differentiated twice with respect to x; we get

$$\frac{d^2}{dx^2}\left(E(x)I(x)\frac{d^2V}{dx^2}\right) + \frac{d^2}{dx^2}(pV) = 0. \tag{9.18}$$

Equation (9.18) is a general buckling equation for Euler column buckling for having only axial compressive load, which includes variation of E, I, and p as a function of x. Using the Galerkin method, the above Eq. (9.18) is discretized as

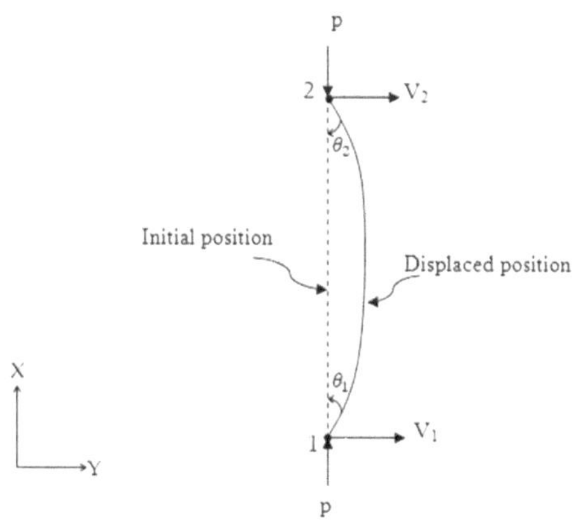


**Fig. 9.3** Column subjected to axial compressive load p

follows:

$$\int_{X_1}^{X_2} \delta V^T \frac{d^2}{dx^2}\left(E(x)I(x)\frac{d^2V}{dx^2}\right)dx + \int_{X_1}^{X_2} \delta v^T \frac{d^2(pV)}{dx^2}dx = 0. \tag{9.19}$$

Integrating the equation by parts results in

$$\begin{aligned} &\delta v^T\left(\frac{d}{dx}\left(E(x)I(x)\frac{d^2V}{dx^2}\right)\right)_{X_1}^{X_2} \\ &\Big| -\delta\left(\frac{dV}{dx}\right)^T E(x)I(x)\left(\frac{d^2V}{dx^2}\right)_{X_1}^{X_2} \\ &\Big| +\delta V^T\left(\frac{d(pV)}{dx}\right)_{X_1}^{X_2}\Big| \int_{X_1}^{X_2} \delta\left(\frac{d^2V}{dx^2}\right)^T E(x)I(x)\left(\frac{d^2V}{dx^2}\right)dx \\ &-\int_{X_1}^{X_2} \delta\left(\frac{dV}{dx}\right)\left(\frac{d(pV)}{dx}\right)dx = 0 \end{aligned} \tag{9.20}$$

At $X_1$ and $X_2$, shear force and bending moments are zero (at nodes 1 and 2). The final equation for constant compressive load p results in

$$\int_{X_1}^{X_2} \delta\left(\frac{d^2V}{dx^2}\right)^T E(x)I(x)\left(\frac{d^2V}{dx^2}\right)dx - p\int_{X_1}^{X_2} \delta\left(\frac{dV}{dx}\right)^T\left(\frac{dV}{dx}\right)dx = 0. \tag{9.21}$$

The transverse displacement at any point within the element can be expressed in terms of nodal transverse displacements and rotations as

$$V(\xi) = N_1V_1 + N_2\theta_1 + N_3V_2 + N_4\theta_2 \tag{9.22}$$

where $N_1,N_2,N_3,N_4$ are shape functions provided in the previous section

$$V = \left\{N_1\ N_2\ N_3\ N_4\right\}\begin{Bmatrix} V_1 \\ \theta_1 \\ V_2 \\ \theta_2 \end{Bmatrix} = \{N\}^T\{q\} \tag{9.23}$$

where

$$\{N\}^T = \left\{N_1\ N_2\ N_3\ N_4\right\} \text{ and } \{q\}^T = \left\{V_1\ \theta_1\ V_2\ \theta_2\right\} \tag{9.24}$$

$$\delta V = \{N\}^T\delta\{q\} \tag{9.25}$$

$$\delta\left(\frac{d^2V}{dx^2}\right)^T = \delta\{q\}^T\frac{d^2\{N\}^T}{dx^2}. \tag{9.26}$$

Similarly

$$\delta\left(\frac{dV}{dx}\right)^T = \delta\{q\}^T\frac{d\{N\}^T}{dx}. \tag{9.27}$$

Substituting Eqs. (9.26) and (9.27) in Eq. (9.21) results is

$$\Delta\{q\}^T\left[\int_{X_1}^{X_2}\frac{d^2\{N\}^T}{dx^2}E(x)I(x)\frac{d^2\{N\}}{dx^2}dx\{q\} - p\int_{X_1}^{X_2}\frac{d\{N\}^T}{dx}\frac{d\{N\}}{dx}dx = 0\right]. \tag{9.28}$$

Since $\delta\{q\}^T$ is an arbitrary virtual displacement, hence $\delta\{q\}^T \neq 0$ results in

$$\left[\int_{X_1}^{X_2}\frac{d^2\{N\}^T}{dx^2}E(x)I(x)\frac{d^2\{N\}}{dx^2}dx\{q\} - p\int_{X_1}^{X_2}\frac{d\{N\}^T}{dx}\frac{d\{N\}}{dx}dx\{q\} = 0\right] \tag{9.29}$$

written in matrix form as

$$[K]\{q\}_i - p_i[K_G]\{q\}_i = 0 \tag{9.30}$$

where [K] and $[K_G]$ are the stiffness matrix and geometrical stiffness matrix, respectively. Solving Eq. (9.30) for eigenvalue provides the eigenpair $(p, \{q\})_i$ corresponding to critical load factor and mode shape vector respectively for the ith mode. Instead of compressive load if the column is subjected to a tensile load, the critical buckling load factor will be negative instead of positive value. This happens due to the conversion of eigenvalue problem provided in Eq. (9.30) into standard form resulting in a negative definite form of the matrix. Eigenvalues of negative definite matrices are negative. From Eq. (9.30), the [K] and $[K_G]$ matrices can be written as

$$[K] = \int_{x_1}^{x_2}\frac{d^2\{N\}^T}{dx^2}E(x)I(x)\frac{d^2\{N\}}{dx^2}dx \tag{9.31}$$

$$[K_G] = \int_{x_1}^{x_2}\frac{d\{N\}^T}{dx}\frac{d\{N\}}{dx}dx. \tag{9.32}$$

Converting integration in Eqs. (9.31) and (9.32) into Gaussian quadrature results in

$$[\mathrm{K}] = \int_{-1}^{1} \frac{\mathrm{d}^2\{\mathrm{N}\}^{\mathrm{T}}}{\mathrm{d}\xi^2}\left(\frac{2}{l}\right)^2 \mathrm{E}(\xi)\mathrm{I}(\xi)\frac{\mathrm{d}^2\{\mathrm{N}\}}{\mathrm{d}\xi^2}\left(\frac{2}{l}\right)^2 \frac{l}{2}\mathrm{d}\xi \tag{9.33}$$

$$[\mathrm{K}] = \frac{8}{l^3}\int_{-1}^{1} \frac{\mathrm{d}^2\{\mathrm{N}\}^{\mathrm{T}}}{\mathrm{d}\xi^2}\mathrm{E}(\xi)\mathrm{I}(\xi)\frac{\mathrm{d}^2\{\mathrm{N}\}}{\mathrm{d}\xi^2}\mathrm{d}\xi. \tag{9.34}$$

Similarly

$$[\mathrm{K_G}] = \frac{2}{l}\int_{-1}^{1} \frac{\mathrm{d}\{\mathrm{N}\}^{\mathrm{T}}}{\mathrm{d}\xi}\frac{\mathrm{d}\{\mathrm{N}\}}{\mathrm{d}\xi}\mathrm{d}\xi. \tag{9.35}$$

Integrating Eqs. (9.34) and (9.35) assuming Young's modulus and moment of inertia to be constant, the following matrices are obtained

$$[\mathrm{K}] = \frac{\mathrm{EI}}{l^3}\begin{bmatrix} 12 & 6l & -12 & 6l \\ 6l & 4l^2 & -6l & 2l^2 \\ -12 & -6l & 12 & -6l \\ 6l & 2l^2 & -6l & 4l^2 \end{bmatrix} \tag{9.36}$$

$$[\mathrm{K_G}] = \frac{1}{30l}\begin{bmatrix} 36 & 3l & -36 & 3l \\ 3l & 4l^2 & -3l & -l^2 \\ -36 & -3l & 36 & -31 \\ 3l & -l^2 & -3l & 4l^2 \end{bmatrix}. \tag{9.37}$$

## 9.4 Finite Element Formulation Using Energy Method

Figure 9.4 shows a pin-ended column having axial compressive load on both ends showing the deformed (buckled) shape. The total potential in the column consists of strain energy and the work done by the axial load p, and the algebraic sum of both is given as

$$\pi = \mathrm{U} - \mathrm{W} \tag{9.38}$$

where U is the strain energy and W is the work potential. Total potential is given as

$$\pi = \frac{1}{2}\int_{\mathrm{V}} \sigma\varepsilon \mathrm{dV} - \int_{\mathrm{X_1}}^{\mathrm{X_2}} \mathrm{p(x)(ds - dx)} \tag{9.39}$$

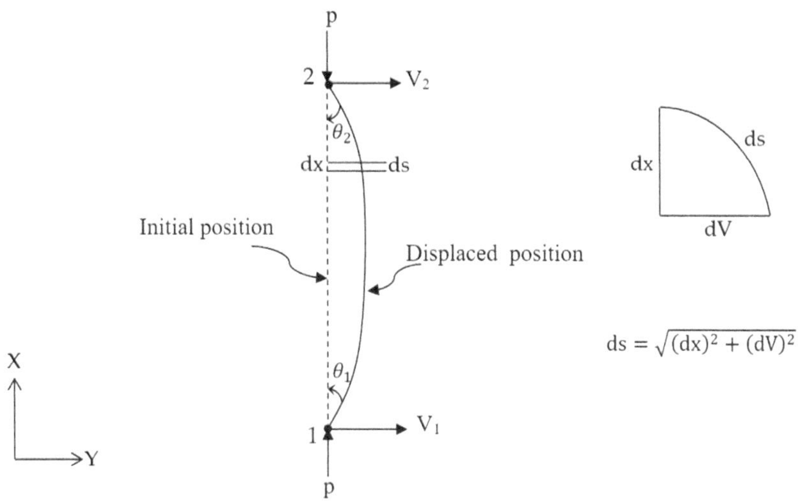


**Fig. 9.4** Column with compressive load showing infinitesimal arc length

$$\sigma\varepsilon = \frac{\sigma^2}{E}. \tag{9.40}$$

Bending stress $\sigma$ is given as

$$\sigma = \frac{My}{I} \tag{9.41}$$

where

$$M = EI\frac{d^2V}{dx^2}. \tag{9.42}$$

Equation (9.40) can be rewritten as

$$\sigma\varepsilon = \frac{\sigma^2}{E} = E\left(\frac{d^2V}{dx^2}\right)^2 y^2. \tag{9.43}$$

Arc-length ds can be written as

$$ds = \sqrt{(dx)^2 + (dV)^2} \tag{9.44}$$

$$ds = dx\left(1 + \left(\frac{dV}{dx}\right)^2\right)^{\frac{1}{2}}. \tag{9.45}$$

Expanding Eq. (9.45) using bi-nominal theorem and neglecting higher order terms can be written as

$$ds = dx\left(1 + \frac{1}{2}\left(\frac{dV}{dx}\right)^2\right) \tag{9.46}$$

$$ds - dx = \frac{1}{2}\left(\frac{dV}{dx}\right)^2 dx. \tag{9.47}$$

Substituting Eqs. (9.43) and (9.47) in Eq. (9.39) can be written for the constant compressive load as

$$\pi = \frac{1}{2}\int_{X_1}^{X_2} E(x)\left(\frac{d^2V}{dx^2}\right)^2 \left[\int_A y^2 dA\right] dx - \frac{p}{2}\int_{X_1}^{X_2}\left(\frac{dV}{dx}\right)^2 dx \tag{9.48}$$

where

$$I(x) = \int_A y^2 dA. \tag{9.49}$$

Total potential after all simplifications for constant compressive load is given as

$$\pi = \frac{1}{2}\int_{X_1}^{X_2} E(x)I(x)\left(\frac{d^2V}{dx^2}\right)^2 dx - \frac{p}{2}\int_{X_1}^{X_2}\left(\frac{dV}{dx}\right)^2 dx. \tag{9.50}$$

Taking variation of $\pi$ with respect to nodal variables results in the same equation that was obtained using the Galerkin Method provided in Eq. (9.21):

$$\delta\pi = \int_{X_1}^{X_2} \delta\left(\frac{d^2V}{dx^2}\right)^T E(x)I(x)\left(\frac{d^2V}{dx^2}\right) dx - p\int_{X_1}^{X_2} \delta\left(\frac{dV}{dx}\right)^T \left(\frac{dV}{dx}\right) dx = 0. \tag{9.51}$$

## 9.5 Closed Form Solutions

The governing differential equation for Euler column buckling for constant E, I, and p is

$$EI\frac{d^2V}{dx^2} + pV = 0. \tag{9.52}$$

This can be rewritten as

$$\frac{d^2V}{dx^2} + \left(\frac{p}{EI}\right)V = 0. \tag{9.53}$$

General solution to Eq. (9.53) is given as

$$V = A\cos\left(\sqrt{\frac{p}{EI}}x\right) + B\sin\left(\sqrt{\frac{p}{EI}}x\right). \tag{9.54}$$

Pinned–Pinned column, with boundary conditions, at x = 0 and l; v = 0 results in

$$A = 0\ \&B = 0 \text{ or } \sqrt{\frac{P}{EI}}l = n\,\pi. \tag{9.55}$$

A = 0 and B = 0 result in a trivial solution. But $\sqrt{\frac{P}{EI}}l = n\,\pi$ provides the critical load factor for different modes. The fundamental critical load is obtained using n = 1

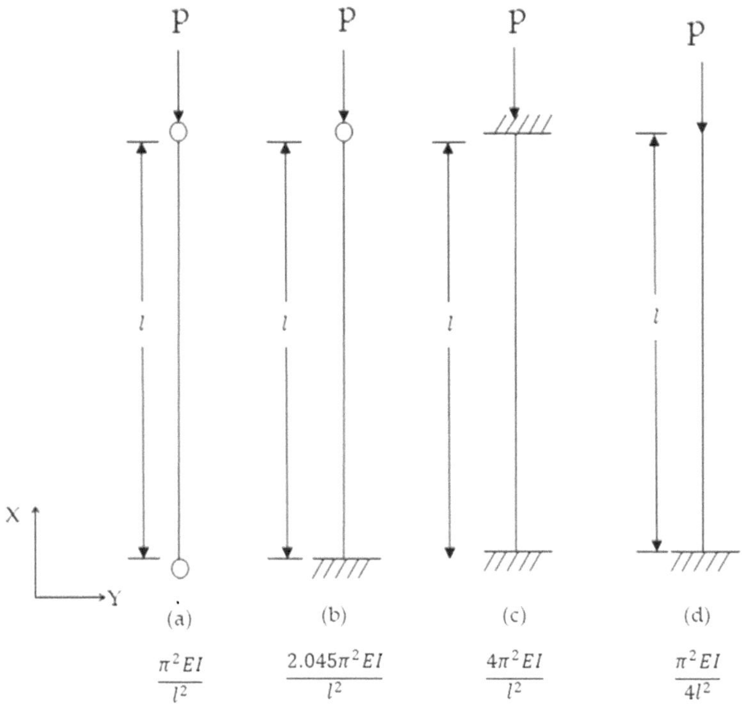


**Fig. 9.5** Euler column with different end conditions and fundamental critical load (a) Pinned–Pinned (b) Pinned–Fixed (c) Fixed–Fixed (d) Fixed-Free

$$p_l = \frac{\pi^2 EI}{l^2}. \tag{9.56}$$

Fig. 9.5 provides the fundamental critical load factor for Euler column for four types of boundary conditions.

## 9.6 Example 1

To study the buckling of Euler column with different boundary conditions, the following properties are considered:

$$E = 2.1 \times 10^{11} \mathrm{N/m^2}$$

$$I = 0.08333 \ \mathrm{m^4}$$

$$l = 100\mathrm{m}$$

For this problem, the solutions obtained from various methods are shown in Table 9.1. The table provides closed form solution for different boundary conditions, solution using the Finite Element Formulation developed in this chapter, and solution from FEAST$^{SMT}$.

Solution provided in Table 9.1 shows good agreement between the closed form solution, present formulation, and FEAST$^{SMT}$. These solutions were obtained with one element for the Fixed–Free boundary condition column and two elements for all other cases.

**Table 9.1** Critical load factor for Euler column buckling

| Boundary Condition | Critical Load Factor × $10^6$ N | | |
|---|---|---|---|
| | Closed form solution | Present FE formulation | FEAST$^{SMT}$ |
| Pinned–Pinned | 17.272 | 17.4016 | 17.3977 |
| Pined–Fixed | 35.32124 | 35.5863 | 36.2212 |
| Fixed–Fixed | 69.088 | 69.606 | 69.913 |
| Fixed–Free | 4.3180 | 4.3504 | 4.3502 |

## 9.7 Example 2

Compute the stiffness matrix for moment of inertia varying linearly within the element and assuming Young's modulus to be constant

$$[\mathrm{K}] = \frac{8\mathrm{E}}{\mathrm{l}^3}\int_{-1}^{1}\frac{\mathrm{d}^2\{\mathrm{N}\}^\mathrm{T}}{\mathrm{d}\xi^2}\mathrm{I}(\xi)\frac{\mathrm{d}^2\{\mathrm{N}\}}{\mathrm{d}\xi^2}\,\mathrm{d}\xi \tag{9.57}$$

where

$$\mathrm{I}(\xi) = \frac{1-\xi}{2}\mathrm{I}_1 + \frac{1+\xi}{2}\mathrm{I}_2 \tag{9.58}$$

where $\mathrm{I}_1, \mathrm{I}_2$ are the moments of inertia at nodes 1 and 2, respectively. The above equation can be rewritten as

$$\mathrm{I}(\xi) = \mathrm{I}_\mathrm{a} + \mathrm{I}_\mathrm{d}\xi \tag{9.59}$$

where

$$\mathrm{I}_\mathrm{a} = \frac{(\mathrm{I}_1 + \mathrm{I}_2)}{2};\ \mathrm{I}_\mathrm{d} = \frac{(\mathrm{I}_2 - \mathrm{I}_1)}{2}. \tag{9.60}$$

Stiffness matrix corresponding to linear variation in the moment of inertia can be written as

$$[\mathrm{K}] = [\mathrm{K1}] + [\mathrm{K2}] \tag{9.61}$$

where

$$[\mathrm{K1}] = \frac{\mathrm{EI}_\mathrm{a}}{\mathrm{l}^3}\begin{bmatrix} 12 & 6\mathrm{l} & -12 & 6\mathrm{l} \\ 6\mathrm{l} & 4\mathrm{l}^2 & -6\mathrm{l} & 2\mathrm{l}^2 \\ -12 & -6\mathrm{l} & 12 & -6\mathrm{l} \\ 6\mathrm{l} & 2\mathrm{l}^2 & -6\mathrm{l} & 4\mathrm{l}^2 \end{bmatrix} \tag{9.62}$$

$$[\mathrm{K2}] = \frac{\mathrm{EI}_\mathrm{d}}{\mathrm{l}^3}\begin{bmatrix} 0 & -2\mathrm{l} & 0 & 2\mathrm{l} \\ -2\mathrm{l} & -2\mathrm{l}^2 & 2\mathrm{l} & 0 \\ 0 & 2\mathrm{l} & 0 & -2\mathrm{l} \\ 2\mathrm{l} & 0 & -2\mathrm{l} & 2\mathrm{l}^2 \end{bmatrix}. \tag{9.63}$$

## 9.8 Example 3

Compute the stiffness matrix assuming the moment of inertia to be a constant and linear variation of Young's modulus within the element

$$[K] = \frac{8I}{l^3}\int_{-1}^{1} \frac{d^2\{N\}^T}{d\xi^2} E(\xi) \frac{d^2\{N\}}{d\xi^2}\, d\xi \tag{9.64}$$

where

$$E(\xi) = \frac{1-\xi}{2}E_1 + \frac{1+\xi}{2}E_2 \tag{9.65}$$

where $E_1, E_2$ are Young's modulus at nodes 1 and 2, respectively. The above equation can be rewritten as

$$E(\xi) = E_a + E_d\xi \tag{9.66}$$

where

$$E_a = \frac{(E_1 + E_2)}{2};\ E_d = \frac{(E_2 - E_1)}{2}. \tag{9.67}$$

Stiffness matrix corresponding to linear variation in Young's modulus can be written as

$$[K] = [K1] + [K2] \tag{9.68}$$

where

$$[K1] = \frac{E_a I}{l^3}\begin{bmatrix} 12 & 6l & -12 & 6l \\ 6l & 4l^2 & -6l & 2l^2 \\ -12 & -6l & 12 & -6l \\ 6l & 2l^2 & -6l & 4l^2 \end{bmatrix} \tag{9.69}$$

$$[K2] = \frac{E_d I}{l^3}\begin{bmatrix} 0 & -2l & 0 & 2l \\ -2l & -2l^2 & 2l & 0 \\ 0 & 2l & 0 & -2l \\ 2l & 0 & -2l & 2l^2 \end{bmatrix}. \tag{9.70}$$

## 9.9 Example 4

Compute the stiffness matrix for linear variation of Young's modulus and moment of inertia within the element

$$[K] = \frac{8}{l^3} \int_{-1}^{1} \frac{d^2\{N\}^T}{d\xi^2} E(\xi) I(\xi) \frac{d^2\{N\}}{d\xi^2}\, d\xi \tag{9.71}$$

where $I(\varepsilon)$, $E(\varepsilon)$ are provided in Eqs. (9.59) and (9.66). The product of these two terms can be written as

$$E(\xi)I(\xi) = E_a I_a + (E_a I_d + E_d I_a)\xi + E_d I_d \xi^2. \tag{9.72}$$

Stiffness matrix corresponding to linear variation in the moment of inertia and Young's modulus is given as

$$[K] = [K1] + [K2] + [K3] \tag{9.73}$$

$$[K1] = \frac{E_a I_a}{l^3} \begin{bmatrix} 12 & 6l & -12 & 6l \\ 6l & 4l^2 & -6l & 2l^2 \\ -12 & -6l & 12 & -6l \\ 6l & 2l^2 & -6l & 4l^2 \end{bmatrix} \tag{9.74}$$

$$[K2] = \frac{(E_a I_d + E_d I_a)}{l^3} \begin{bmatrix} 0 & -2l & 0 & 2l \\ -2l & -2l^2 & 2l & 0 \\ 0 & 2l & 0 & -2l \\ 2l & 0 & -2l & 2l^2 \end{bmatrix} \tag{9.75}$$

$$[K3] = \frac{E_d I_d}{l^3} \begin{bmatrix} 7.2 & 3.6l & -7.2 & 3.6l \\ 3.6l & 2.133l^2 & -3.6l & 1.4667l^2 \\ -7.2 & -3.6l & 7.2 & -3.6l \\ 3.6l & 1.4667l^2 & -3.6l & 2.1333l^2 \end{bmatrix}. \tag{9.76}$$

## 9.10 Knockdown Factor

Structural configurations that are used in Finite Element analysis for buckling are perfectly modeled and imposed boundary conditions at appropriate nodal locations. Due to this reason, the buckling load factor is overestimated. The same configuration, when experimentally tested, buckles well ahead of Finite Element predictions. The ratio between experimental prediction and Finite Element estimate is known as the

**Knockdown** factor. This factor is always less than one and is multiplied with the Finite Element estimate for all practical purposes.

## 9.11 Exercise

9.11.1 Consider a column of length 10 m long with Young's modulus assumed as E $= 2.0 \times 10^{11}$ N/m$^2$. Solve this column to estimate the buckling load factor for the following boundary conditions and compare it with the closed form solutions. Assume 10 elements to obtain the solutions. Boundary conditions for the columns may be assumed as (a) Pinned–Pinned, (b) Pinned–Fixed, (c) Fixed–Free, and (d) Fixed–Fixed:

(i) Plot the buckling load factor for a circular cross section where the radius ranges from 0.1 m to 1 m with an increment of 0.1 m.
(ii) Plot the buckling load factor for a square cross section where breadth and width range from 0.1 $\sqrt{\pi}$ m to $\sqrt{\pi}$ m with an increment of 0.1 $\sqrt{\pi}$ m.
(iii) Inference from both the problem. Plot the buckling load factor versus sectional area in the same graph to have a clear picture of the influence of sectional properties on the buckling load factor.

9.11.2 Solve problem 9.8.1 assuming the length of the column as 15 m. Write your inference based on both the problems on the buckling load factor of the column.

9.11.3 Estimate the buckling load factor for all four types of boundary conditions for the configuration shown in Fig. 9.6 with l = 1 m and EI can be assumed conveniently.

9.11.4 Predict the buckling plane for the rectangular column as shown in Fig. 9.7 for the cases (i) a > b and (ii) a < b. Justify your answer.

9.11.5 A circular column of diameter 1 m at one end which reduces to 0.5 m gradually at the other end having a length of 10 m is discretized with five elements having 6 nodes. Moments of inertia for nodes from one to six in m$^4$ are given as 0.0491, 0.03934, 0.0307878, 0.021482, 0.0122744, and 0.003068, respectively. Young's modulus is assumed as $2.0 \times 10^{11}$ N/m$^2$. Moment of inertia is varying linearly within the element. The following boundary conditions are assumed: (i) Fixed–Free, (ii) Pinned–pinned, (iii)

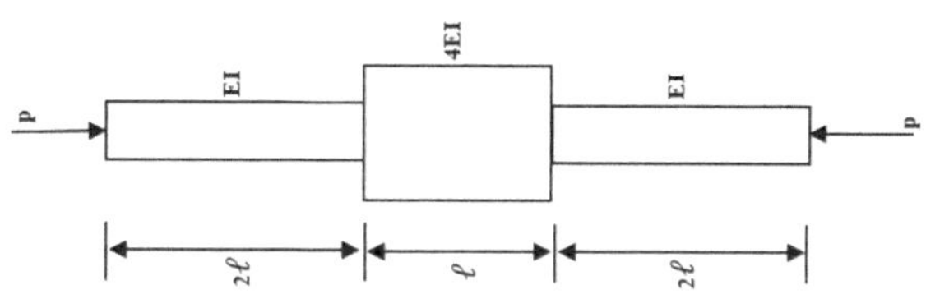


**Fig. 9.6** Column having non-uniform cross-sectional properties

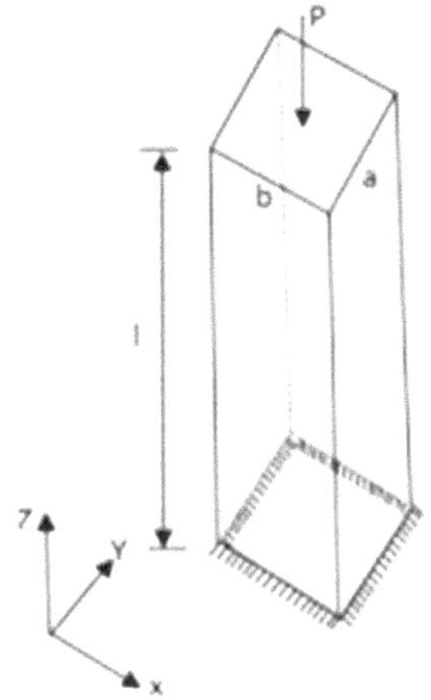


**Fig. 9.7** Fixed–Free rectangular column

Fixed–Pinned, and (iv) Fixed–Fixed. Predict the buckling load factor for (a) Assuming a fixed end at the node with a maximum moment of inertia and (b) Assuming a fixed end at the node with a minimum moment of inertia. Write your observations based on the results.

9.11.6 A circular column of diameter 1 m with length 10 m is discretized with five elements having 6 nodes. Young's modulus is varying from ($2.0 \times 10^{11}$ to $1.0 \times 10^{10}$) N/m$^2$. Young's modulus for nodes one to six is given as (20, 16.2, 12.4, 8.6, and 4.8,1) $\times$ $10^{10}$ N/m$^2$, respectively. The following boundary conditions are assumed: (i) Fixed–Free, (ii) Pinned–pinned, (iii) Fixed–Pinned, and (iv) Fixed–Fixed. Predict the buckling load factor for (a) Assuming a fixed end at the node with maximum Young's modulus and (b) Assuming a fixed end at the node with minimum Young's modulus. Write your observation based on the results.

9.11.7 A circular column of diameter 1 m at one end which reduces to 0.5 m at the other and with length of 10 m is discretized with five elements having 6 nodes. The moment of inertia for nodes from one to six in m$^4$ is given as 0.0491, 0.03934, 0.0307878, 0.021482, 0.0122744, and 0.003068. Young's modulus is varying from $2.0 \times 10^{11}$ to $1.0 \times 10^{10}$ N/m$^2$. Young's modulus for nodes one to six is given as (20, 16.2, 12.4, 8.6, and 4.8, 1) $\times$ $10^{10}$ N/m$^2$. The following boundary conditions are assumed: (i) Fixed–Free, (ii) Pinned–pinned, (iii) Fixed–Pinned, and (iv) Fixed–Fixed. Predict the buckling load factor for (a) Assuming a fixed end at the node with a maximum moment of inertia and Young's modulus and (b) Assuming a fixed end at the node with a minimum moment of inertia and Young's modulus. Write your observation based on the results.

## Bibliography

Cook RD, Malkus DS, Plesha ME, Witt RJ (2007) Concepts and applications of finite element analysis. Wiley (Asia) Pte. Ltd., ISBN-13 978–81–265–1336–9

MacNeal RH, Harder RL (1985) A proposed standard set of problem to test finite element accuracy. Finite Element in Analysis and Design 1(1):3–20

Megson THG (2007) Aircraft Structures for Engineering Students, Elsevier, First edition 2007. ISBN-13:978-0-75066-7395

Tirupathi R, Chandrapatla R, Ashok D Belegundu (2007) Introduction to finite elements in engineering. 4th edn, Prentice Hall, ISBN 13:978-0132162746

Zienkiewicz OC, Taylor RL (2000) The finite element method, Vol 1, The Basis, Fifth Edition, Butterworth–Heinemann, ISBN 0 7506 5049 4

# Chapter 10
# Features of FEAST$^{SMT}$ Software

**P. V. Anil Kumar**

*Knowing is not enough; we must apply. Being willing is not enough; we must do—Leonardo da Vinci*

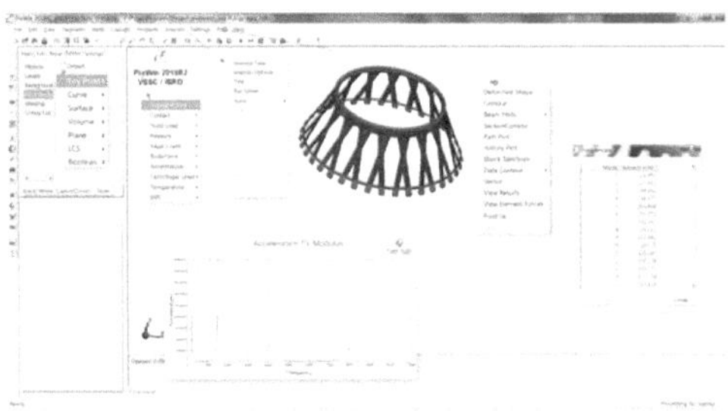

## 10.1 Introduction

Finite element analysis involves pre-processing, solving, and post-processing phases. The goals of pre-processing are to develop an appropriate finite element mesh, assign suitable material properties, and apply boundary conditions in the form of restraints and loads. Developing the mesh is usually the most time-consuming task in FEA. Knowledge, experience, and engineering judgment are very important in modeling the geometry of a system. In many cases, finely detailed geometrical features play only an aesthetic role and have negligible effects on the performance of the engineering system. These features can be deleted, ignored or simplified, though this may not be true in some cases, where a fine geometrical change can give rise to a significant difference in the simulation results.

P. V. A. Kumar (✉)
Former DH, FSDD/SDAG/STR, Veli, VSSC, Thiruvananthapuram 695022, India
e-mail: pvanilkumar1@gmail.com

 
S. U. Nair and S. Somanath (eds.), *Introduction to Finite Element Analysis*,
https://doi.org/10.1007/978-981-19-7989-7_10

FEAST$^{SMT}$ (acronym for Finite Element Analysis of Structures) is a structural analysis software developed by VSSC/ISRO utilizing the expertise of generations of ISRO scientists and engineers. This software has been extensively used by the structural design community in all ISRO centers across India apart from academia and other institutions. It has conventional capabilities to carry out structural and thermal engineering analysis as expected from the software of a similar type, apart from meeting certain specific needs of the aerospace community. Static, stability, and dynamic analysis capabilities are broad areas where finite element analysis is applied, which are incorporated into FEAST$^{SMT}$. The FEAST$^{SMT}$ software is continually updated based on feedback from the user community as well as by adding new features that might be required to solve a specific problem encountered. PreWin is an interactive Pre and Post processor that can be used to create finite element models and the associated input file for FEAST$^{SMT}$ and also for viewing the results obtained from FEAST$^{SMT}$. The Solver is seamlessly integrated with PreWin as a single application. Updated versions of FEAST$^{SMT}$ are released every year with more features.

A trial version of FEAST$^{SMT}$ can be freely downloaded from the site https://feast.vssc.gov.in.

## 10.2 Analysis Capabilities

FEAST$^{SMT}$ is capable of handling many types of analyses with both metallic and composite material models. Some of the capabilities are listed below:

- Linear static
- Free vibration
- Buckling
- Viscoelasticity
- Thermoelasticity
- Heat transfer
- Transient response
- Frequency response
- Random response
- Shock response
- Electro-static
- Magneto-static.

## 10.3 Graphical User Interface

PreWin provides graphic interfaces (Fig. 10.1) to help in the creation and manipulation of the finite element analysis data, in both menu-driven and manual command-entry approaches. Context-sensitive help is available for all operations. A log of all operations used in a model-making session is maintained in a file that may be re-run

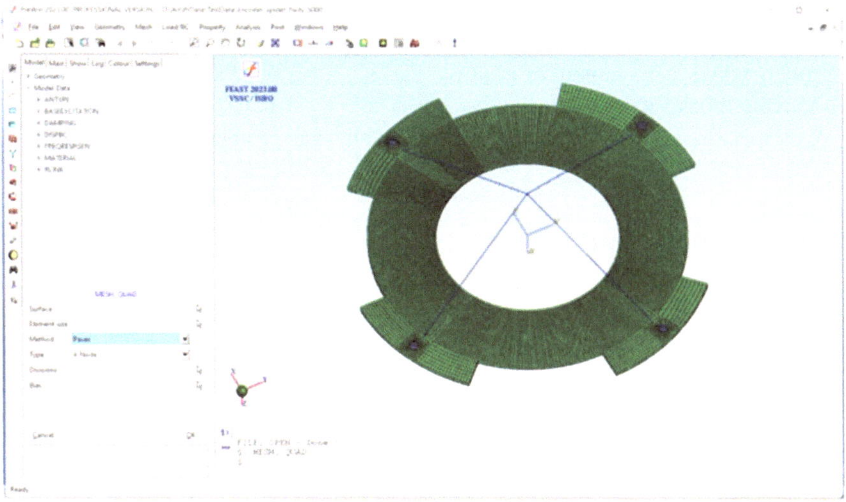

**Fig. 10.1** PreWin graphical user interface

with necessary changes to update the model. A multilevel UNDO/REDO feature is provided to recover from previous changes.

## 10.4 Model Viewing and Displaying

PreWin can show or mask all geometric entities, nodes, elements, loads, boundary conditions, and coordinate systems (Fig. 10.2) individually with a range of navigational options. The user may rotate, zoom or pan models with mouse and key controls. There are also features such as:

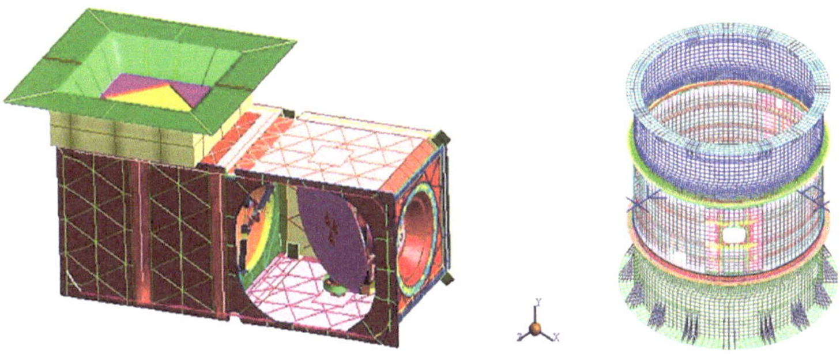

**Fig. 10.2** Attribute-based shading

- Choose standard or user-definable views
- Wireframe, hidden, filled, and shaded views
- Independent management of multiple views
- Moving cross-sectional plots
- Color shading based on material properties, physical properties, and element type.

## 10.5 Selection Options

It requires multiple options to select a set of nodes, elements or element faces from a complex FE model for an operation such as applying loads. PreWin has options to select all objects that are either fully or partially within a rectangular, circular or polygonal region. There are also options to pick the topmost objects from the current view of the 3D model based on element attributes, geometric associativity, and previously defined groups. The selected sets may be combined using Boolean operations—intersection, union or subtraction

## 10.6 Building Geometry Model

Real structures, components or domains are in general very complex and have to be reduced to a manageable geometry. Curved parts of the geometry and its boundary can be modeled using curves and curved surfaces (Fig. 10.3). PreWin provides graphic interfaces to help in the creation and manipulation of geometrical objects. Points can be created simply by keying in the coordinates. Lines and curves can be created by connecting the points or nodes. Surfaces can be created by connecting, rotating or translating the existing lines or curves; and solids can be created by connecting, rotating or translating the existing surfaces. Points, lines and curves, surfaces, and solids can be translated, rotated or reflected to form new ones.

There are numerous computer-aided design (CAD) software packages used for engineering design that can produce files containing the geometry of the designed engineering system. A common approach is to develop the mesh directly on the CAD geometry. PreWin can read the geometry definitions produced in this manner

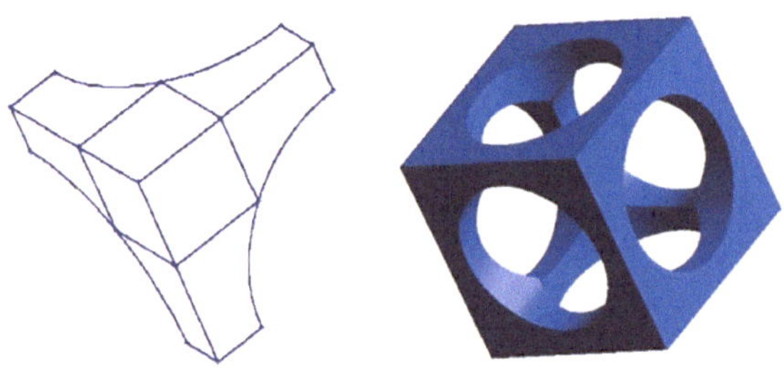

**Fig. 10.3** Geometric modelling

in neutral geometry formats, viz. IGES (Initial Graphics Exchange Specification) and STEP (Standard for the Exchange of Product model data). This can significantly save time when creating the geometry of the models.

It should be noted that the geometry is eventually represented by a collection of elements, and the curves and curved surfaces are approximated by piecewise straight lines or flat surfaces, if linear elements are used. Consequently, in many cases, complex objects read directly from a CAD file may need to be modified and simplified before performing meshing or discretization. PreWin provides several commands to perform the cleanup operation such as removing small holes and merging small faces and small edges.

PreWin provides a large set of modeling tools mimicking CAD software. A geometric model is created using points, curves, surfaces, and volumes and expanded through the following transformations:

- Translation,
- Rotation,
- Reflection,
- Scaling,
- Sweeping,
- Extrusion,
- Projection.

## 10.7 Mesh Generation

The geometry meshes with a mapping algorithm or an automatic free-meshing algorithm. The algorithm first maps a rectangular grid onto a geometric region, which must, therefore, have the correct number of sides. Free-meshing automatically subdivides meshing regions into elements, with the advantages of fast meshing and easy mesh-size transitioning for a denser mesh in regions of a large gradient. Disadvantages include the generation of huge models and the generation of distorted elements (Fig. 10.4).

PreWin supports mapped and automatic meshing to generate the following types of elements:

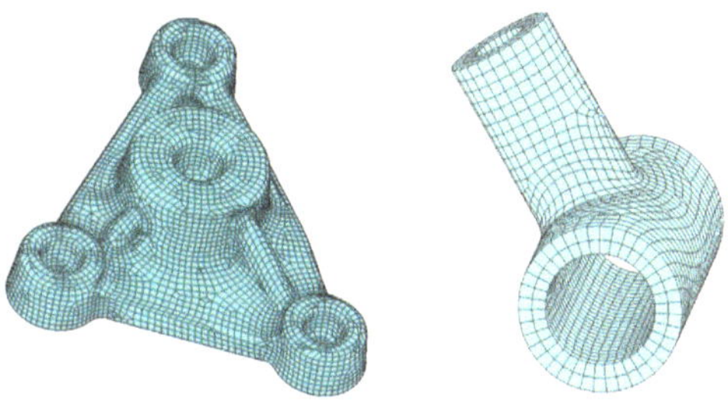

**Fig. 10.4** Mesh on surfaces and solids

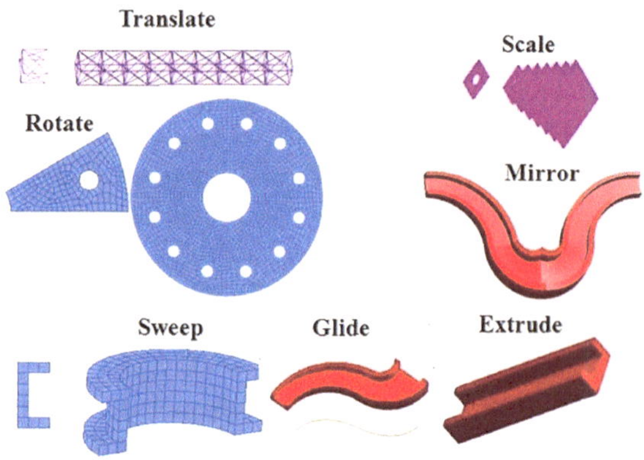


**Fig. 10.5** Transformation of FE Models

- BAR2, BAR3,
- QUAD4, QUAD8,
- TRIA3, TRIA6,
- HEXA8, HEXA20,
- TETRA.

Initial FE mesh is generated on the geometry and may be further manipulated through the following transformations (Fig. 10.5):

- Translation,
- Rotation,
- Reflection,
- Scale,
- Sweep,
- Extrude,
- Extrude to surface.

Other conversions possible on elements are as follows:

- Project to surfaces,
- Convert among various types,
- Break quadrilateral elements to refine the mesh,
- Offset 2D elements along normal direction to form 3D elements,
- Making circular cut-outs on a regular surface mesh.

## 10.8 Model Checking

Element quality is very important for a solver to provide a reasonable result. It is always important to check elemental distortion prior to the solution. A badly distorted element will cause a matrix singularity, killing the solution. A less distorted element may solve but can deliver very poor answers. Element quality is most often defined as a deviation from a perfect element. Some of the basic element quality checks:

- Highlight free edges,
- Identify duplicate elements,
- Visualize plate warpage,
- Check aspect ratio,
- Determine bounds of interior angles of elements.

## 10.9 Element Library

FEAST$^{SMT}$ has an element library to model a wider class of engineering problems—beam, truss, shell, solid, axisymmetric, plane stress/strain, mass, gap, glue, spring/scalar element, and rigid link.

## 10.10 Material Data

Many engineering systems consist of more than one material. Property of materials can be defined either for a group of elements or each individual element if needed. For different phenomena to be simulated, different sets of material properties are required. For example, Young's modulus and Poisson's ratio are required for the stress analysis of structures, whereas the thermal conductivity coefficient will be required for thermal analysis. Inputting material properties into a pre-processor is usually straightforward; all the analyst needs to do is to key in the data on material properties and specify either to which region of the geometry or which elements the data applies. However, obtaining these properties is not always easy. There are material databases to choose from, but experiments are usually required to accurately determine the property of materials to be used in the system. FEAST$^{SMT}$ supports multiple material types to represent different material behavior suitable for linear analyses—isotropic, orthotropic, incompressible, and layered orthotropic.

## 10.11 Solution Schemes

Efficient solution methods are provided for static and dynamic structural problems—Linear equation solvers are Cholesky solver, Multifrontal solver, and Pre-conditioned conjugate gradient solver. For eigenvalue solutions, Lanczos and Subspace Iteration methods are available. Analysis jobs run in parallel on multi-core processors for faster simulation of results. Solver exploits substructuring technique to create multiple independent jobs. Static substructuring is based on the static condensation technique for stress analysis and dynamic substructuring is based on Craig–Bampton method.

## 10.12 Loads and Boundary Conditions

In order to prevent the component from moving around, a constraint is needed, similar to the rigid clamp of a bending beam. Further boundary conditions define the external forces on the component, since without them, no deflection would occur. Loads like pressure or force have an effect on parts of the component, which are defined by the user, who also has to assign numerical values to them. It is preferable to apply boundary conditions to the CAD geometry, with the FEA package transferring them to the underlying model, to allow for the simpler application of adaptive and optimization algorithms. It is worth noting that the largest error in the entire process is often in the boundary conditions. All common forms of loads and constraints are available in FEAST$^{SMT}$, which are discussed as follows:

- Single point constraints,
- Multi-point constraint equations,
- Prescribed displacements,
- Nodal forces and moments,
- Pressure on the plate and solid elements,
- Nodal temperature.

Load application is possible in user-specified rectangular, spherical or cylindrical coordinate systems. It is also possible to define material and loading properties as functions of spatial coordinates, temperature, and time.

## 10.13 Post-processing

The result generated after solving the system equation is usually a vast volume of digital data. The results have to be visualized in such a way that they are easy to interpret, analyze, and present. The visualization is performed through a post processor. PreWin extracts the analysis results from FEAST$^{SMT}$ and displays the results, with appropriate conversions, in various graphical forms (Fig. 10.6):

- Visualization of deformed geometry

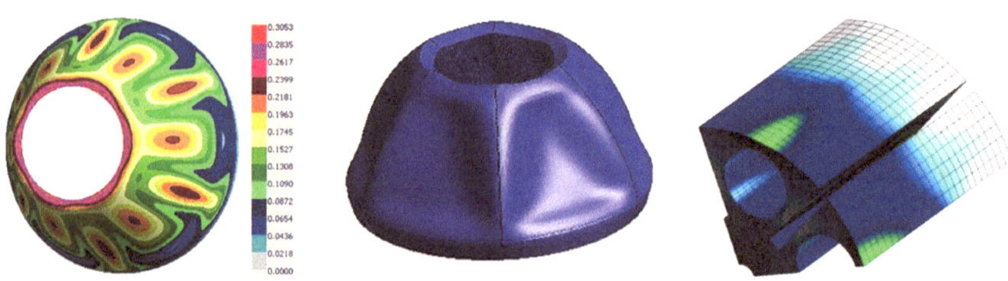


**Fig. 10.6** Post processor graphics

- Contour color-coded plots of field variables like displacements and nodal temperatures
- Animation of dynamic model behavior
- X–Y plot
- Arrow plot of vector quantities
- Tabulated display of computed nodal quantities.

Principal stresses and combined stress forms such as von Mises, and octahedral shear stresses are computed by the post processor. The computed values are presented in different coordinate systems such as cartesian, cylindrical, spherical, and other user-defined coordinate systems.

## 10.14 Structural Deflection/Mode Shapes

The display of deformed configuration enables a better understanding of the stiffness of the structure and acceptability of the finite element model. The deformed and undeformed shapes are plotted together or separately with user-specified features. PreWin scales the displacements so that the form of the displaced shape relative to the original shape of the structure is clearly visible. An option is provided to re-scale the displacements with user-specified scale factor. This enables the display of the true magnitude of the displacements.

## 10.15 Contours

The distribution of membrane or shear stresses or a combined quantity such as the von Mises stress can be illustrated graphically using a contour map. PreWin provides the options to plot the contours in the form of discrete lines joining points of equal stress and in the form of a continuous shaded image. The use of colors with legend enhances the visual impact of the contours. The option is provided to mark labels on the contour lines to help distinguish the lines in a black & white hard copy. The contour levels, equispaced over the range of computed values of the parameter, are automatically calculated. It is possible to change the contour levels to have a refinement in a particular interval. Arbitrary contour levels can also be specified to display regions having stresses above a certain level using a single color.

The contour map indicates the distribution of a single variable on a surface or on the visible surfaces of a solid object. This is sufficient in the context of stress analysis, in which the maximum stresses tend to occur at the surfaces. However, there is a need to visualize the distribution in the interior of a solid object for cases such as in thermal analysis, in which important phenomena might occur at interior points. For such cases PreWin allows the definition of a cutting plane and the display of the distribution on it.

## 10.16 Arrow Plots

Vector quantities such as displacement, principal stresses, and principal strains can be visualized in the form of arrows. The length of the arrow varies proportionally between the minimum and maximum of the magnitude of the item being displayed.

## 10.17 X–Y Plots

It is sometimes useful to produce X–Y plot of the variable along a particular curve in space such as in the case of the distribution of stress along a line approaching a stress concentration. In PreWin, the curve is defined as a path of nodes and the variable is plotted against the distance of nodes in the path from a node at one end of the path.

## 10.18 Animation

PreWin is capable of showing animations involving displaced shapes and vibration modes. The presentation of the results of transient analyses introduces the factor of time into output data. PreWin shows animation sequences of contours illustrating spatial variation of parameters as they change with time.

For animating the deflection or mode shape, a specified number of linearly scaled instances are plotted, the images are stored and then the frames are displayed with a preset time lag. In the transient analysis cases, the frames are created for discrete time steps.

Substructured and Multithreaded implementation of the solver ensures high performance by exploiting the multi-core architecture of modern computing platforms. With advanced solution algorithms, the solver is able to handle large-order problems of structural engineering. The software is designed to work on personal computers and workstations running Windows or LINUX. It is packaged with various licensing options for academic and commercial usage.

# Chapter 11
# Modeling Techniques and Interpretation of Results

**T. J. Raj Thilak**

*An experiment is a question which science poses to Nature, and a measurement is the recording of Nature's answer— Max Planck..*

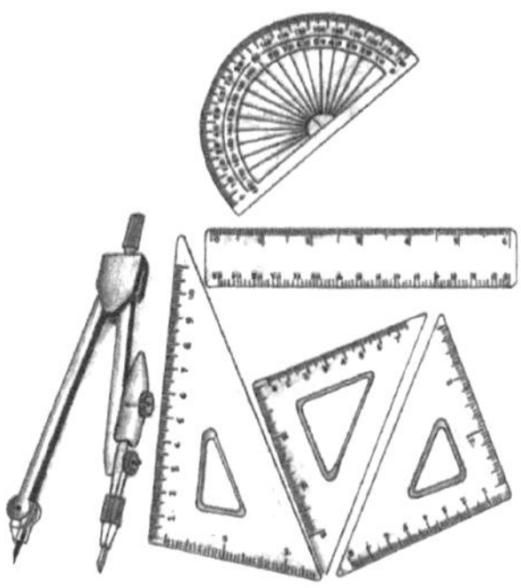

## 11.1 Introduction

In real life, all structures are three dimensional, and it will be appropriate to discretize the model using three-dimensional solid elements. Solid element is a more practical way of modeling as it can capture all finer details like fillets and corners. In a solid mesh, all outputs such as displacement and stress plots can be viewed realistically. Figure 11.1 shows the finite element discretization of a gear hub in which all finer details of the component, like fillets, are exactly captured. The CAD model was imported in FEAST$^{SMT}$ software and is meshed with solid elements. However, in a practical situation, it is very difficult to mesh all models using solid elements and it

T. J. R. Thilak (✉)
FSDD/SDAG/STR, VSSC, Veli, Thiruvananthapuram, Kerala 695022, India
e-mail: tj_rajthilak@vssc.gov.in

 
S. U. Nair and S. Somanath (eds.), *Introduction to Finite Element Analysis*,
https://doi.org/10.1007/978-981-19-7989-7_11

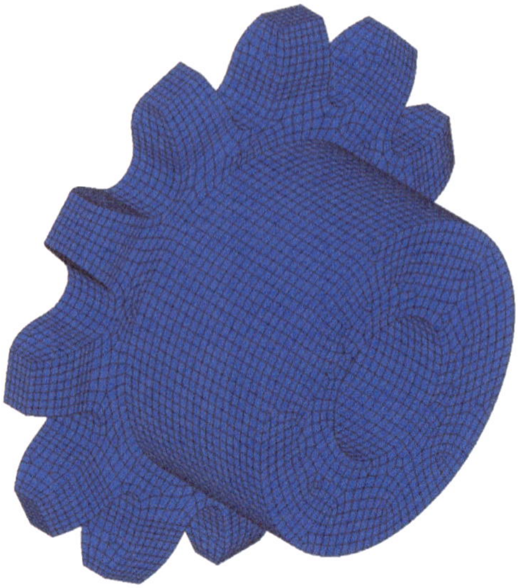

**Fig. 11.1** Gear hub meshed with solid elements

increases the number of degrees of freedom (DOF) of the model. Since the solution time is directly proportional to the square of the number of degrees in the model, the execution time will also increase. The following example illustrates use of different elements to discretize the same problem and to find end deflection of an aluminum cantilever beam subjected to end point load.

**Problem Description**:

Length—100 mm; Cross Section—Square of 5 mm; Material—Aluminum subjected to a load of P—10N.

The closed form solution gives a deflection value of $\frac{PL^3}{3EI} = 0.915$ mm.

In the finite element approach, the cantilever beam can be modeled using beam element, shell element, or solid element as shown in Fig. 11.2 to determine the displacement.

It can be observed from Table 11.1 that the beam element discretization is the better option as it has lesser DOF resulting in a lesser number of equations to be solved and still gives the most accurate result. To get the same accuracy for the displacement, a much larger number of shell and solid elements are needed which results in a greater number of DOF. Even though the number of equations to be solved for shell and solid elements are same, bandwidth of the stiffness matrix formed using solid element is usually high resulting in a larger solution time. If the beam under consideration is a stepped beam with a fillet in the step and the result that is to be predicted is stress at the fillet, it is very difficult, if not impossible, to model this structural component with beam element and the natural choice would be solid element. Suppose that the model under consideration is a cylinder made of thin sheet. To discretize the cylinder with solid elements will result in a large number of elements and hence a more practical choice would be to discretize the cylinder with shell elements.

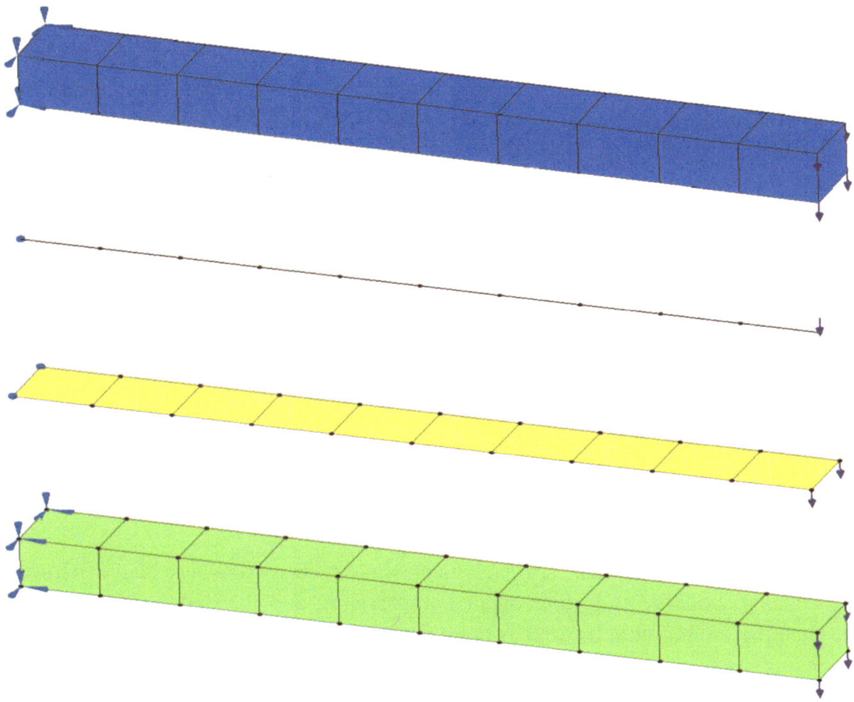

**Fig. 11.2** Cantilever beam subjected to end load with different element discretizations

**Table 11.1** Displacement values for different discretization

| Element | Mesh | No. of DOF | Displacement, mm |
|---|---|---|---|
| Beam | 5 | 30 | 0.916 |
| Shell | 5 × 1 | 60 | 0.894 |
| | 10 × 1 | 120 | 0.908 |
| | 20 × 1 | 240 | 0.913 |
| | 40 × 2 | 720 | 0.914 |
| Solid | 5 × 1 × 1 | 60 | 0.889 |
| | 10 × 1 × 1 | 120 | 0.904 |
| | 20 × 1 × 1 | 240 | 0.91 |
| | 40 × 2 × 2 | 720 | 0.912 |

## 11.2 Usage of Elements

In real life, all structures are three-dimensional and the user has to decide the right type of elements for discretizing the structure. Depending on the structure, the choice could be a single element type or a combination of different types of elements for discretization. Proper selection of element for discretization needs a proper understanding of the elements available in the software. In FEAST$^{SMT}$ software, dimension of elements is based on number of coordinates in the shape function used for interpolating the field variable within the element. Even though beam element can be used for discretizing a three-dimensional structure, the shape function used for interpolating field variable uses a single coordinate and hence it is categorized as a one-dimensional element. The features that are available in FEAST$^{SMT}$ 2023 are discussed in this book . For the sake of simplicity, only static analysis is discussed.

### *11.2.1 One-dimensional elements in FEAST$^{SMT}$*

i. Truss Element (TRUSS)
ii. Beam Element (BEAM3D)

**TRUSS Element**:

**Modeling considerations**: Truss element should be used where the structure is expected to undergo only tension or compression, and the end condition should be such that no moment transfer is possible between adjacent elements. If two truss elements are used to discretize a single member, then it is similar to introducing a pinned joint within the structure. Hence only one element should be used to discretize a structural member between two joints. For a constant cross section, the stress in the element is constant and hence the results are converged results according to the patch test. Examples include truss structures in buildings, truss girders in bridges, tie rods, thrust frames in launch vehicles, etc.

**Applicable DOF per node**: Translational DOF along the axis of the element when only area is specified as property (option-I), Translational and Rotational DOF along the axis of element if area and torsional constant is specified as properties (option-II).

**Strain**: Axial strain along the element for option-I Axial and shear strain for option-II.

**Stress**: Axial stress along the element for option-I Axial and shear stress for option-II.

**Input required**: Material properties, area of cross section for option-I, material properties, area of cross section, and torsional constant for option -II.

**Element type in FEAST$^{SMT}$**: TRUSS3D.

**Minimum number of DOF to be constrained**: All three translations DOF of the structure. Arrest rotational DOF in any one node to make the stiffness matrix nonsingular even though rotational DOF is not applicable for option-I. All three translational and rotational DOF of the structure for option-II.

**Applicable analysis types**: Linear static, geometric nonlinear static, buckling, free vibration, transient response, frequency response, random response, shock response.

**Applicable material types**: Isotropic

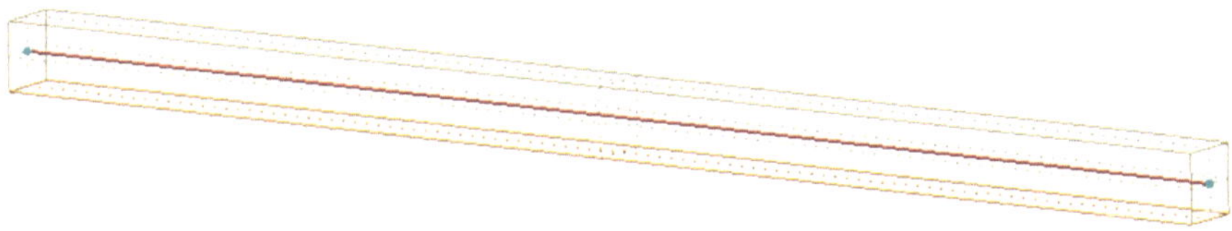

**Fig. 11.3** Beam element idealized through the cross-sectional centroid of beam structure

**Applicable load types**: Point, edge, acceleration, body force, centrifugal, temperature.

**Variants**: 2-Node element.

**3D BEAM Element**:

**Modeling considerations**: If one dimension is very large compared to other two dimensions and the load transfer is also through bending action, then beam element is used to discretize the structure. The largest dimension is the length of the beam and the shape formed by the other two dimensions is the cross section of the beam. The beam element passes through the centroid of the cross section as shown in Fig. 11.3. If the centroid is not aligned with the beam element, corresponding centroid offsets can be provided as an input to the element. The orientation of the cross section has to be provided either through a third node or through a reference direction. The third node or the reference direction along with the two nodes of beam element defines the X-Y plane of the cross section which in turn defines the local coordinate system of the element. Examples where beam element is used include chassis of an automobile, framed structures, etc.

**Applicable DOF per node**: Three translational and three rotational DOF.

**Strain**: Axial, bending, shear.

**Stress**: Axial, bending, shear.

**Input required**: Material properties, cross section shape or cross section properties, centroid offset, reference node or reference direction.

**Element type in FEAST$^{SMT}$**: BEAM3D.

**Minimum number of DOF to be constrained**: All three translations and rotations to be arrested for the structure.

**Applicable analysis types**: Linear static, geometric nonlinear static, buckling, free vibration, transient response, frequency response, random response, shock response.

**Applicable material types**: Isotropic, thermal.

**Applicable load types**: Point, edge, acceleration, body force, centrifugal, temperature.

**Variants**: 2-Node element.

### *11.2.2 Two-Dimensional elements*

i. Plane Strain Element (PSTRAIN, HPSTRAIN)
ii. Plane Stress Element (PSTRESS)
iii. Axisymmetric solid or solid of revolution Element (AXISYM, HAXISYM)
iv. Shell Element (SHELL, CSHELL)

**Plane Strain Element**:

**Modeling considerations**: If cross section of the structure does not vary along longitudinal direction, and if loading and field variable constraints act throughout the longitudinal direction, then a section of the structure along the longitudinal direction can be considered for analysis with appropriate loads and boundary conditions. Examples of plane strain analysis include gravity dam, a long cylinder subjected to internal pressure, etc.

Figure 11.4 shows a gravity dam subjected to gravity and water pressure load, only the hatched portion of the dam is sufficient to represent the dam. Since the structure is very long, the strain in the longitudinal direction is negligible or close to zero. However, the stress in the longitudinal direction cannot be neglected because of the Poisson effect. The geometry of the structure in the X–Y plane has to be discretized and the Z-coordinate for all the nodes must be 0.

**Applicable DOF per node**: Two translational DOF in X–Y plane.

**Strain**: $\varepsilon_x, \varepsilon_y, \gamma_{xy}$

**Stress**: $\sigma_x, \sigma_y, \sigma_z, \tau_{xy}$

**Input required**: Material properties.

**Element type in FEAST$^{SMT}$**: PSTRAIN, HPSTRAIN (Mixed formulation).

**Minimum number of DOF to be constrained**: Two translations and rotation with respect to Z-plane have to be arrested. The Z-plane rotation can be arrested by arresting the translation DOF at two geometrically non-coincident nodes.

**Applicable analysis types**: Linear static, geometric nonlinear static, viscoelastic, fracture

**Applicable material types**: Isotropic, orthotropic, anisotropic, viscoelastic, layered

**Fig. 11.4** Earth dam subjected to gravity load and water pressure load

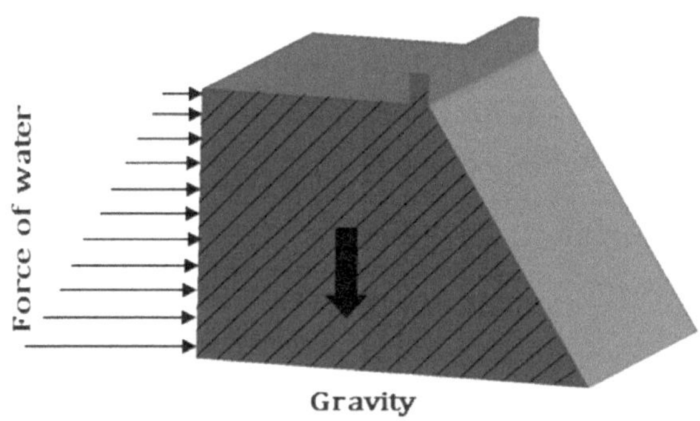

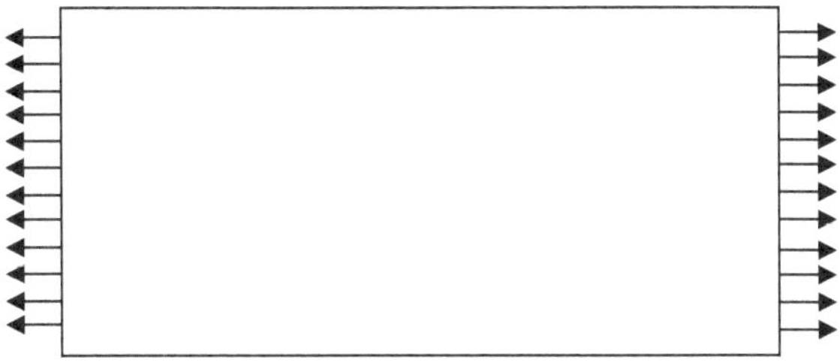

**Fig. 11.5** Thin plate with in-plane load

**Applicable load types**: Point, edge, acceleration, body force, centrifugal, temperature.

**Variants**: 3-node triangle, 6-node triangle, 4-node quadrilateral, 8-node quadrilateral.

**Plane Stress Element**:

**Modeling considerations**: If the structure under consideration is very thin compared to other two dimensions and if the loading and boundary conditions are constant throughout the thickness direction, then it is sufficient to model only a section of the structure in the thickness direction. Figure 11.5 shows an example of plane stress problem where a thin plate is subjected to a uniform edge load applied throughout the thickness of the structure. Since the structure is very thin, the stress in the thickness direction is negligible or close to zero. The strain in the thickness direction cannot be neglected because of the Poisson effect.

**Applicable DOF per node**: Two translational DOF in X–Y plane.

**Strain**: $\varepsilon_x, \varepsilon_y, \varepsilon_z, \gamma_{xy}$

**Stress**: $\sigma_x, \sigma_y, \tau_{xy}$

**Input required**: Material Properties, Thickness.

**Element type in FEAST$^{SMT}$**: PSTRESS.

**Minimum number of DOF to be constrained**: Two translations and rotation with respect to Z-plane has to be arrested by arresting the translation DOF at two geometrically non-coincident nodes.

**Applicable analysis types**: Linear static, geometric nonlinear static, fracture, steady state heat transfer, transient heat transfer.

**Applicable material types**: Isotropic, orthotropic, anisotropic, layered, thermal.

**Applicable load types**: Point, edge, acceleration, body force, centrifugal, temperature, convection, heat flux.

**Variants**: 3-node triangle, 6-node triangle, 4-node quadrilateral, 8-node quadrilateral.

**Axisymmetric Solid Element**:

**Modeling considerations**: If the structure, loading, and its boundary conditions are symmetric with respect to an axis then a section of structure along the hoop direction can be considered for analysis. A thick cylinder with internal or external pressure and a rocket nozzle as shown in Fig. 11.6 are typical examples of axisymmetric analysis.

**Fig. 11.6** Typical rocket nozzle with internal pressure

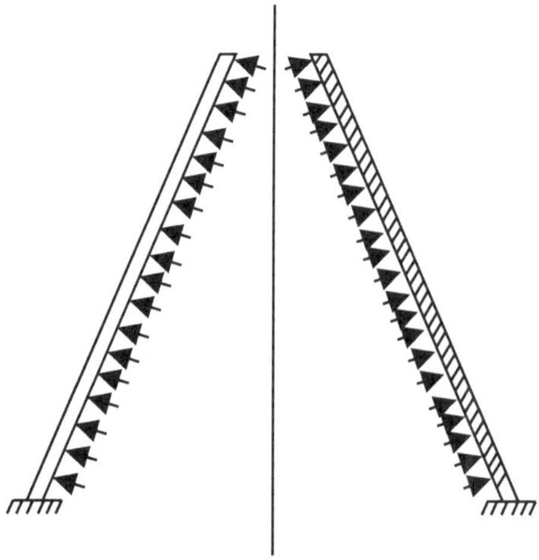

In FEAST[SMT] software Y-axis is the axis of revolution and the structure has to be modeled in first and/or fourth quadrant of X–Y plane, where X-axis is the radial direction, Y-axis is the longitudinal direction, and the Z-axis is the hoop direction for this element.

**Applicable DOF per node**: Two translational DOF in X–Y plane.

**Strain**: $\varepsilon_x, \varepsilon_y, \varepsilon_z, \gamma_{xy}$

**Stress**: $\sigma_x, \sigma_y, \sigma_z, \tau_{xy}$

**Input required**: Material Properties.

**Element type in FEAST[SMT]**: AXISYM, HAXISYM (Mixed Formulation).

**Minimum number of DOF to be constrained**: Y Translation DOF to be arrested.

**Applicable analysis types**: Linear static, gometric nonlinear static, viscoelastic, steady state heat transfer, transient heat transfer

**Applicable material types**: Isotropic, orthotropic, anisotropic, layered, thermal

**Applicable load types**: Point, edge, acceleration, body force, centrifugal, temperature, convection, heat flux.

**Variants**: 3-node triangle, 6-node triangle, 4-node quadrilateral, 8-node quadrilateral.

**Shell Element**:

**Modeling considerations**: If the thickness of a structure is small compared to the other two dimensions and the structure resists load through bending as well as membrane action, then shell element can be used to discretize the structure. Only a plane of the structure is modeled while the thickness of the structure is an input for the analysis. Usually, the mid-plane will be considered for the analysis even though other planes can also be modeled with mid-plane offset as an additional input.

**Applicable DOF per node**: Three translations and three rotations in global coordinate system (rotation DOF with respect to element normal is absent in element coordinate system).

**Strain**: $\varepsilon_x, \varepsilon_y, \varepsilon_z, \gamma_{xy}, \gamma_{xz}, \gamma_{yz}$

**Stress**: $\sigma_x, \sigma_y, \sigma_z, \tau_{xy}, \tau_{xz}, \tau_{yz}$ (Normal stress in the thickness direction is 0 in element coordinate system).

**Input required**: Material Properties, Thickness, Mid-plane offset. For layered element, no. of layers, fibre orientation, and layup sequence.

**Element type in FEAST$^{SMT}$**: SHELL, CSHELL (For layered properties).

**Minimum number of DOF to be constrained**: All the six DOF of the structure has to be arrested.

**Applicable analysis types**: Linear static, geometric nonlinear static, buckling, free vibration, transient response, frequency response, random response, shock response

**Applicable material types**: Isotropic, orthotropic, anisotropic, layered

**Applicable load types**: Point, edge, pressure, acceleration, body force, centrifugal, temperature.

**Variants**: 3-node triangle, 6-node triangle, 4-node quadrilateral, 8-node quadrilateral.

### *11.2.3 Three-Dimensional elements*

i. Solid Element

**Solid Element**:

**Modeling considerations**: Practically any structure can be modeled using solid element. Due consideration must be given to execution time, result visualization and its interpretation.

**Applicable DOF per node**: Three translations.

**Strain**: $\varepsilon_x, \varepsilon_y, \varepsilon_z, \gamma_{xy}, \gamma_{xz}, \gamma_{yz}$

**Stress**: $\sigma_x, \sigma_y, \sigma_z, \tau_{xy}, \tau_{xz}, \tau_{yz}$

**Input required** Material Properties.

**Element type in FEAST$^{SMT}$**: BRICK, HBRICK (Mixed Formulation).

**Minimum number of DOF to be constrained**: Three translation DOFs to be arrested, three rotations of the structure have to be arrested by arresting translation DOF at two geometrically non-coincident nodes.

**Applicable analysis types**: Linear static, geometric nonlinear static, buckling, free vibration, transient response, frequency response, random response, shock response, viscoelastic, steady state heat transfer, transient heat transfer.

**Applicable material types**: Isotropic, orthotropic, anisotropic, vicoelastic, thermal.

**Applicable load types**: Point, edge, pressure, acceleration, body force, centrifugal, temperature, convection, heat flux.

**Variants**: 4-node tetrahedral, 10-node tetrahedral, 6-node wedge, 15-node wedge, 8-node hexahedral, 20-node hexahedral.

### 11.2.4 Zero-Dimensional elements

There exists a set of elements in FEAST$^{SMT}$ software, where the usage of shape function is not required for deriving the characteristic matrices. Those elements are termed as zero-dimensional elements.

i. Spring Element
ii. Point Mass Element
iii. Tying Element
iv. Rigid Link Element

**Spring Element**:

**Modeling considerations**: Spring stiffness can be provided as an input between DOF of two nodes. Any structure whose stiffness alone has to be simulated without representing the actual geometry can be modeled using spring element. If the structure can be defined in terms of stiffness associated with its degrees of freedom, then it can be modeled using spring element.

**Applicable DOF per node**: Depending on the connected nodes where the nodes are part of other element types.

**Strain**: Not applicable.

**Stress**: Not applicable.

**Input required** Stiffness properties and a local coordinate system if the nodes are coincident. If the nodes are not coincident then only stiffness is required and the vector connecting node 1 to node 2 will be taken as local x-axis.

**Element type in FEAST$^{SMT}$**: SPRING, SPRING1.

**Minimum number of DOF to be constrained**: Appropriate DOF to arrest the singularity of the structure.

**Applicable analysis types**: Linear static, free vibration, transient response, frequency response, random response, shock response

**Applicable material types**: Not applicable.

**Applicable load types**: Point.

**Variants**: Not applicable.

**Point Mass Element**:

**Modeling considerations**: Mass properties are to be assigned at a single node. Any structure whose mass alone has to be simulated without representing the actual geometry can be modeled using mass element.

**Applicable DOF per node**: Depending on the input provided.

**Strain**: Not applicable.

**Stress**: Not applicable.

**Input required**: Mass properties which comprise scalar mass in three directions and mass moment of inertia components are defined in the global coordinate system.

**Element type in FEAST$^{SMT}$**: MASS.

**Minimum number of DOF to be constrained**: Appropriate DOF to arrest the singularity of the structure.

**Applicable analysis types**: Linear static, free vibration, transient response, frequency response, random response, shock response

**Applicable material types**: Not applicable.

**Applicable load types**: Acceleration.

**Variants**: Not applicable*

*—In addition to the mass element, mass can be given as nonstructural mass property that can be applied on edge, surface or on a volume.

**Tying Element**:

**Modeling considerations**: Even though tying or coupled displacement is a constraint, in FEAST$^{SMT}$ software it is considered as an element to couple the DOF between two nodes.

**Applicable DOF per node**: Depending on the input provided.

**Strain**: Not applicable.

**Stress**: Not applicable.

**Input required**: Two coincident or non-coincident node ids whose DOF has to be coupled, type of DOF to be coupled, local coordinate system if the coupling of DOF is with respect to a coordinate system.

**Element type in FEAST$^{SMT}$**: TYING.

**Minimum number of DOF to be constrained**: Appropriate DOF to arrest the singularity of the structure.

**Applicable analysis types**: Linear static, buckling, free vibration, transient response, frequency response, random response, shock response, viscoelastic

**Applicable material types**: Not applicable.

**Applicable load types**: Not applicable.

**Variants**: Not applicable.

**Rigid Link Element**:

**Modeling considerations**: Even though rigid link is a constraint, it is considered as an element in FEAST$^{SMT}$ software. The relative rotation between the nodes that are connected with rigid link depends on the length of the rigid link. If the length of the rigid link is zero, then it is equivalent to TYING element.

**Applicable DOF per node**: Depending on the input provided.

**Strain**: Not applicable.

**Stress**: Not applicable.

**Input required**: Master node and one or more slave nodes with the applicable DOF, local coordinate system if the coupling of DOF is with respect to a coordinate system.

**Element type in FEAST$^{SMT}$**: RLINK.

**Minimum number of DOF to be constrained**: Appropriate DOF to arrest the singularity of the structure.

**Applicable analysis types**: Linear static, buckling, free vibration, transient response, frequency response, random response, shock response, viscoelastic

**Applicable material types**: Not applicable.

**Applicable load types**: Point.

**Variants**: Not applicable.

## 11.3 Element Quality

The ideal shape for a quadrilateral element is square, and for triangle, it is an equilateral triangle. But it is not always possible to have an ideal shape for the element while discretizing a complicated structure and there will be deviations from the ideal shape for elements in a mesh of real-life structures. To get a meaningful result, deviation from the ideal shape of the element should be kept to a minimum or in other words the element quality must be maintained within an acceptable level. In critical regions where the results are so crucial for the design, the element shape must be as close to the ideal shape as possible. Deviation from the ideal shape causes inaccuracies in the computation of the Jacobian matrix which defines the transformation for the characteristic matrix and load vectors from parent element to the physical element in the model. Deviations in the element shape result in inaccuracy in the calculation of the determinant of the Jacobian matrix, which is the ratio of the area or volume of the parent element to the area or volume of the model element. The deviation also causes inaccuracies in unique mapping of every point from the parent element to the physical element. In particular, the results are poor for derived quantities when compared to the field variables. The acceptable deviation from the ideal shape of the elements in the critical region is discussed in this section.

**Aspect Ratio**: It is the ratio of the longest element edge to the shortest element edge as shown in Fig. 11.7. Consider, one side of the element is very large compared to other side of the element, the Gauss point in the direction of shorter side almost coincides with each other and the real characteristics of the element cannot be captured. Otherwise, the physical element cannot be mapped properly to the parent element. For this reason, the aspect ratio must be limited to 5 in the overall structure and in the critical regions where stress is major concern, it is limited to 2. However, the ideal shape of a quadrilateral element is a square and for triangle it is an equilateral triangle.

**Included angle**: The included angle between two edges of a quadrilateral element should be greater than 60° less than 135°

**Skew**: The acute angle formed between the lines connecting center of two opposite edges defines skewness of the element. The skew angle should not be less than 30° as shown in Fig. 11.8.

**Warping index**: It defines the out-of-plane deviation of the quadrilateral element with respect to the mean plane. Warping index must be less than 0.2 as shown in Fig. 11.9.

$$\frac{h}{\sqrt{A}} < 0.2$$

**Moving mid-side nodes**: The mid node at the edge of quadratic element should not be moved unless otherwise it is demanded. The maximum movement of the node should be less than one-fourth of the element edge length.

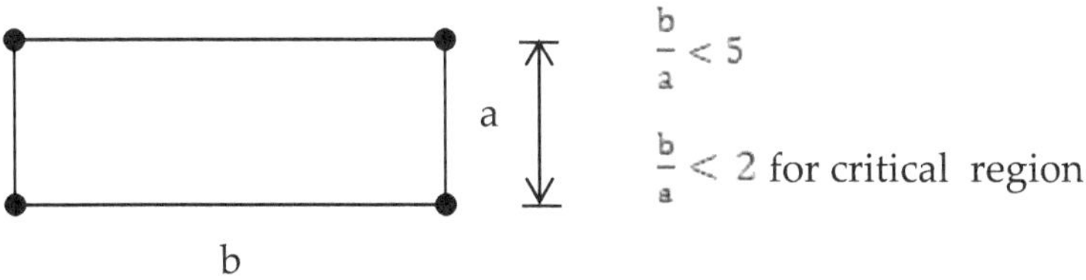


**Fig. 11.7** Element quality check for aspect ratio

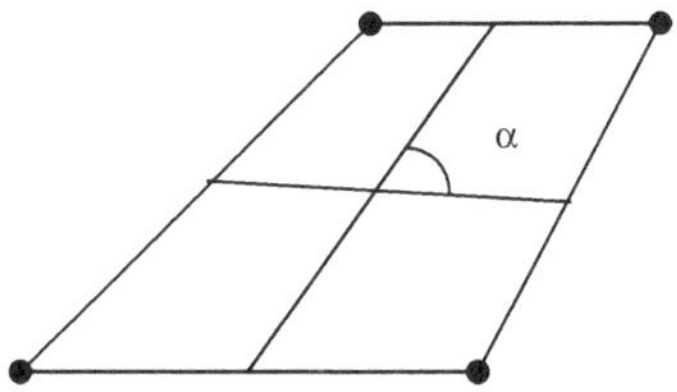


**Fig. 11.8** Element quality check for skew angle

**Unacceptable mesh shapes**:

Figure 11.10 shows few unacceptable mesh shapes that have to be avoided in the model.

**Incompatible meshing**: As far as possible, elements of the same order must be used throughout the model. Linear and quadratic elements should not be combined within a single model. If at all required, they must be connected using proper constraints. The edge of an element must be completely shared by edge of another element. It should not be shared by two or more elements. For field variables to be continuous through an edge of the element in a finite element model, the variation of the field variable along the edge must be same for the elements sharing the edge. For a linear element, the field variable varies linearly, whereas for a quadrilateral element it is a parabolic variation, which will result in a discontinuity in the field variable across the edge connecting two elements. Figure 11.11 shows typical incompatible meshes to be avoided in finite element modeling.

## 11.4 Boundary Conditions

Boundary conditions are the support conditions existing at boundaries of the structure. These support conditions have to be modeled exactly to get correct results. For example, hinged boundary conditions can be simulated by allowing rotations free at the node, if all the rotations are arrested it becomes a fixed support. But, for a solid element, where there is only translation DOF how to simulate hinged boundary condition? Before starting an analysis, a proper plan has to be worked out for the analysis.

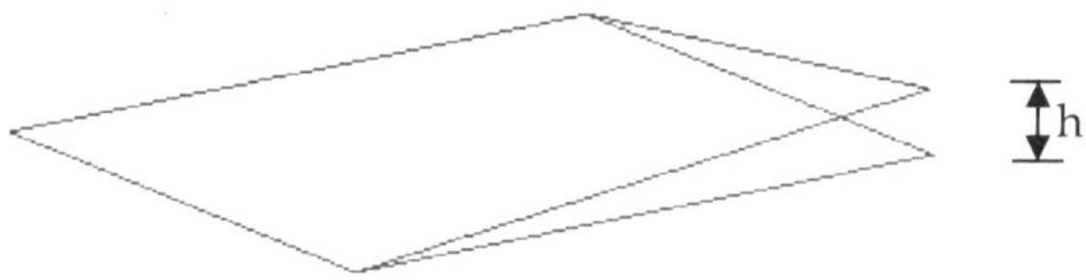


$$\frac{h}{\sqrt{A}} < 0.2$$

**Fig. 11.9** Element quality check for warping angle

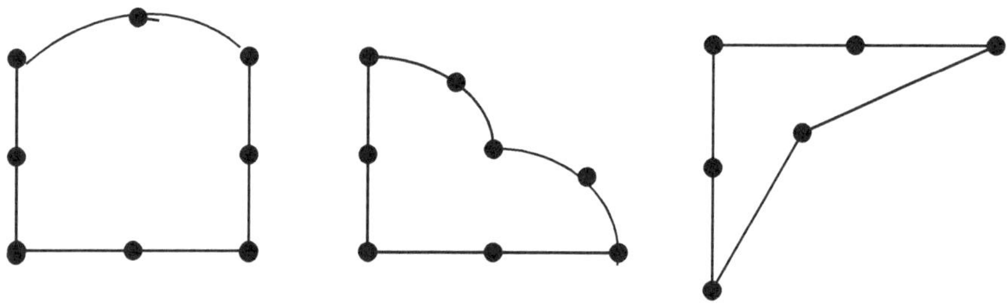

**Fig. 11.10** Unacceptable mesh shapes

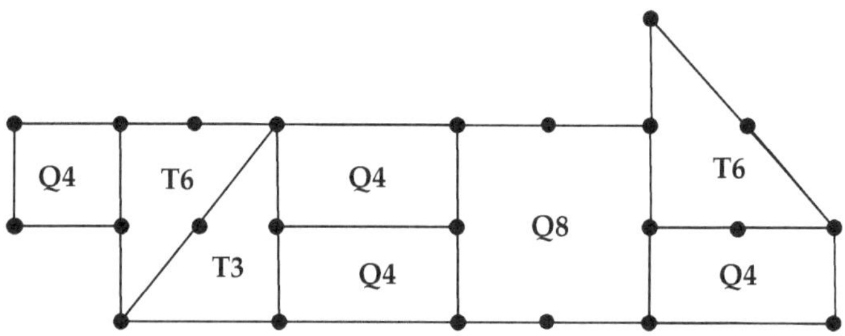


**Fig. 11.11** Incompatible meshing

Model has to be meshed with appropriate element types and appropriate number of elements to capture results at the critical region. Even boundary conditions dictate the number of elements to be meshed in a particular structure. To simulate hinged boundary conditions while using a solid element for bending loads, to allow rotation at the boundary, odd number of nodes must be used along the boundary in the model. The translational DOF at node exactly at the mid of boundary must be arrested to simulate the hinged conditions as shown in Fig. 11.12.

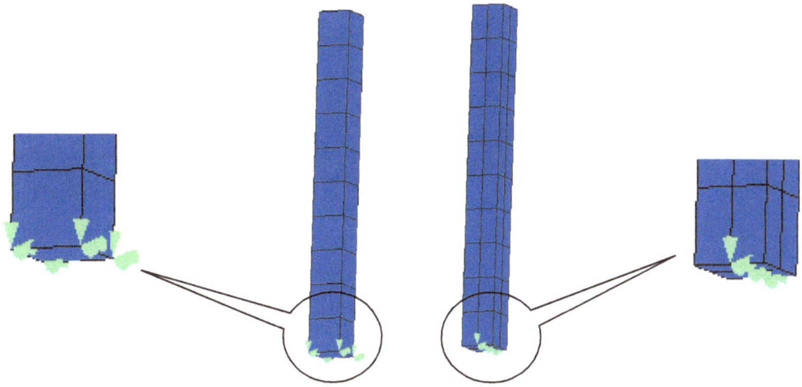

**Fig. 11.12** Simply supported boundary condition for model discretized with solid elements

### *11.4.1 Exploiting Symmetries*

**Mirror Symmetry**: If a structure with its loads and boundary conditions are symmetric with respect to a plane, then the resulting deformation pattern and stress distributions will also be symmetric with respect to the plane. Therefore, it is sufficient to model one-half of the structure. The other half of the structure is simulated by applying symmetric boundary condition on the plane of symmetry. The advantage of modeling part of the structure is to reduce the number of DOF, reducing the time to prepare the mesh and hence reducing the overall execution time. If the structure is symmetric with respect to two planes, a quarter of the structure can be simulated with proper boundary conditions applied on both the planes. Figure 11.13 shows a plate simply supported at all edges and point load is applied at the center of the plate. Table 11.2 shows the results obtained by using different idealizations. From the results, it can be observed that results are identical for different idealizations by simulating proper boundary conditions.

When a structure is symmetric with respect to a plane but the loading is anti-symmetric with respect to that plane then it is sufficient to model one-half of the structure anti-symmetric boundary condition has to be applied on the plane of anti-symmetry.

When a structure and its boundary condition are symmetric but the applied load is not symmetric with respect to a plane, as shown in Fig. 11.14 then for a linear analysis, principle of superposition holds and a symmetric portion of structure is sufficient for modeling the structure. Initially, the structure has to be analyzed for symmetric boundary condition and a second analysis has to be repeated with anti-symmetric boundary condition. The summation of the individual result gives the result for the entire structure. Tables 11.3 and 11.4 show boundary conditions to be applied for symmetric and anti-symmetric boundary conditions with symmetric geometry. u, v, w represent the translational DOF with respect to x, y, z axes respectively. whereas,

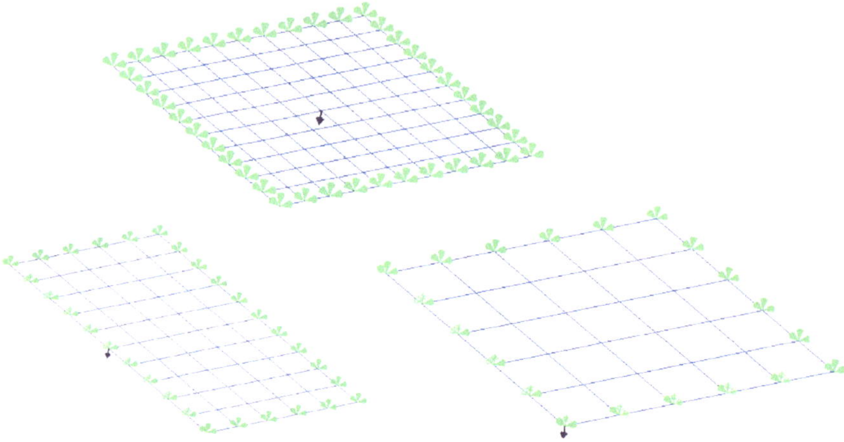

**Fig. 11.13** Simply supported plate—full model, half model, quarter model

**Table 11.2** Result for a symmetric plate with different idealization

| | Displacement, mm |
|---|---|
| Full Model | 0.22 |
| Half Model | 0.22 |
| Quarter Model | 0. 22 |

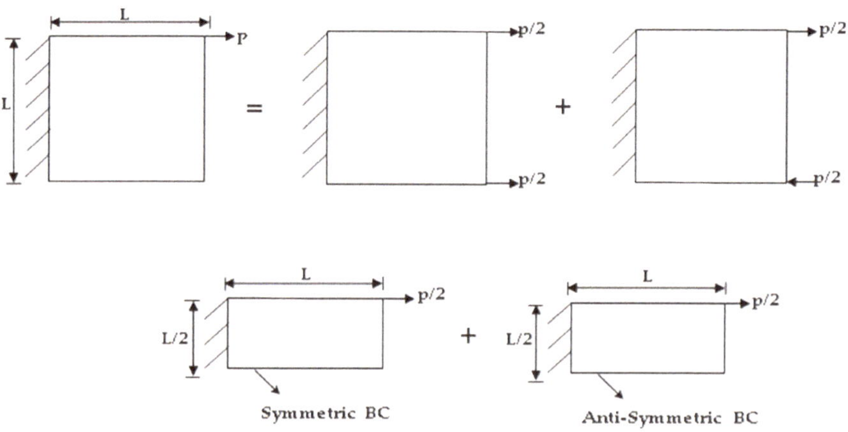


**Fig. 11.14** Symmetric structure with unsymmetrical loading

**Table 11.3** Symmetric boundary conditions for symmetric structure with symmetric loading

| Plane of Symmetry | u | v | w | $\theta_x$ | $\theta_y$ | $\theta_z$ |
|---|---|---|---|---|---|---|
| xy | Free | Free | Arrest | Arrest | Arrest | Free |
| yz | Arrest | Free | Free | Free | Arrest | Arrest |
| *zx* | *Free* | *Arrest* | *Free* | *Arrest* | *Free* | *Arrest* |

**Table 11.4** Anti-symmetric boundary conditions for symmetric structure with anti-symmetric loading

| Plane of symmetry | u | v | w | $\theta_x$ | $\theta_y$ | $\theta_z$ |
|---|---|---|---|---|---|---|
| xy | Arrest | Arrest | Free | Free | Free | Arrest |
| yz | Free | Arrest | Arrest | Arrcst | Free | Free |
| zx | Arrest | Free | Arrest | Free | Arrest | Free |

$\theta_x, \theta_y, \theta_z$ represent the corresponding rotations. In the table, free indicates the corresponding DOF is free to move while arrest indicates boundary condition for the corresponding DOF.

**Symmetry in Dynamic analysis**: Dynamic analysis is a global phenomenon in which the mode shape can be symmetric or anti-symmetric with respect to a plane of symmetry of the structure. Therefore, it is advisable to model the entire structure for the dynamic analysis to get a clear picture of various mode shapes. Nevertheless, symmetry can be exploited by performing $2^n$ analysis where n is the number of planes of symmetries. Figure 11.15 shows a plate simply supported at all the edges. The plate has two symmetric planes. The full model of the plate is analyzed first, and the results are compared with a quarter symmetric model with four sets of different boundary conditions. First set (Sym-Sym), symmetric boundary conditions in both plane-1 and plane-2. Second set (Asym-Asym), anti-symmetric boundary conditions in both plane-1 and plane-2. Third set (Sym-Asym), symmetric boundary condition in plane-1 and anti-symmetric boundary condition in plane-2. Fourth set (Asym-Sym), anti-symmetric boundary condition in plane-1 and symmetric boundary condition in plane-2. Table 11.5 shows summary of results for first 10 mode shapes of the plate.

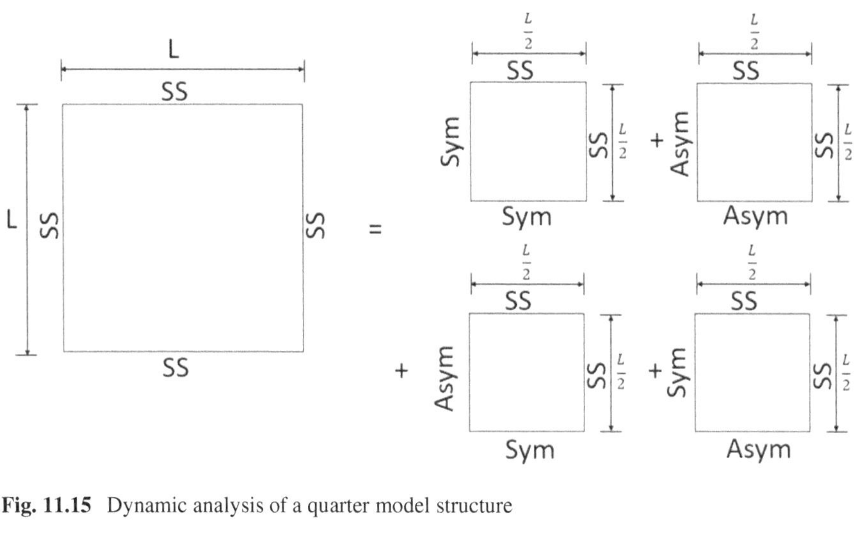


**Fig. 11.15** Dynamic analysis of a quarter model structure

**Table 11.5** Exploiting symmetry for dynamic analysis

| Mode number | Frequency (Hz) | | | | |
|---|---|---|---|---|---|
| | Full plate | Quarter plate | | | |
| | | Sym-Sym | Asym-Asym | Sym-Asym | Asym-Sym |
| 1 | 9.43855 | 9.43855 | | | |
| 2 | 23.69 | | | 23.69 | |
| 3 | 23.69 | | | | 23.69 |
| 4 | 37.0293 | | 37.0293 | | |
| 5 | 48.3323 | 48.3323 | | | |
| 6 | 48.3323 | 48.3323 | | | |
| 7 | 60.1551 | | | | 60.1551 |
| 8 | 60.1551 | | | 60.1551 | |
| 9 | 80.7405 | 80.7405 | | | |
| 10 | 85.0767 | | | 85.0767 | |

**Cyclic Symmetry**: If a structure is having cyclic symmetry, it is sufficient to model a sector of the structure and symmetry boundary conditions can be applied as shown in Fig. 11.16.

Rotating turbo machineries are very good examples of cyclic symmetry. It is sufficient to model only one blade for predicting the displacements, strains, stresses, and temperatures.

**Repetitive Symmetry**: If a structure can be idealized by repeating a unit cell, then it is sufficient to model only one unit cell and symmetry boundary condition can be

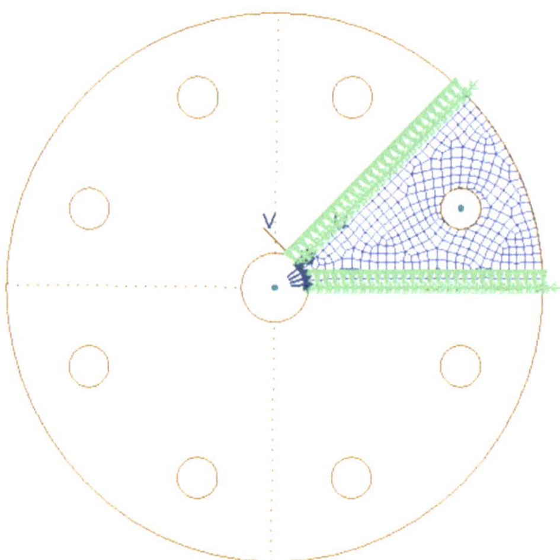

**Fig. 11.16** Structure with cyclic symmetry

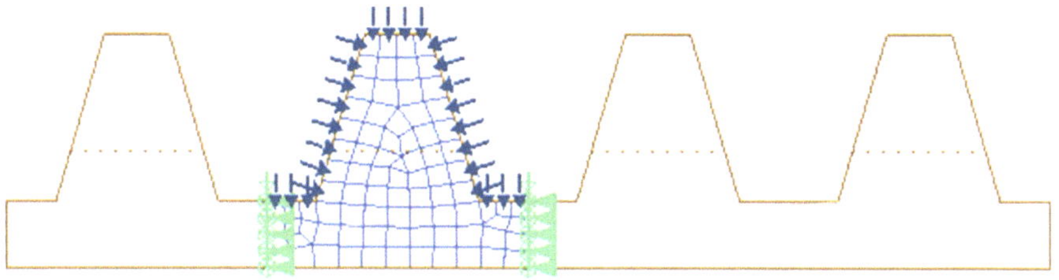

**Fig. 11.17** Structure with repetitive symmetry

given to the boundaries of the unit cell. For example, for designing railway tracks it is sufficient to model a single rail with appropriate boundary conditions or to design teeth of a rack and pinion mechanism, it is sufficient to model a single tooth of a rack as shown in Fig. 11.17.

## 11.5 Stress Concentration

When there is an abrupt change in the cross section of a structure, possibility of stress peaking is maximum at that region. These are termed as the critical regions of the structure. To avoid stress peaking, abrupt changes must be avoided instead a fillet must be provided to smoothly change the cross section. Stress pattern must be carefully studied around these critical regions. Mesh must be sufficiently refined to capture the stress exactly around these regions, but smaller element size increases the number of DOFs of the structure and hence the computational time. Therefore, while modeling, these small features are ignored and the resulting stress peaking is also ignored for the initial modeling. For consecutive modeling, otherwise known as sub-model, the region around the critical region is modeled in detail with all the small features that were ignored in the initial run. Displacement results obtained from the initial execution are given as boundary condition for the second execution to obtain stress pattern in detail.

In stress analysis, the most critical regions are around corners, cutouts, near to the fillets, etc. In critical regions, it is better to have quadrilateral element instead of triangular elements in case of a 2D mesh and hexahedral elements instead of tetrahedral elements in case of a 3D mesh. The reason being,

Shape function for linear triangular element $u = a_0 + a_1x + a_2y$.

Shape function for linear tetrahedral element $u = a_0 + a_1x + a_2y + a_3z$.

Strain for linear triangular element $\varepsilon_x = \frac{\partial u}{\partial x} = a_1 = \text{constant}$.

Strain for linear tetrahedral element $\varepsilon_x = \frac{\partial u}{\partial x} = a_1 = \text{constant}$.

Shape function for linear quadrilateral element $u = a_0 + a_1x + a_2y + a_3xy$.

Shape function for linear hexahedral element $u = a_0 + a_1x + a_2y + a_3z + a_4xy + a_5yz + a_6xz + a_7xyz$.

Strain for linear quadrilateral element $\varepsilon_x = \frac{\partial u}{\partial x} = a_1 + a_3y = \text{linear function of } y$.

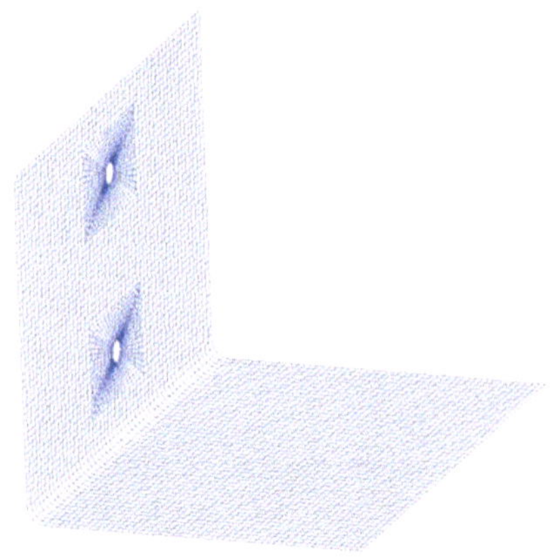

**Fig. 11.18** Meshing of a structure with fillets and holes

Strain for linear hexahedral element $\varepsilon_x = \frac{\partial u}{\partial x} = a_1 + a_4y + a_6z + a_7yz =$ linear function of y and z.

For linear triangular or tetrahedral elements, the shape functions are linear, hence the derivative of the displacement, i.e. strains, and subsequently the stress will be constant throughout the element which is not always correct. Therefore, capturing a complex state of strain using "constant strain elements" requires a refined mesh. Whereas, for a linear quadrilateral element the displacement variation is bilinear and the strain and stress variations are linear. For a linear hexahedral element, the displacement variation is trilinear and the strain and stress variations are bilinear. Therefore, a relatively coarser mesh of these elements can capture the complex state of strain. For the figure shown in Fig. 11.18, the fillet should be meshed with minimum three elements to represent the geometry better and to get the results with reasonable accuracy. Around the cutouts, there should be washer mesh (A washer mesh is a layer of circular meshes around the cut out) for two layers before the mesh gets transition to the remaining regions.

## 11.6 Good Finite Element Modeling Practices

**Consistent Units**: Since finite element is independent of units, user should be conscious about the units used. When geometry is modeled in meter and force is applied in Newton then Young's modulus must be provided in N/m$^2$ and the density in kg/m$^3$ (Ns$^2$/m$^4$).

**Choosing Elements**: Right type of elements must be chosen for proper representation of the structure under consideration. For example, shell elements must be used instead of solid elements while modeling sheet metal parts. For a thick bracket, it is preferred to use solid elements instead of shell elements.

**Geometry Representation**: Mesh should exactly represent the geometry. For example, a cylinder should not be meshed with 4 elements along its circumference.

It will be no longer a cylinder but a square tube. A minimum of 36 elements must be used along the circumferential direction to represent a cylinder or otherwise to model atleast one element must be used for every 10°. If the region is a critical region, then it is better to have element for every 5°. Unnecessary details in model must be avoided. Holes of smaller diameter and fillets away from critical region can be avoided which will unnecessarily increase the number of elements in the model and hence the solution time.

For shell elements, element normal must be consistent. If the element normals are of opposite sign for adjacent elements, the bottom and top planes will be different which leads to discontinuous stress contours. Free edges of the geometry must be checked before submitting for analysis. It ensures that there are no duplicate nodes in the mesh. Duplicate elements can be checked by comparing the mass of the model with the mass of the structure. A free vibration analysis can be performed initially to ensure appropriate rigid body modes. More number of rigid body modes indicate that all the elements in the model are not connected together.

**Element Quality**: The geometry should be meshed with good quality elements around the critical region. Triangular and tetrahedral elements must be avoided around the critical region. Element quality check must be performed on the model to find bad quality elements and if there are any around the critical regions, those regions must be re-meshed with good quality elements.

**Checking Duplicate Nodes**: It is possible to have duplicate nodes in the models resulting in discontinuities in the mesh. For example, while meshing adjacent geometries at different instants adjacent elements will have separate nodes at the interface. The interface can be a point in the case of line elements, a line in case of surface elements and a surface in case of solid elements. Before invoking the solver, the model has to be checked for free edges. If unwanted free edges are seen, it indicates discontinuous mesh and duplicates nodes at those interfaces have to be merged together so that the mesh is continuous.

**Checking Duplicate Elements**: If a structure is inadvertently meshed twice or more it causes duplicate elements in the model which results in a heavier and stiffer structure. If the mass density is an input for the model, then it can be identified by verifying the mass of the structure, if not element labels can be switched on and the software shows overlapped labels, which is an indication of existence of duplicate elements.

**Missing Elements**: While modeling the structure, there is a possibility of deleting an element by mistake which results in a hole in the structure. If the model is in wireframe view, then it is not possible to identify the missing element. Therefore, before invoking the solver, the model has to be viewed in shaded mode to verify any such mistakes.

## 11.7 Exercise Problems

11.7.1. Which is the appropriate element to model a car windshield if the objective of analysis is to estimate flexibility of windshield?

11.7.2. What kind of elements have to be used for finding the fundamental frequency of a heavy vehicle chassis?

11.7.3. For the same chassis, if the objective is to find the stress distribution at the intersection of two members, which element will be appropriate?

11.7.4. Which elements are appropriate to analyze a thick cylinder under internal pressure for finding stress distribution?

11.7.5. How the following joints can be represented in a finite element model?

a. Hinged Joint
b. Universal Joint
c. Spherical Joint
d. Prismatic Joint
e. Cylindrical Joint

11.7.6. Is it necessary to model fillets and rivets in a ship to find its natural frequencies?

11.7.7. A wall mounted bookshelf has to be designed to carry 20 books weighing 500 g each. How the loads due to book can be represented without applying it as a point or surface load? Which elements are suitable for modeling the clamps?

11.7.8. Identify the errors in the plate model shown below

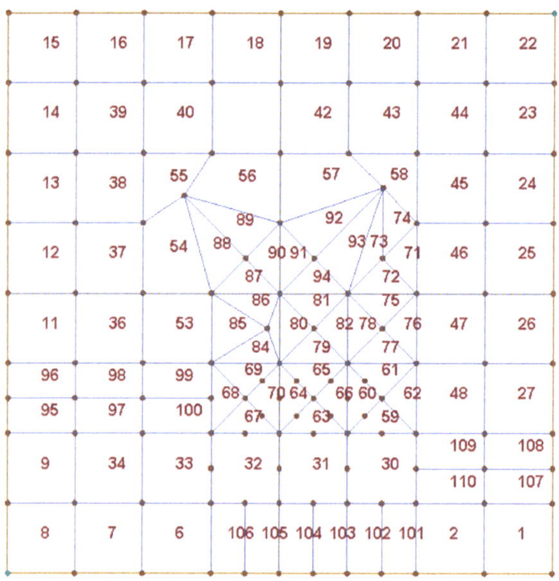

11.7.9. A plate of dimension 1 m X 1 m X 0.001 m which is constrained at one edge and pulled uniformly at the other edge has to be modeled. A novice engineer chose plane stress element to model the plate and is shown in the figure below. However, he has made few mistakes in the analysis. List it out.

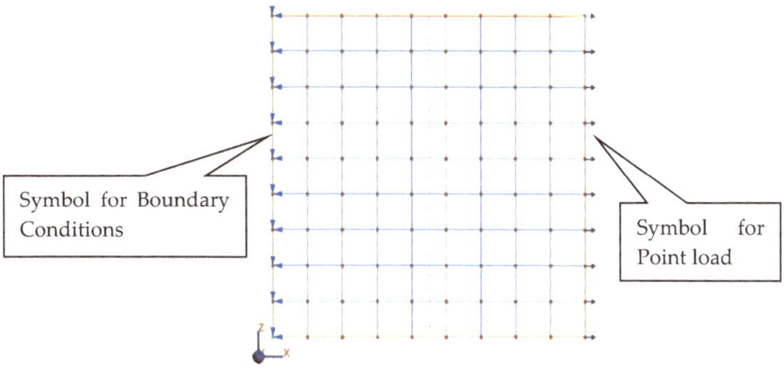


11.7.10. What are the boundary conditions to be applied for a plate pulled uniformly at both the ends?

## Bibliography

Vince A (2006) How to Manage Finite Element Analysis in the Design Process", NAFEMS Ltd., ISBN 1874376123

Beattie GA (1995) Management of Finite Element Analysis : Guidelines to Best Practice", NAFEMS Ltd., R0033

Baguley D, Hose DR (1994) How to plan a Finite Element Analysis", NAFEMS Ltd., ASIN B002HX0JGS

Baguley D, Hose DR (1994) Why do Finite Element Analysis", NAFEMS Ltd.

Prinja NK (1994) Use of Finite Element Analysis in the Design Process", NAFEMS Ltd., 0075

Stress Analysis by the Finite Element Method for Practicing Engineers", Lexington Books, ASIN B005S0SYT6, 1975

Gokhale NS, Deshpande SS, Bedekar SV, Thite AN (2008) Practical finite element analysis. Finite To Infinite, ISBN-13 978–8190619509

Cook RD, Malkus DS, Plesha ME, Witt RJ (2007) Concepts and applications of finite element analysis. Wiley (Asia) Pte. Ltd., ISBN-13 978–81–265–1336–9

# Chapter 12
# Linear Static Analysis Using FEAST$^{SMT}$ Software

**T. J. Raj Thilak**

*I learned very early the difference between knowing the name of something and knowing something.*
*—Richard P. Feynman*

## 12.1 Static Analysis of an Earth Dam

An earth dam is subjected to a gravitational force and lateral pressure by the water stored in the reservoir. The objective of this problem is to find the maximum displacement of the dam due to the applied forces and the stress distribution on the dam. Since the dam is very long and the cross section of the dam, loading, and boundary conditions are not varying across the length of the dam, a representative cross section along the length of the dam can be considered for modeling. Plane strain element can be used to discretize the model. Material properties of concrete are considered for idealizing the dam as shown in Fig. 12.1.

Element Type

Plane Strain

Properties

Modulus of Elasticity = 27.38 GPa.

Poisson's Ratio = 0.15

Mass density = 2400 kg/m$^3$

For plane strain element, thickness is considered as 1 unit by the software.

T. J. R. Thilak (✉)
FSDD/SDAG/STR, Veli, VSSC, Thiruvananthapuram, Kerala 695022, India
e-mail: tj_rajthilak@vssc.gov.in; rajthilaks@gmail.com



S. U. Nair and S. Somanath (eds.), *Introduction to Finite Element Analysis*,
https://doi.org/10.1007/978-981-19-7989-7_12

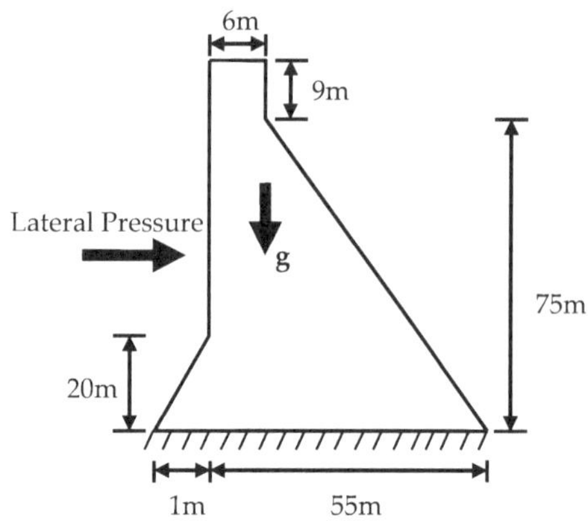


**Fig. 12.1** Static analysis of an earth dam

## *12.1.1 Pre-processor*

**Create geometric coordinates (0:0:0), (56:0:0), (1:20:0), (1:84:0), (7:84:0), (7:75:0).**

Command : POINT, ADD
Menu : Geometry→ Key point→ Create→Add
Parameters : (To be filled by the user)

| Point Data | 0/0/0 |
|---|---|

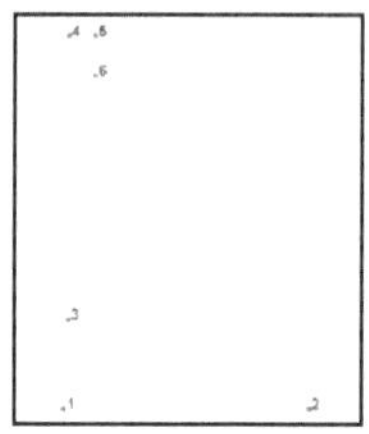

**Note**:

After entering the coordinate data click 'Apply' or press 'Cntrl+Enter'. If the AutoExec box is checked, then the command will be automatically executed by filling minimum number of parameters required to create the entity. Similarly other coordinates are created using the above command or GUI options. The 'Entity ID' in the parameter box is the label ID of the entity created. The label of the first entity is by default 1 and it increases by 1 for following similar entities. User can give a different Entity ID if required.

**Create a line**

Command : CURVE, LINE
Menu : Geometry → Curve → Create → Line

| End points | Use Mouse to pick the points |
|---|---|

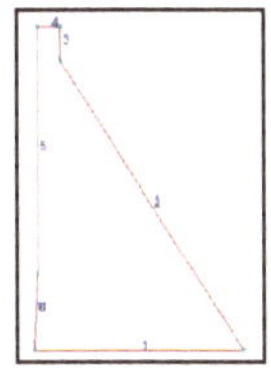

**Create surface using six curves**

Menu : Geometry → Surface → Create → On Geometry→ Bounded
Command : SURFACE, BOUNDED
Parameters : (To be filled by the user)

| Boundary Curves | Use Mouse to select the curves |
|---|---|

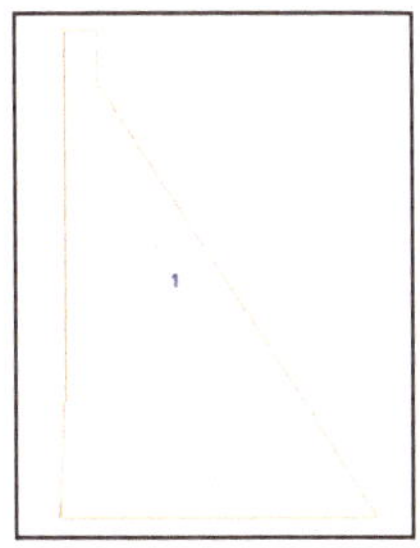

**Generate mesh on the surface**

Commands : MESH, QUAD
Menu : Mesh → Mesh Gen→ QUAD
Parameters : (To be filled by the user)

| Surface | 1 |
|---|---|
| Element Size | 1.5 |
| Method | Paver |
| Type | 4-Node |
| Divisions | |
| Bias | |

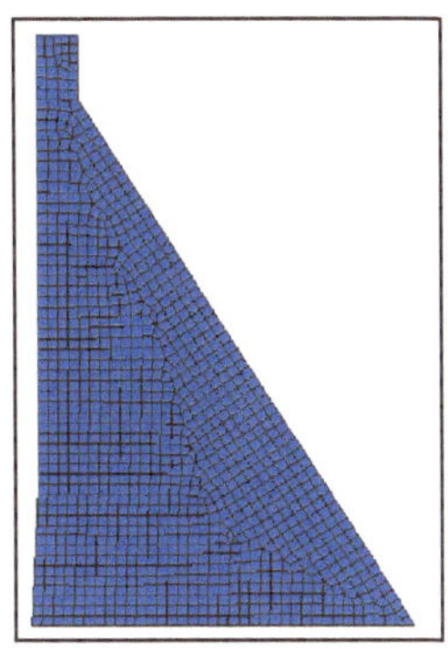

**Note**:

The initial size gives an approximate edge length of a representative element. According to the initial size, the software divides the edge into number of sub-divisions. The user can increase or decrease the number of sub-divisions by right click/left click of the mouse pointer and placing the mouse pointer on the surface edge. The user can also alter the bias of the edge (Refined mesh on one side of the edge and coarser on the other side by maintaining same number of divisions) by right click/left click of the mouse pointer and placing mouse pointer on an edge of the surface.

**Set the element type**

Command : ELEMENT,TYPE
Menu : Mesh → Element → Modify → Type
Parameters : (To be filled by the user)

| **Elements** | All[0] |
|---|---|
| **Type** | Plane Strain |

**Note**:

Default element type refers to one-dimensional elements—beam elements, two-dimensional elements—shell elements, and three-dimensional elements—displacement-based solid elements.

**Specify displacement constraints**

Command : DispBC,ADD
Menu : Load/BC → Structural → DispBC
Parameters : (To be filled by the user)

| **Nodes** | Select the nodes on the bottom boundary edges |
|---|---|
| **DispBC** | 0/0//// |
| **LCS** | 0 |
| **Label** | |

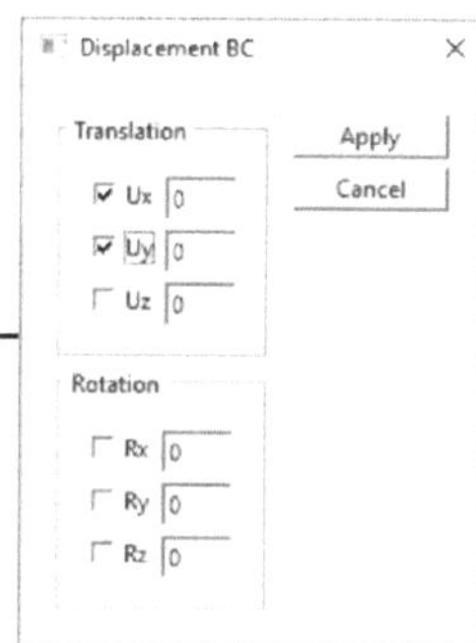


**Note**:

By default, the specified displacements for each DOF are zero. User can define specified displacements in place of zero. The Set ID is similar to the Entity ID starts from 1 and increments by 1 automatically. User can define a different Set ID if required.

**Specify material properties**

Command : MATERIAL,ISOTROPIC
Menu : Property→ Material→ Structural→ Isotropic
Parameters : (To be filled by the user)

| **Elements** | All[0] |
|---|---|
| **Young's Modulus** | 2.783E+10 |
| **Nu** | 0.15 |
| **Density** | 2400 |
| **Alpha** | 0 |
| **Label** | |

**Specify function**

Command : FUNCTION,EXPRESSION
Menu : Property → Function → Expression

Parameters : (To be filled by the user)

| **Expression** | 1000*9.81*(75-Y) |
|---|---|
| **Label** | |

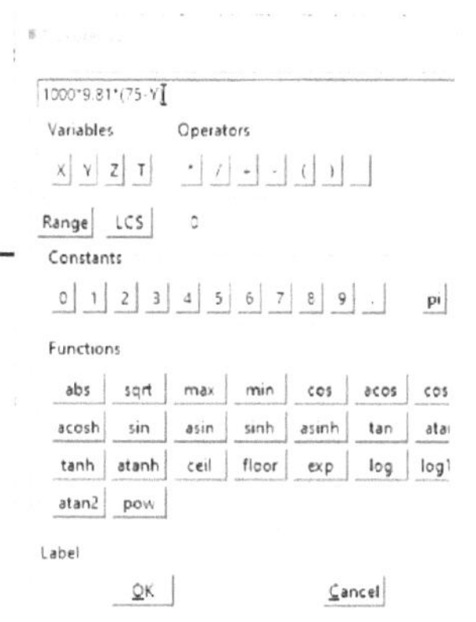

**Specify edge load**

Command : EDGELOAD, ADD
Menu : Load/BC → Structural → Edge Loads
Parameters : (To be filled by the user)

| | |
|---|---|
| **Edges** | Use Mouse to select the element edges |
| **Magnitude** | F1 |
| **Direction** | Along x-axis |
| **LCS** | 0 |
| **Label** | |

**Specify acceleration**

Command : GRAVITY, ADD
Menu : Load/BC → Structural → Gravity
Parameters : (To be filled by the user)

| | |
|---|---|
| **Elements** | ALL[0] |
| **Acceleration** | -9.81 |
| **Direction** | Y-Axis |
| **LCS** | 0 |
| **Label** | |

### *12.1.2 Solution*

**Set analysis type**

Commands : ANTYPE, ADD
Menu : Analysis → Analysis Type
Parameters : (To be filled by the user)

| | |
|---|---|
| **Analysis Types** | Static |

**Set analysis general**

| | | |
|---|---|---|
| Commands | : | STATICGEN,ADD |
| Menu | : | Analysis→ Static→ General |
| Parameters | : | (To be filled by the user) |

| **Stress Output** | YES |
|---|---|

Save the job and submit the job into FEAST$^{SMT}$

**Note**:

For solving user can click [!] icon in right end of tool bar.

**Note**:

By invoking the solver, the software automatically creates data file and submits it to solver. The user can also create data file without invoking the solver using GUI through File → Create Data file menu. The summary of the finite element information is as follows.

| | |
|---|---|
| **Number of Elements** | 1069 |
| **Number of Nodes** | 1152 |
| **Number of Active DOF** | 2228 |

### *12.1.3 Post-processing*

**Deformed shape**

| | | |
|---|---|---|
| Command | : | POST, DEFLECTION |
| Menu | : | Post→ Deflection |

**Displacement/Stress/Strain Contour**

| | | |
|---|---|---|
| Command | : | POST, CONTOUR |
| Menu | : | Post → Contour |

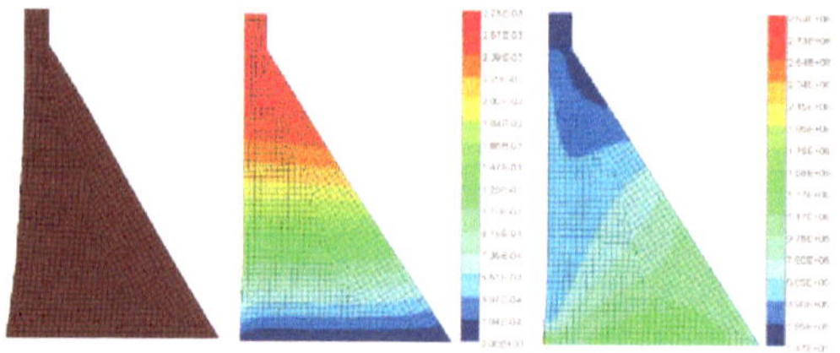

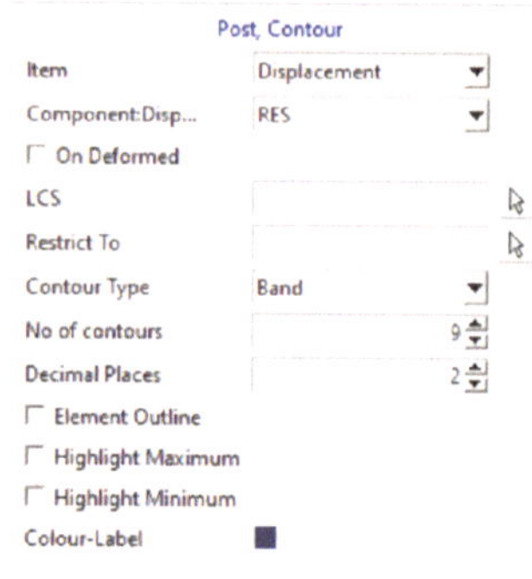


__Note__:

To plot the stress–strain contour from the 'item' dialogue box select stress and strains. Maximum displacement and von Mises Stress obtained from the above analysis are **2.75 mm** and **2.92 MPa** respectively. The maximum stress is concentrated at the lower left corner of the structure. Since the variation of stress within the element is ranging from 2.92 to 1.17 MPa, an average value of 2.045 MPa can be taken as the maximum value for von Mises stress.

## 12.2 Static Analysis of a Portal Frame

A portal frame containing three structural elements is designed to carry a load of 210 kN at the middle of the top member. The bottom side of the vertical members is arrested for all degrees of freedom as shown in Fig. 12.2. The cross section of the members of the portal frame is 10 cm × 20 cm rectangular cross section. The portal frame is made of aluminum material. The objective of the analysis is to estimate the deflection, shear force, and bending moment diagrams for all the structural members. Since the cross section of the structure is very small compared to length of the structure, beam element is used to idealize the structure.

__Properties__

Modulus of Elasticity = 70 GPa

Poisson's Ratio = 0.3

Mass density = 2800 kg/m$^3$

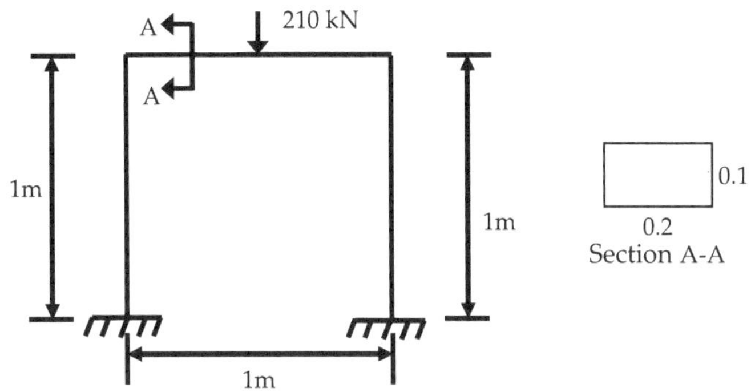


**Fig. 12.2** Static analysis of a portal frame

## *12.2.1 Pre-processor*

**Create geometric coordinates (0:0:0), (1:0:0), (0:1:0), (1:1:0).**

Commands : POINT, ADD
Menu : Geometry → Key point → Create → Add
Parameters : (To be filled by the user)

| **Point Data** | 0:0:0 |
|---|---|

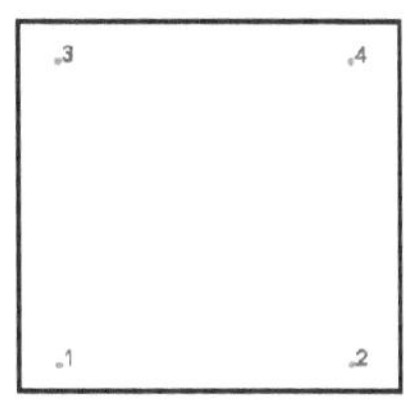

After entering the coordinate data click 'Apply' or press 'Cntrl+Enter'. Likewise, each coordinate is created using the above command options.

**Create a line**

Commands : CURVE, LINE
Menu : Geometry → Curve → Create → Line

| **End points** | Use Mouse to pick the points |
|---|---|

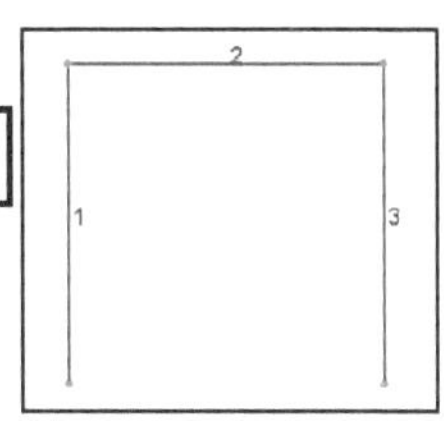

**Meshing using beam elements**

Commands : MESH, BAR
Menu : Mesh →MeshGen→ Bar
Parameters : (To be filled by the user)

| **Curve** | Use Mouse to select the curves |
|---|---|
| **Element Size** | 0.1 |
| **Type** | 2-node |
| **Divisions** | 10 |
| **Bias** | |

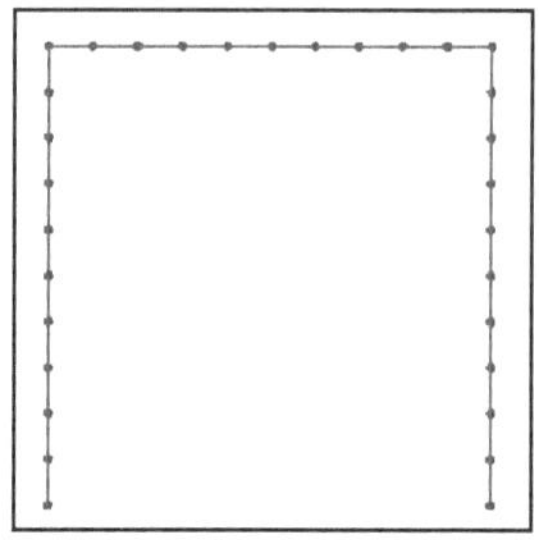

**Note**:

Right click/left click mouse point to alter sub-divisions. Bar is a name given for one-dimensional meshing, whereas Beam and Truss are different types of one-dimensional elements.

**Specify material properties**

Commands : MATERIAL, ISOTROPIC
Menu : Property → Material → Structural→ Isotropic
Parameters : (To be filled by the user)

| **Elements** | All[0] |
|---|---|
| **Young's Modulus** | 7E+10 |
| **Nu** | 0.3 |
| **Density** | 0 |
| **Alpha** | |
| **Label** | |

**Note**:

It is possible for the user to add material properties from the material library provided and user also can load the material properties to the existing library.

**Specify beam properties**

Command : BEAMPROP, ADD
Menu : Property → Physical → Beam Properties → Cross Section
Parameters : (To be filled by the user)

| Elements | All[0] |
|---|---|
| Cross section Shape | RECT/0.2/0.1 |
| Centroid Offset | 0/0 |
| Label | |

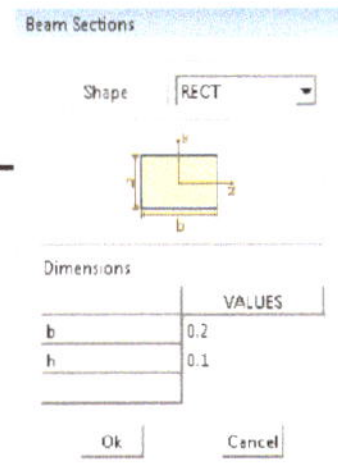


**Specify reference node for beam**

Commands : BEAMLCS, ADD
Menu : Property → Physical → Beam Properties → BeamLCS
Parameters : (To be filled by the user)

| Elements | 11T20[0] |
|---|---|
| Type | Direction |
| Vector | 0:1:0 |
| LCS | |
| Label | |

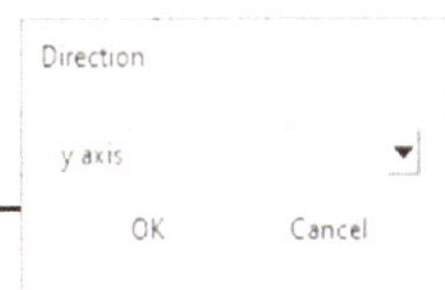


**Note**:

The element is idealized as the structure passes through the centroid of the cross section. If the beam element modeled does not pass through the centroid of the beam element, then the distance between the cross-sectional centroid and the element must be given as Node offsets in the Y-axis and Z-axis.

The direction connecting node-1 and node-2 is the local x-axis of the beam element. While assigning the cross section for the beam element, local y and z-axes are shown in the cross-sectional diagram. These y and z-axes can be oriented in any direction. Therefore, the direction of the y-axis should be supplied to the software through either a reference node or a reference direction. By default local y-axis is aligned with the global y-axis. If the beam element itself is oriented toward global y-axis, then global z-axis is considered as the local y-axis. This reference direction is required for properly extracting the two area moment of inertia of the cross section, I-yy and I-zz.

**Specify displacement constraints**

Commands : DispBC,ADD
Menu : Load/BC → Structural→ DispBC
Parameters : (To be filled by the user)
Fixed support

| **Node IDs** | 1/33 |
|---|---|
| **DispBC** | 0/0/0/0/0/0 |
| **LCS** | 0 |
| **Label** | |

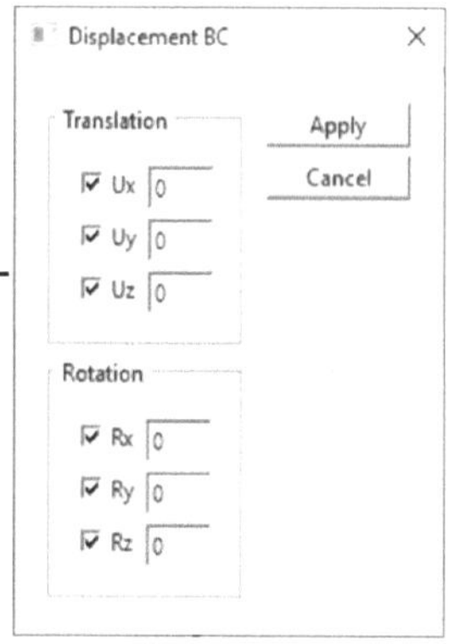


**Specify point load**

Command : PONITLOAD, ADD
Menu : Load/BC → Structural → Point Load
Parameters : (To be filled by the user)

| **Node IDs** | 17 |
|---|---|
| **Magnitude** | -210000 |
| **Component** | Fy |
| **LCS** | 0 |
| **Label** | |

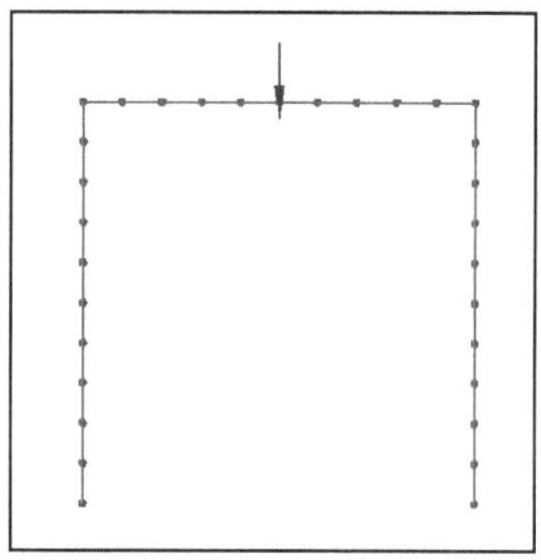

### 12.2.2 Solution

**Set analysis type**

Commands : ANTYPE, ADD
Menu : Analysis → Analysis Type
Parameters : (To be filled by the user)

| Analysis Types | Static |
|---|---|

**Set analysis general**

Commands : STATGEN,ADD
Menu : Analysis → Static → General
Parameters : (To be filled by the user)

| Stress Output | YES |
|---|---|

**Save the job and submit the job into FEAST$^{SMT}$**

<u>Note</u>:

For solving user can click [ ! ] icon in right end of tool bar.

### 12.2.3 Post-processing

**Deformed shape**

Command : POST,Deflection
Menu : Post → Deflection

Figure 12.3 represents deformed configuration plot of the frame structure for the applied loads.

**Fig. 12.3** Deformed configuration plot

**Stress contour**

Command : POST, CONTOUR
Menu : Post → Contour

| | |
|---|---|
| **Item** | Beam Stress |
| **Component** | BENDING-XY |
| **Restrict To** | ALL[0] |
| **Contour Type** | Band |
| **No. of contours** | 9 |
| **Decimal Places** | 2 |

Figure 12.4 represents stress contour of the frame structure, where the strain and stresses are derived from nodal displacements.

Figure 12.5 represents the bending moment diagram of the frame structure for the given configuration and loads.

**Note**:

Additional options to view **AXIAL, BENDING-XZ, TORISIONAL** are also available under Beam Strain and Beam Stress item.

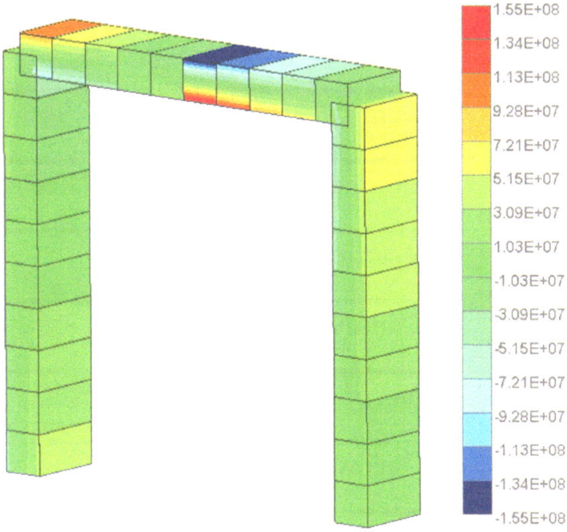


**Fig. 12.4** Realistic stress plot

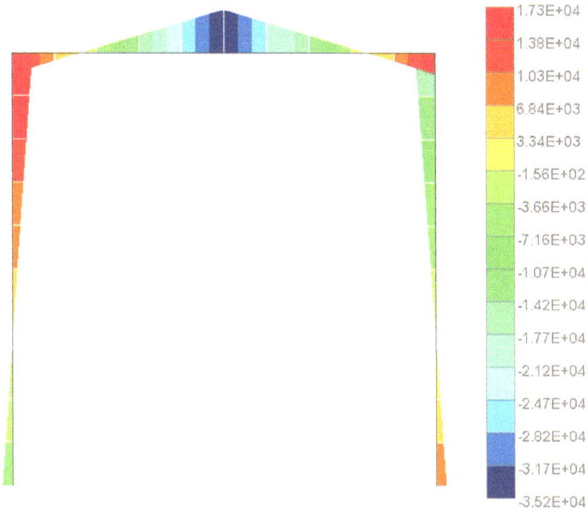


**Fig. 12.5** Bending moment diagram

**Strain contour**

Command : POST, CONTOUR
Menu : Post→ Contour

| Item | Beam Strain |
|---|---|
| Component | BENDING-XY |
| Restrict To | All[0] |
| Contour Type | Band |
| No. of contours | 9 |
| Decimal Places | 2 |

Note:

Additional options to view AXIAL, Torque, Bending Moments are also available under Element Forces item. SHEAR1 and SHEAR2 define the shear forces in two planes. BM1 and BM2 define the bending moments in two planes. The Plane parameter represents the plane in which the result has to be displayed.

## 12.3 Static Analysis of a Truss Structure

A planar truss structure is modeled using Truss element. The truss structure with its loading and displacement constraints is shown in Fig. 12.6. The cross section is 5 mm circular and the material is aluminum. The objective is to find the maximum displacement of the structure and the element forces acting in each truss member.

Properties:

Area = 7.85e−05 $m^2$

Modulus of Elasticity = 70 GPa

Poisson's Ratio = 0.3

Mass density = 2800 $kg/m^3$

**Fig. 12.6** Planar truss structure

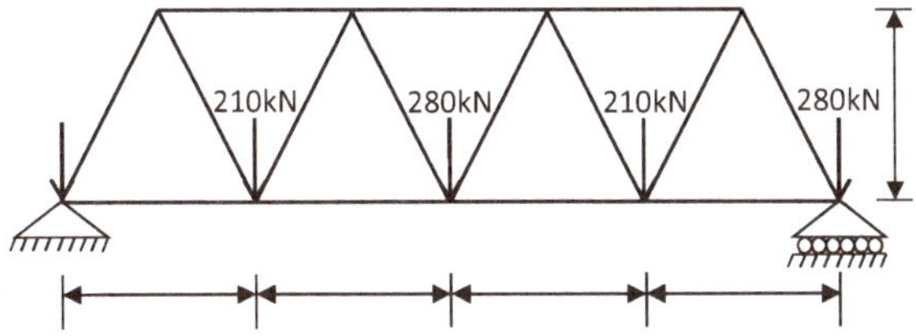


### *12.3.1 Pre-processor*

**Create geometric coordinates-(0:0:0), (1:0:0), (0.5:1:0), (2:0:0), (1.5:1,:0), (3:0:0), (2.5:1:0), (4:0:0), (3.5:1:0).**

Commands : POINT, ADD
Menu : Geometry → Key point → Create → Add
Parameters : (To be filled by the user)

| **Point Data** | 0:0:0 |
|---|---|

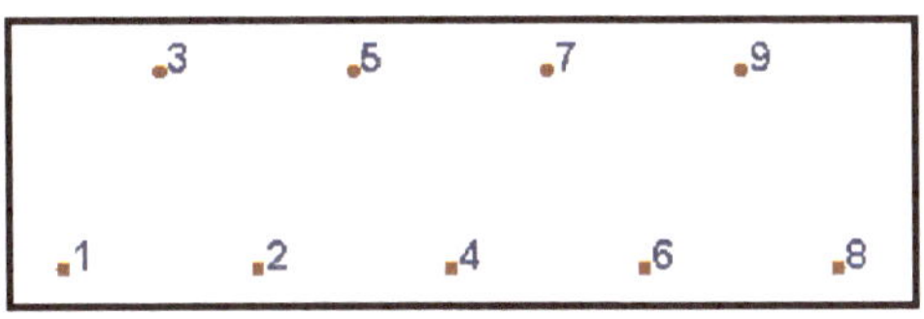


After entering the coordinate data click 'Apply' or press 'Cntrl+Enter'. Likewise, each coordinate is created using the above command options.

**Generate nodes for each point IDs created**

Commands : NODE, ADD
Menu : Mesh → Node → Create → Add
Parameters : (To be filled by the user)

Click on  button above the parameter box in order to select the points created in the above step and generate corresponding nodes at locations.

| Coordinates | P1/ P2/ P3/ P4/ P5/ P6/ P7/ P8/ P9 |
|---|---|
| Type | Cartesian |
| Entity ID | |

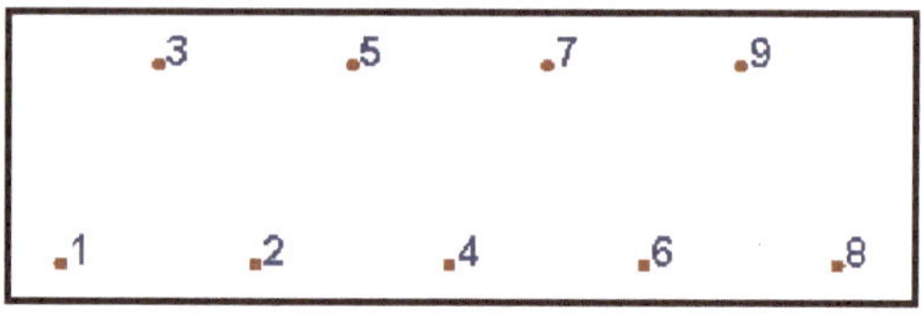


**Note**:

*Erase the points created initially to make the nodes visible by the [POINT, ERASE, ALL] command option.

**Create bar elements**

Commands : ELEMENT, ADD
Menu : Mesh → Element → Create → Add
Parameters : (To be filled by the user)

| Nodes | 1/2 |
|---|---|
| Dimension | 1 |

Selecting the element type as 'Bar2', 9 bar elements are created by selecting appropriate connecting nodes. After performing the above operations, a total number of 9 nodes and 15 elements are displayed.

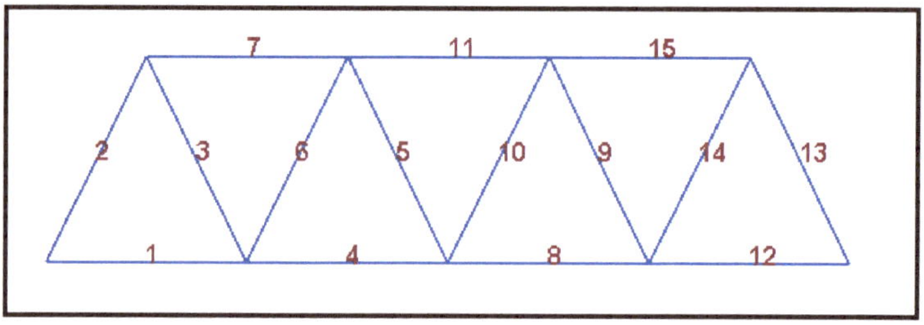

**Set the element type**

Commands : ELEMENT,TYPE
Menu : Mesh → Element → Modify → Type
Parameters : (To be filled by the user)

| Elements | All |
|---|---|
| Type | Truss |

**Define the element properties**

Commands : RODPROP,ADD
Menu : Property → Physical → Rod Property → Add
Parameters : (To be filled by the user)

| Element IDs | All |
|---|---|
| Area of cross-section | 7.85E-05 |
| Torsional constant | |
| Set ID | 1 |
| Label | |

Note:

If the area of cross-sectional parameter alone is filled, then the element is assigned only translational DOF. If the torsional constant is also specified along with area of cross section, then translational and rotational DOF are assigned for each node of the element.

**Specify displacement constraints**

Commands : DispBC,ADD
Menu : Load/BC→ Structural→ DispBC
Parameters : (To be filled by the user)

(i) Pinned support

| **Nodes** | 1 |
|---|---|
| **DispBC** | 0/0/0/// |
| **LCS** | 0 |
| **Label** | |

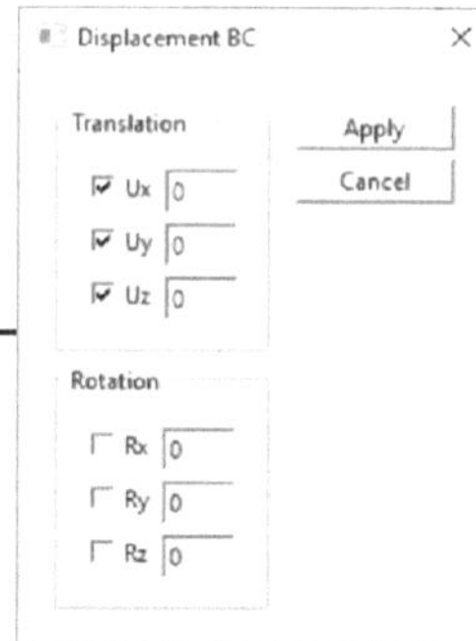


(ii) Roller support

| **Nodes** | 8 |
|---|---|
| **DispBC** | /0/0/// |
| **LCS** | 0 |
| **Label** | |

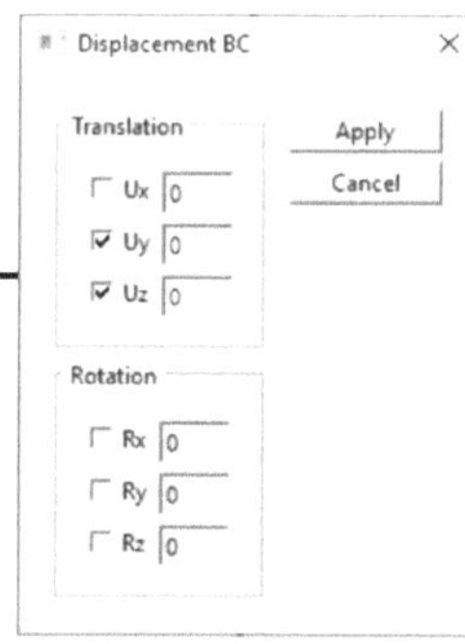


(iii) Out-of-plane DOF is arrested at all the nodes to avoid singularity in the stiffness matrix

| **Nodes** | ALL |
|---|---|
| **DispBC** | //0/// |
| **LCS** | 0 |
| **Label** | |

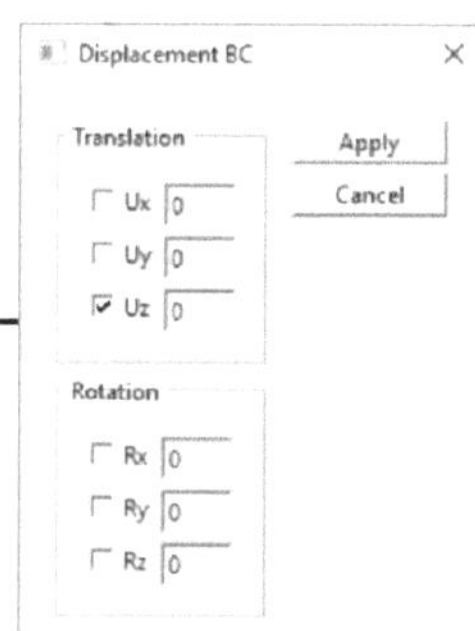

**Specify loads applied**

Commands : POINT,ADD
Menu : Load/BC → Structural→ Point Load
Parameters : (To be filled by the user)

(i) 280 kN at node id 1, 4 & 8

| **Nodes** | 1/4/8 |
|---|---|
| **Magnitude** | -280000 |
| **Component** | Fy |
| **LCS ID** | 0 |
| **Label** | |

(ii) 210 kN at node id 2 & 6

| **Nodes** | 2/6 |
|---|---|
| **Magnitude** | -210000 |
| **Component** | Fy |
| **LCS ID** | 0 |
| **Label** | |

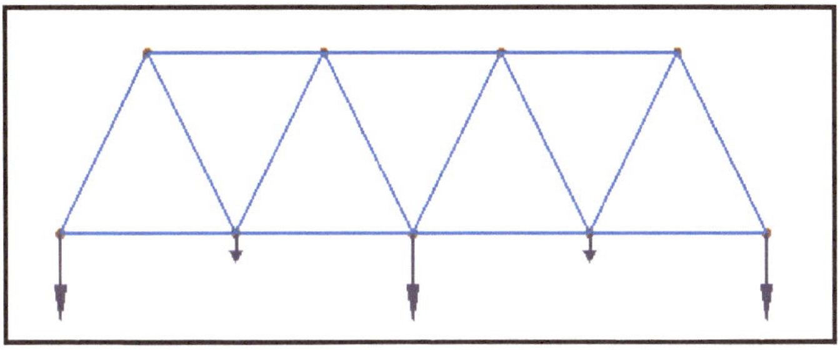

**Specify material properties**

Commands : MATERIAL, ISOTROPIC
Menu : Property → Material → Structural→ Isotropic
Parameters : (To be filled by the user)

| | |
|---|---|
| **Elements** | All |
| **Young's Modulus** | 7E+10 |
| **Nu** | 0.3 |
| **Density** | 0 |
| **Alpha** | 0 |
| **Label** | |

**Note**:

It is possible for the user to add material properties from the material library provided.

### *12.3.2 Solution*

**Set analysis type**

Commands : ANTYPE, ADD
Menu : Analysis → Analysis Type
Parameters : (To be filled by the user)

| | |
|---|---|
| Analysis Type | Static |

**Set analysis general**

Commands : STATGEN,ADD
Menu : Analysis → Static → General
Parameters : (To be filled by the user)

| | |
|---|---|
| Stress Output | YES |

**Save the job and submit the job into FEAST$^{SMT}$**

**Note**:

For solving user can click [ ! ] icon in right end of tool bar.

### 12.3.3 *Post-processing*

**Deformed shape**

| | | |
|---|---|---|
| Command | : | POST,DEFLECTION |
| Menu | : | Post → Deflection |

Note that the value for the Maximum Displacement at node 4 is listed as **0.5435** with deformed shape.

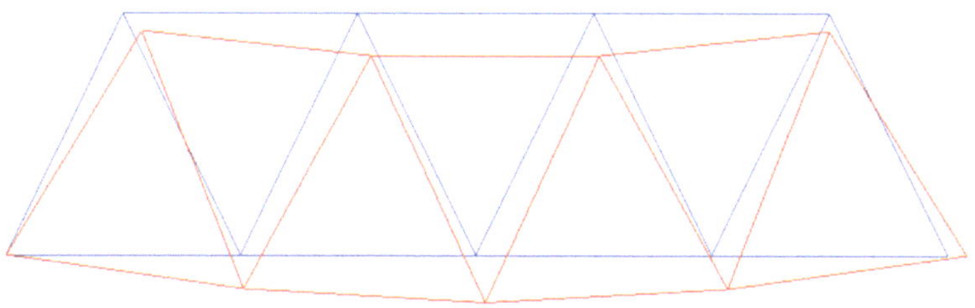

(i) **Element force**

| | | |
|---|---|---|
| Command | : | POST, Table View |
| Menu | : | Post → View Table |
| Parameters | : | (To be filled by the user) |

| **Item** | **TRUSS3D** |
|---|---|
| **Elements** | ALL[0] |
| **Components** | |

**Note**:

Results are shown in table format

Element Forces

| Element ID | Node ID | FX | FY | FZ | MX | MY |
|---|---|---|---|---|---|---|
| 1 | 1 | -175000 | 0 | 0 | 0 | 0 |
| 1 | 2 | 175000 | 0 | 0 | 0 | 0 |
| 2 | 1 | 391312 | 0 | 0 | 0 | 0 |
| 2 | 3 | -391312 | 0 | 0 | 0 | 0 |
| 3 | 2 | -391312 | 0 | 0 | 0 | 0 |
| 3 | 3 | 391312 | 0 | 0 | 0 | 0 |
| 4 | 2 | -420000 | 0 | 0 | 0 | 0 |
| 4 | 4 | 420000 | 0 | 0 | 0 | 0 |
| 5 | 4 | -156525 | 0 | 0 | 0 | 0 |
| 5 | 5 | 156525 | 0 | 0 | 0 | 0 |
| 6 | 2 | 156525 | 0 | 0 | 0 | 0 |
| 6 | 5 | -156525 | 0 | 0 | 0 | 0 |
| 7 | 5 | 350000 | 0 | 0 | 0 | 0 |
| 7 | 3 | -350000 | 0 | 0 | 0 | 0 |

Copy Close

(ii) **Reaction force**

| | | |
|---|---|---|
| Command | : | POST, TableView |
| Menu | : | Post → View Table |
| Parameters | : | (To be filled by the user) |

| **Item** | Reaction Force |
|---|---|
| **Components** | ... |
| **Nodes** | All |

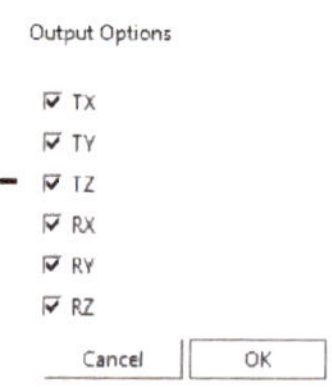


**Note**:

Results are shown in table format

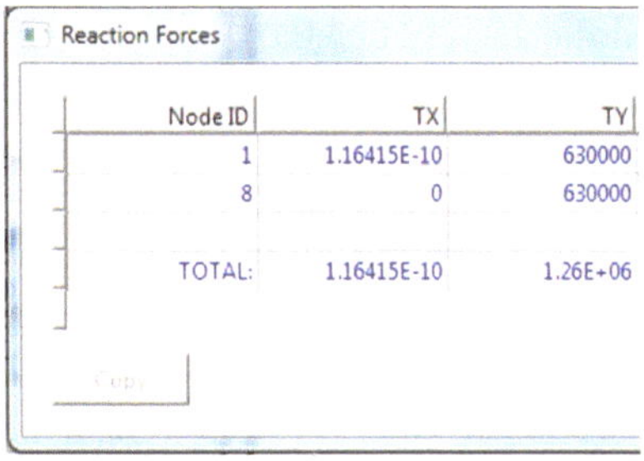

Reaction Forces

| Node ID | TX | TY |
|---|---|---|
| 1 | 1.16415E-10 | 630000 |
| 8 | 0 | 630000 |
| TOTAL: | 1.16415E-10 | 1.26E+06 |

## 12.4 Static Analysis of a Rotating Disk

A stepped disk with dimensions shown in Fig. 12.7 attached to a shaft is rotating at a speed of 2000 rpm with its base in contact with a frictionless rotating platform. The disk is free to deform in radial direction and is made of lead material. The objective of this analysis is to find the maximum deflection and the maximum stress in the disk. Since the geometry, loading, and boundary conditions are symmetric with respect to an axis, only a section of the structure is modeled and it is discretized with axisymmetric element. In FEAST$^{SMT}$ software, the axis of symmetry is global Y-axis and the structure is modeled in the first quadrant of the global Cartesian coordinate system.

Properties:

Modulus of Elasticity = 14GPa

Poisson's Ratio = 0.15.

Mass density = 11,340 kg/m$^3$

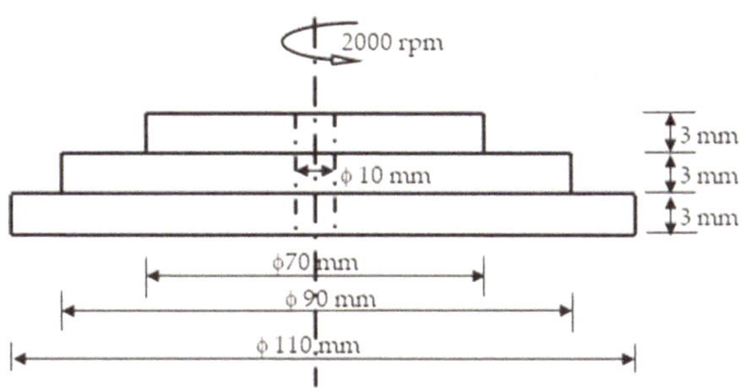


**Fig. 12.7** Static analysis of a stepped plate

### 12.4.1 Pre-processor

Create geometric coordinates (0.005:0:0), (0.005:0.009:0), (0.035:0.009:0), (0.035:0.006:0), (0.045:0.006:0), (0.045:0.003:0), (0.055:0.003:0), (0.055:0:0).

| | | |
|---|---|---|
| Commands | : | POINT, ADD |
| Menu | : | Geometry → Key point → Create → Add |
| Parameters | : | (To be filled by the user) |

| Point Data | 0.005:0:0 |
|---|---|

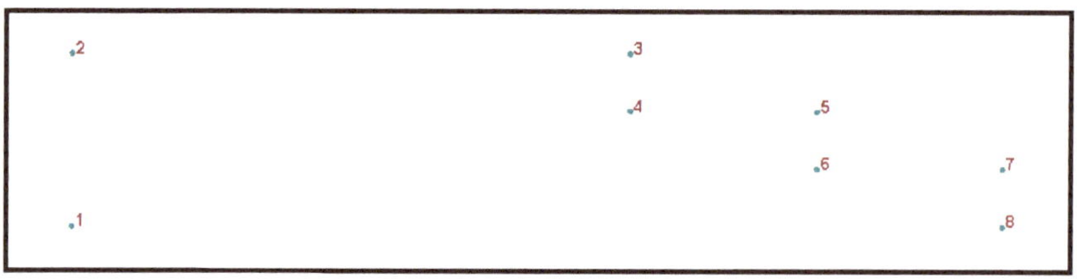


After entering the coordinate data click 'Apply' or press 'Cntrl+Enter'. Likewise each coordinate is created using the above command options.

**Create a line**

| | | |
|---|---|---|
| Commands | : | CURVE, LINE |
| Menu | : | Geometry → Curve → Create → Line |

| End points | Use Mouse to pick the points |
|---|---|

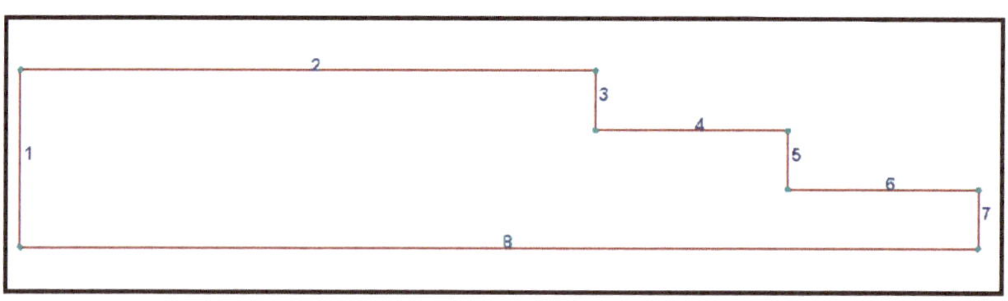

**Create surface using curves**

| | | |
|---|---|---|
| Menu | : | Geometry → Surface → Create → On Geometry→ Bounded |
| Command | : | SURFACE, BOUNDED |
| Parameters | : | (To be filled by the user) |

| **Boundary Curves** | Use Mouse to select the curves |
|---|---|

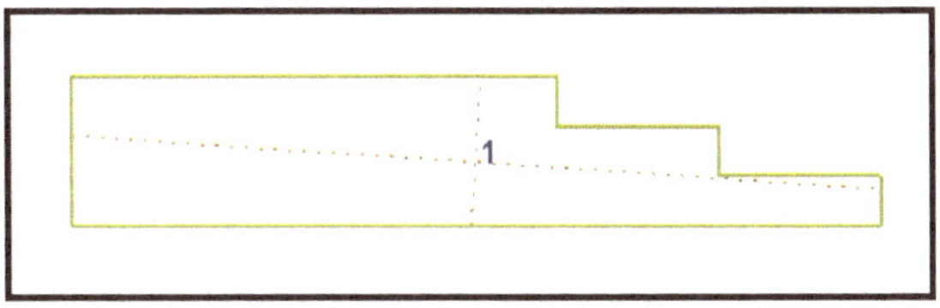

**Note**:

Break the surfaces for meshing elements.

| | | |
|---|---|---|
| Menu | : | Geometry → Surface → Modify→ Break |
| Command | : | SURFACE, BREAKP |
| Parameters | : | (To be filled by the user) |

| **Surface** | 1 |
|---|---|
| **Plane** | X axis |
| **Location** | P3 |

**Note**:

A cutting plane has to be defined to break the surface. The cutting plane is identified by specifying a plane parallel to the global plane and a key point in the plane is mentioned to specify the cutting plane.

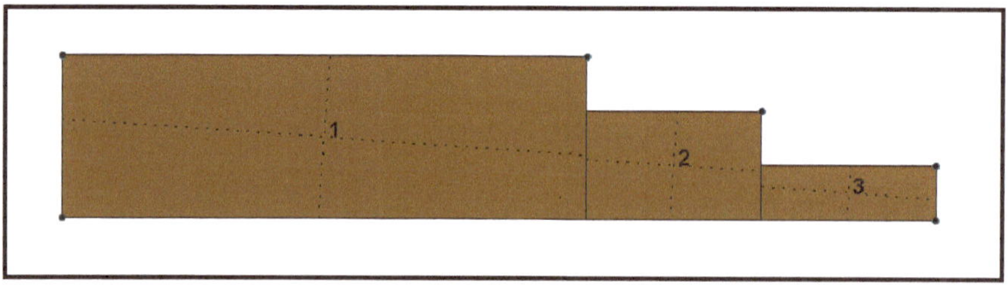


**Create elements using quadrilateral elements**

Commands : MESH, QUAD

Menu : Mesh→ MeshGen→ QUAD

Parameters : (To be filled by the user)

| | |
|---|---|
| **Surface** | 1T3 |
| **Element Size** | 0.001 |
| **Method** | Mapped |
| **Type** | 4-Node |
| **Divisions** | |
| **Bias** | |

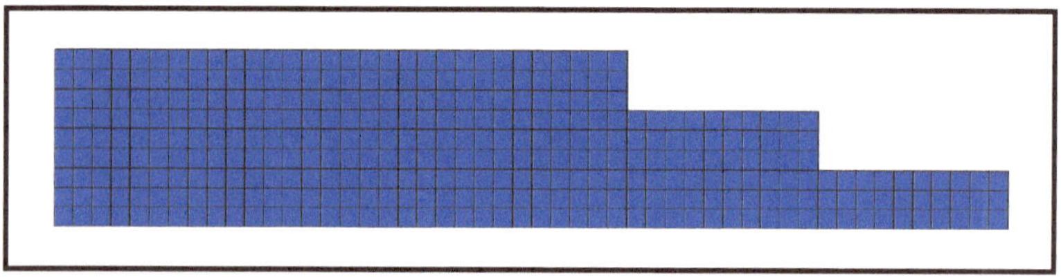

**Note**:

Right click/left click mouse point to alter sub-divisions and Bias.

**Set the element type**

Commands : ELEMENT, TYPE
Menu : Mesh → Element → Modify → Type
Parameters : (To be filled by the user)

| **Elements** | All[0] |
|---|---|
| **Type** | Axisymmetric |

**Specify displacement constraints**

Commands : DISPBC, ADD
Menu : Load/BC→ Structural→ DispBC
Parameters : (To be filled by the user)

| Node IDs | Select the bottom nodes |
|---|---|
| BC Value | /0//// ··· |
| LCS ID | 0 |
| Label | |

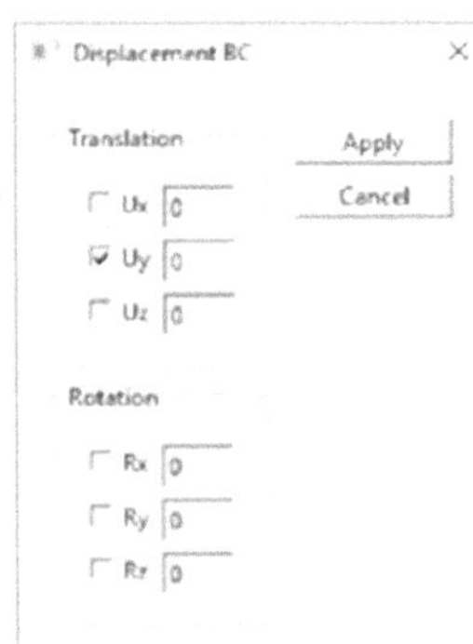


**Specify loads applied**

Commands : CENTRIFUGAL,ADD
Menu : Load/BC → Structural→ Centrifugal Load
Parameters : (To be filled by the user)

| | |
|---|---|
| **ElementS** | All[0] |
| **Angular Velocity** | 333.33 |
| **Axis** | YAXIS |
| **LCS ID** | 0 |
| **Label** | |

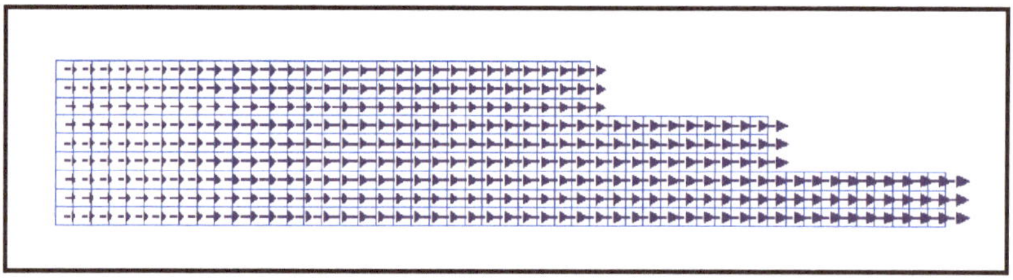

**Specify material properties**

Commands : MATERIAL,ISOTROPIC
Menu : Property → Material → Structural→ Isotropic
Parameters : (To be filled by the user)

| | |
|---|---|
| **Elements** | All |
| **Young's Modulus** | 1.4E+10 |
| **Nu** | 0.15 |
| **Density** | 11340 |
| **Alpha** | 0 |
| **Label** | |

**Note**:

It is possible for the user to add material properties from material library provided.

### 12.4.2 Solution

**Set analysis type**

Commands : ANTYPE, ADD
Menu : Analysis → Analysis Type
Parameters : (To be filled by the user)

| Analysis Types | Static |
|---|---|

**Set analysis general**

Commands : STATGEN,ADD
Menu : Analysis → Static → General → Add
Parameters : (To be filled by the user)

| Stress Output | YES |
|---|---|

**Save the job and submit the job into FEAST$^{SMT}$**

**Note**:

For solving user can click [ ! ] icon in right end of tool bar.

### 12.4.3 Post-processing

**Deformed shape**

Command : POST,Deflection
Menu : Post → Deflection

**Displacement/Stress/Strain Contour**

Command : POST, CONTOUR
Menu : Post→ Contour

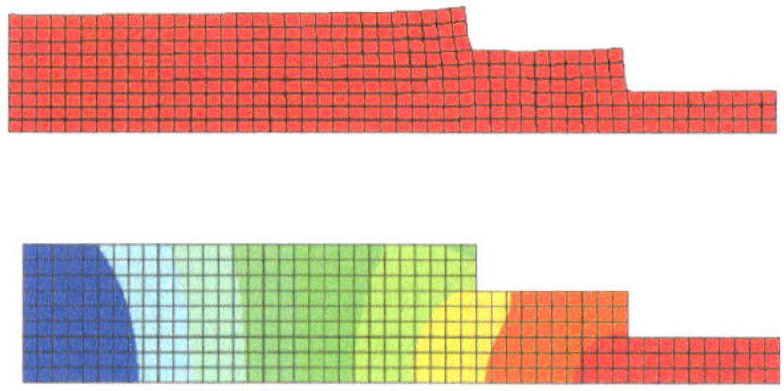

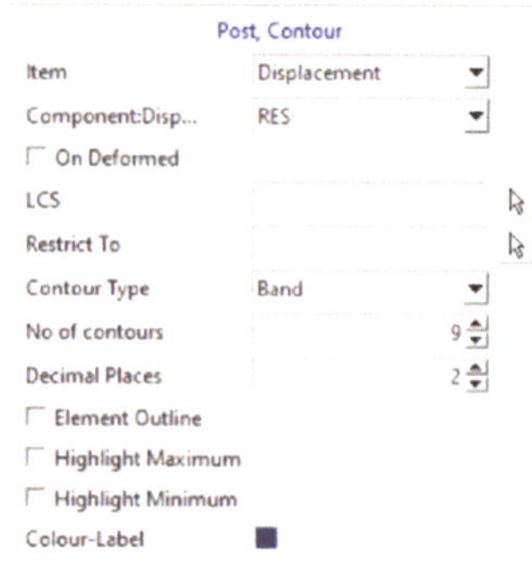


**Note**:

To plot the stress/strain contour, from the 'item' dialogue box select stress and strains. Maximum displacement and von Mises Stress obtained from the above analysis are **2.02E−06 m** and **22 MPa** respectively.

## 12.5 Exercise Problems

12.5.1 The figure below shows a biaxial test specimen pin loaded* at the holes. The magnitude of the load at each hole is 1 MPa. Assume the material for the specimen as aluminum. Model the specimen with appropriate elements, apply appropriate boundary conditions, and find the stress distribution at the gauge area.

*—Pin load—The holes are hooked onto a pin and the pin is pulled to apply the load. Therefore, the loads act only on one-half of the perimeter of the hole.

Hint: Exploit the symmetry of the structure.

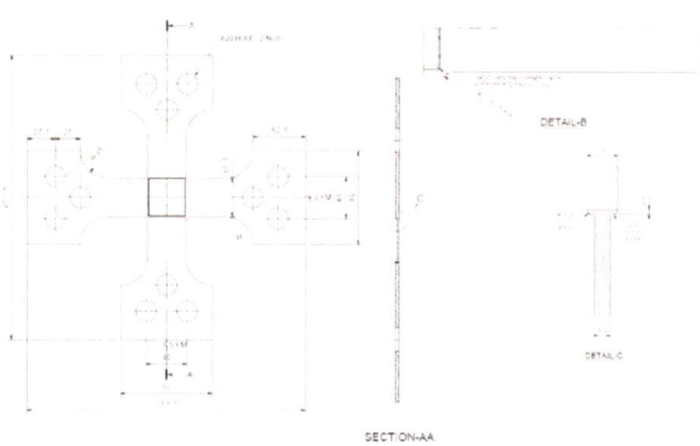

12.5.2 Assume that the gauge area is also having the same thickness as the arm area. Find the distribution of the stress pattern and compare the results with the previous problem.

12.5.3 Analyze a steel cylinder of inner radius 20 cm with thickness 5 mm and height 100 cm closed at both the ends. The internal pressure of the cylinder is 50 MPa. Find the volume change of the cylinder due to the internal load, estimate the maximum stress, and compare it with the closed form solution.

12.5.4 A swing is suspended from a hook as shown below. The hook is made of stainless steel with yield strength of 500 MPa. What is the maximum weight it can withstand without yielding? Try to find the weight in a single analysis. All dimensions are in mm.

**Hint**: Apply unit load and it is a linear static analysis.

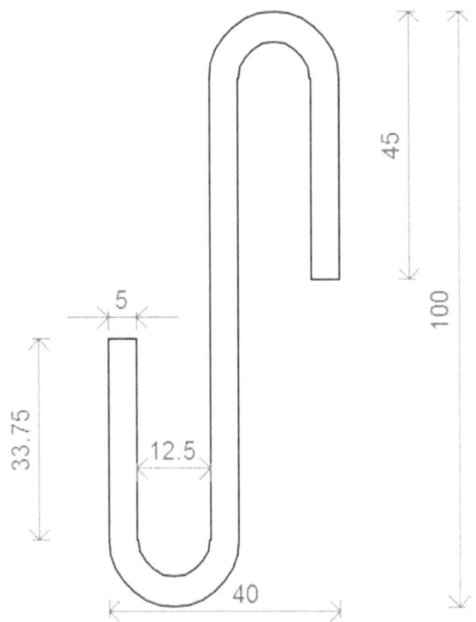


12.5.5 A composite plate of dimension 1 m × 3 m with nine layers of lamina is fixed at all the ends. A uniform pressure of 10 MPa is applied on the plate. Find the central deflection of the plate. The lamina sequence is 0/90/45/90/45/90/45/90/0. The material property of each layer is $E_L$ = 279E+03 MPa; $E_{LT}$ = 5952 MPa; $E_{LZ}$ = 100 MPa; $\upsilon_{LT} = 0.346$; $\upsilon_{Lz} = 0$; $\upsilon_{TZ} = 0$; $G_{LT}$ = 4413 MPa; $G_{LZ}$ = 10 MPa; $G_{TZ}$ = 10 MPa. The thickness of each lamina is 0.1 mm.

# Chapter 13
# Heat Transfer Analysis Using FEAST$^{SMT}$ Software

**T. J. Raj Thilak**

*I never teach my pupils. I only attempt to provide the conditions in which they can learn.*
*—Albert Einstein*

## 13.1 Rocket Nozzle Heat Transfer

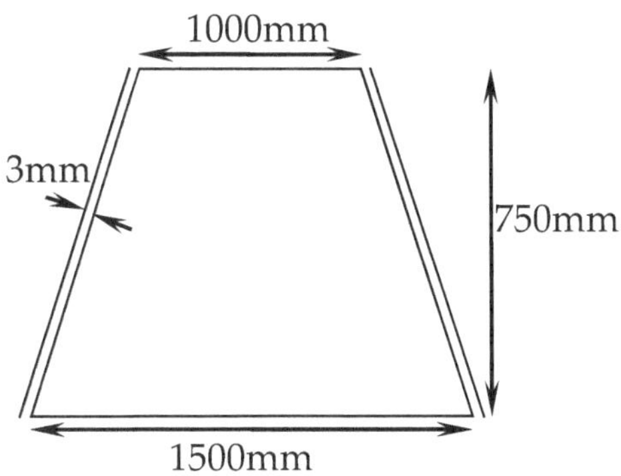


**Problem definition**: The nozzle of a rocket engine has a convergent-divergent section to accelerate the flow to supersonic conditions at the exit of the nozzle. The divergent portion of the nozzle is heated to very high temperatures during its operation. One way to maintain the temperature is to remove heat from the nozzle by the process of radiation. The typical materials used for making nozzle are high temperature steels such as Stellite and Columbium. Consider the following example.

T. J. R. Thilak (✉)
FSDD/SDAG/STR, Veli, VSSC, Thiruvananthapuram, Kerala 695022, India
e-mail: tj_rajthilak@vssc.gov.in; rajthilaks@gmail.com



S. U. Nair and S. Somanath (eds.), *Introduction to Finite Element Analysis*,
https://doi.org/10.1007/978-981-19-7989-7_13

A 3 mm thick conical nozzle divergent made of Stellite is radiatively cooled during its operation. Determine its steady state temperature if emissivity of nozzle is 0.8 and Stellite thermal conductivity is as given below.

| Temperature (K) | Thermal Conductivity (W/(m–K)) |
|---|---|
| 293 | 9.8 |
| 373 | 11.2 |
| 473 | 13 |
| 573 | 14.7 |
| 673 | 16.6 |
| 773 | 18.5 |
| 873 | 20.4 |
| 973 | 22.4 |
| 1073 | 24.4 |
| 1173 | 26.5 |
| 1273 | 28.42 |

Assume that during engine firing, internal heat transfer coefficient from the combustion products h = 100 W/m$^2$K and the gas temperature is 3000 K. The aim of this problem is to find the temperature distribution along the nozzle.

### *13.1.1 Pre-processor*

**Create geometric coordinates (0.75:0:0), (0.753:0:0), (0.5:0.75:0), (0.503:0.75:0).**

Commands : POINT, ADD
Menu : Geometry → Key point → Create → Add
Parameters : (To be filled by the user)

| **Point Data** | 0.75:0:0 |
|---|---|

After entering the coordinate data click 'Apply' or press 'Cntrl + Enter'. Likewise, each coordinate is created using the above command options.

**Create a line**

| | | |
|---|---|---|
| Commands | : | CURVE, LINE |
| Menu | : | Geometry → Curve → Create → Line |

| | |
|---|---|
| **End points** | Use Mouse to pick the points |

**Create surface using four curves**

| | | |
|---|---|---|
| Menu | : | Geometry → Surface → Create → On Geometry→ Bounded |
| Command | : | SURFACE, BOUNDED |
| Parameters | : | (To be filled by the user) |

| | |
|---|---|
| **Boundary Curves** | Use Mouse to select the curves |

**Create elements using quadrilateral elements**

| | | |
|---|---|---|
| Commands | : | MESH, QUAD |
| Menu | : | Mesh → MeshGen→ QUAD |
| Parameters | : | (To be filled by the user) |

| | |
|---|---|
| **Surface** | 1 |
| **Element Size** | 0.001 |
| **Method** | Mapped |
| **Type** | 4-Node |
| **Divisions** | |
| **Bias** | |

**Note**:

Right click/left click mouse point to alter sub-divisions and Bias.

**Set the element type**

Command : ELEMENT, TYPE
Menu : Mesh → Element → Modify → Type
Parameters : (To be filled by the user)

| **Element** | All[0] |
|---|---|
| **Type** | Axisymmetric |

*Default Type* represents for
One Dimension—Beam Element
Two Dimension—Shell Element
Three Dimension—Displacement formulation-based Solid Element

**Create the function using real table**

Commands : FUNCTION, TABLESCALAR
Menu : Property → Function → Scalar Table
Parameters : (To be filled by the user)

| **Table Data** | 293/9.8/373/11.2/473/13/573/14. |
|---|---|

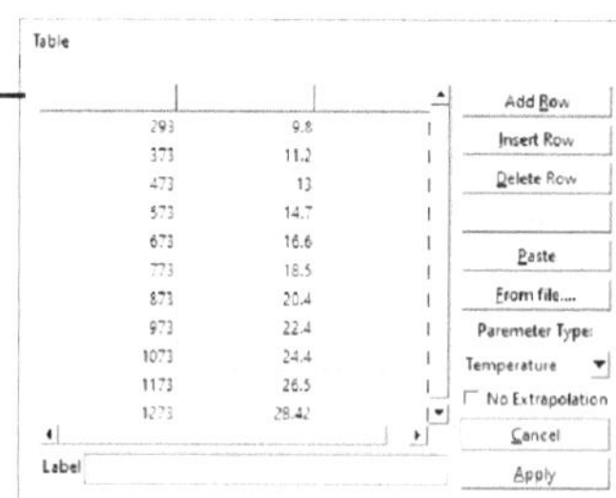

**Create material property**

Command : MATERIAL, HTISOTROPIC
Menu : Property → Material → Thermal→ Isotropic
Parameters : (To be filled by the user)

| | |
|---|---|
| **Elements** | ALL[0] |
| **Thermal Conductivity** | F1 |
| **Density** | 1 |
| **Specific Heat** | |
| **Label** | |

**Create the Heat Transfer Boundary Conditions**

Command : HTTEMP, ADD
Menu : Load/BC → Thermal → Temperature
Parameters : (To be filled by the user)

| | |
|---|---|
| **Node IDs** | Use Mouse to select the Nodes |
| **Temperature** | 3000 |
| **Label** | 1 |

Command : HTCONV, ADD
Menu : Load/BC → Thermal → Convection
Parameters : (To be filled by the user)

| **Entity Type** | Edge |
|---|---|
| **Entity Edge** | Use Mouse to select the Element edges/faces |
| **Film coefficient** | 100 |
| **Ambient temperature** | 3000 |
| **Label** | 1 |

Command : HTRADSIMPLE, ADD
Menu : Load/BC → Heat Transfer → Radiation
Parameters : (To be filled by the user)

| **Entity Type** | Edge |
|---|---|
| **Entity Edge** | Use Mouse to select the Element edges/faces |
| **Emissivity** | 0.8 |
| **Temperature of surrounding** | 293 |
| **Stefan–Boltzmann constant** | 5.670373×10−8 |
| **Label** | 1 |

### 13.1.2 *Solution*

**Set analysis type**

| | | |
|---|---|---|
| Commands | : | ANTYPE, ADD |
| Menu | : | Analysis → Analysis Type |
| Parameters | : | (To be filled by the user) |

| **Analysis Type** | HT-Steady State |
|---|---|

**Save the job and submit the job into FEAST$^{SMT}$**

**Note**:

For solving user can click ! icon in right end of tool bar.

### 13.1.3 *Post-processing*

**Temperature Contour**

| | | |
|---|---|---|
| Command | : | POST, CONTOUR |
| Menu | : | Post → Contour |

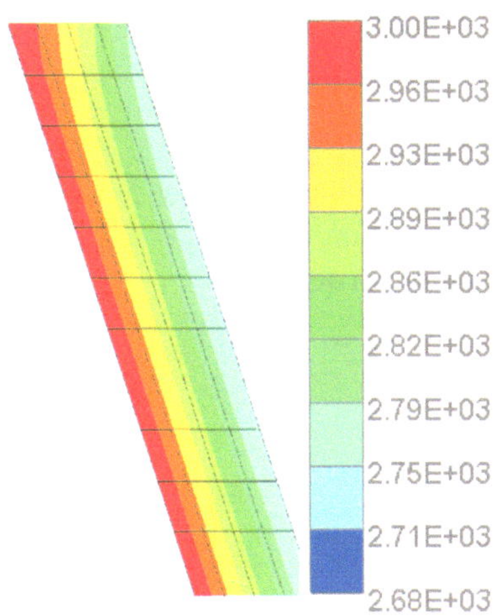


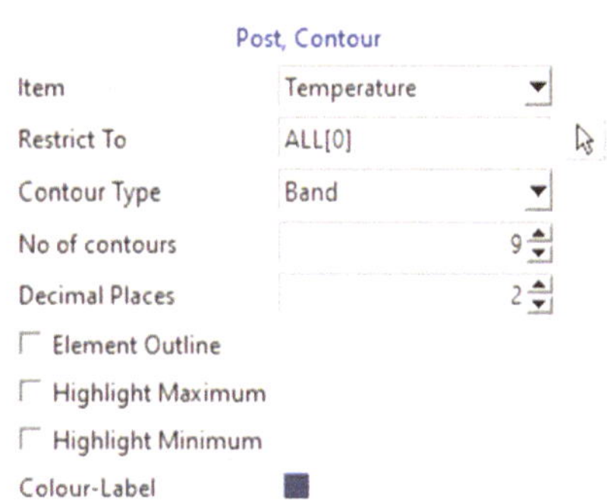

Nozzle divergent is radiatively cooled to ensure that the temperatures are below 1000–1100 °C. This ensures the structural integrity of the nozzle during its operation.

## 13.2 Heat Transfer Analysis of a Printed Circuit Board

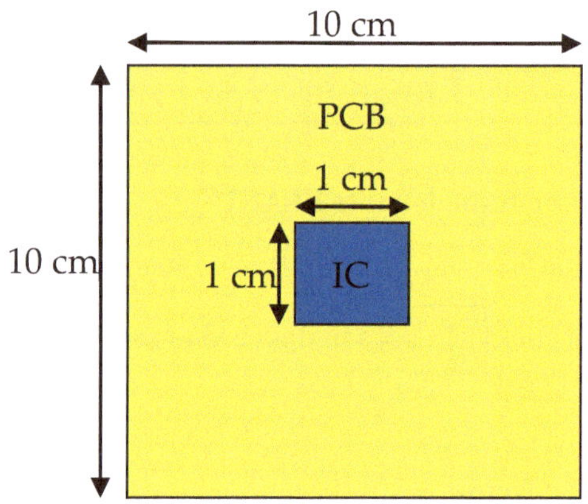


**Problem definition:** Printed circuit boards (PCB) are widely used in avionics packages. The electronic components mounted onto the PCB will generate heat during its operation. Good thermal design ensures that heat is properly dissipated to surroundings without creating hot spots in the PCB.

A PCB with 1 mm thick board of size 10 cm × 10 cm made of FR4 material with thermal conductivity 0.7 W/mK has an electronic chip of size 1 cm × 1 cm in the center that dissipates 100 mW. Determine the steady state temperature distribution of the board if the board is having free convection to an ambient of 30 °C from both sides.

The PCB board can be modeled with 2D elements in which the center portion under the IC can be specified as the source of heat. Convection boundary condition shall be specified over the entire board with h = 5 W/m$^2$K and Tgas = 303 K.

The objective is to find the temperature distribution on the PCB due to the IC heat dissipation.

### 13.2.1 Pre-processor

**Create geometric coordinates (0:0:0), (0.1:0:0), (0:0.1:0), (0.1:0.1:0).**

Commands : POINT, ADD
Menu : Geometry → Key point → Create → Add
Parameters : (To be filled by the user)

| **Point Data** | 0:0:0 |
|---|---|

After entering the coordinate data click 'Apply' or press 'Cntrl + Enter'. Likewise, each coordinate is created using the above command options.

**Create a line**

Commands : CURVE, LINE
Menu : Geometry → Curve → Create → Line

| **End points** | Use Mouse to pick the points |
|---|---|

**Create surface using four curves**

Menu : Geometry → Surface → Create → On Geometry→ Bounded
Command : SURFACE, BOUNDED
Parameters : (To be filled by the user)

| **Boundary Curves** | Use Mouse to select the curves |
|---|---|

**Create elements using quadrilateral elements**

Commands : MESH, QUAD
Menu : Mesh → MeshGen→ QUAD
Parameters : (To be filled by the user)

| Surface | 1 |
|---|---|
| Element Size | 0.005 |
| Method | Mapped |
| Type | 4-Node |
| Divisions | |
| Bias | |

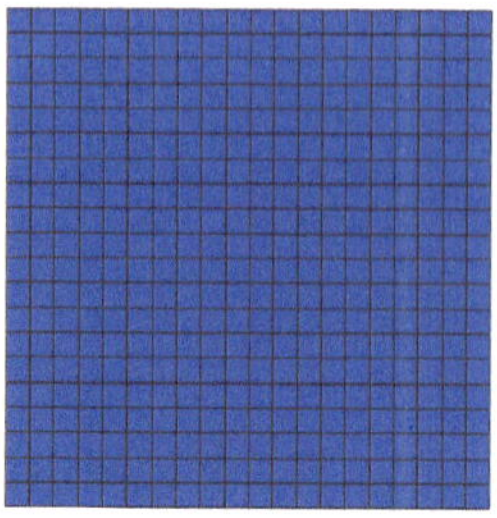

**Note**:

Right click/left click mouse point to alter sub-divisions and Bias.

**Extrude the shell element**

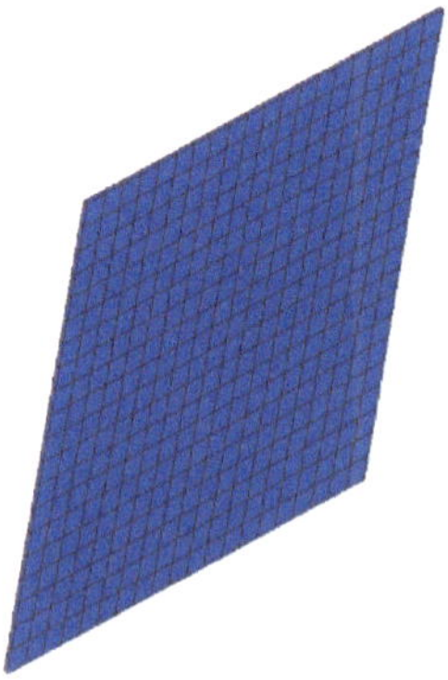

Command : ELEMENT, EXTRUDE
Menu : Mesh → Element → Create →
Extrude
Parameters : (To be filled by the user)

| List of Elements | ALL[0] |
|---|---|
| Extrusion Vector | 0:0:0.001 |
| Twist | |
| Scale | |
| Anchor | |
| Segments | 1 |

**Create material property**

Command : MATERIAL, HTISOTROPIC
Menu : Property → Material → Thermal→ Isotropic
Parameters : (To be filled by the user)

| **Elements** | ALL |
|---|---|
| **Thermal Conductivity** | 0.0007 |
| **Density** | 0 |
| **Specific Heat** | 0 |
| **Label** | |

**Create the Heat Transfer Boundary Conditions**

Command : HTGEN, ADD
Menu : Load/BC → Thermal → Heat Generation
Parameters : (To be filled by the user)

| **Element IDs** | Use Mouse to select the Elements |
|---|---|
| **Heat generation** | 10 |
| **Label** | |

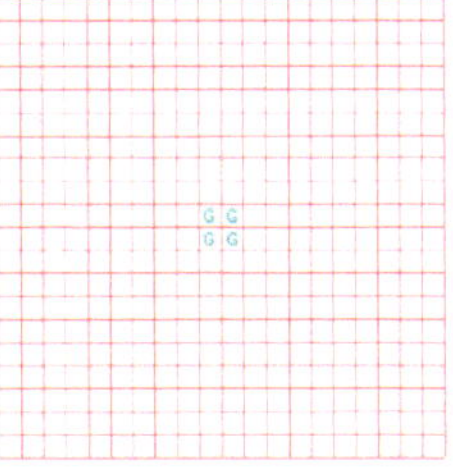

Command : HTCONV, ADD
Menu : Load/BC → Thermal → Convection
Parameters : (To be filled by the user)

| **Entity Type** | Edge |
|---|---|
| **Entity Edge** | Use Mouse to select the Element edges/faces |
| **Film coefficient** | 5 |
| **Ambient temperature** | 303 |
| **Label** | |

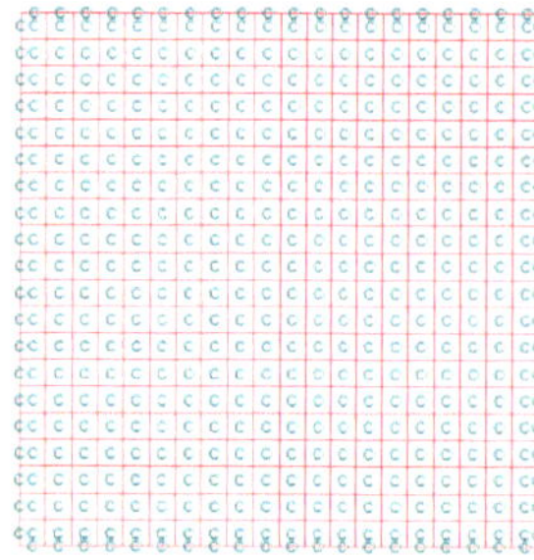

## *13.2.2 Solution*

**Set analysis type**

Commands : ANTYPE, ADD
Menu : Analysis → Analysis Type
Parameters : (To be filled by the user)

| **Analysis Type** | HT-Steady State |
|---|---|

**Save the job and submit the job into FEAST[SMT]**

**Note**:

For solving user can click  icon in right end of tool bar.

### *13.2.3 Processing*

**Temperature Contour**

| | | |
|---|---|---|
| Command | : | POST, CONTOUR |
| Menu | : | Post Contour |

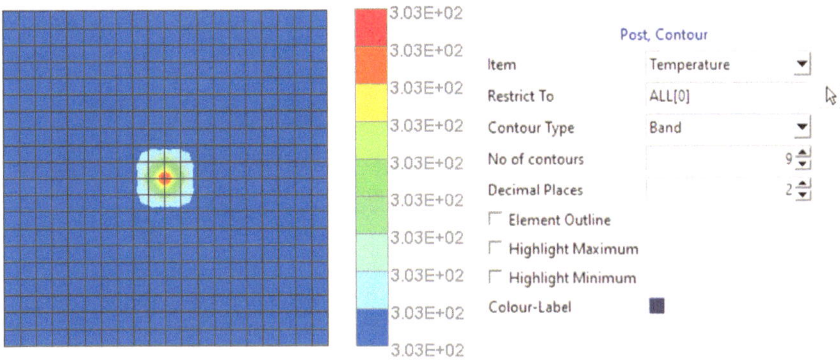


**Comments on results**

Maximum steady temperature of PCB at chip location is 329 K or 56 °C. Knowing the resistance of the chip $\Theta_{jc}$, which is specified by the manufacturer, one can work out the junction temperature for the given heat dissipation. For safe operation of the chip, the junction temperature has to be typically below 100 °C.

## 13.3 Heat Transfer Analysis of Cooling Fins

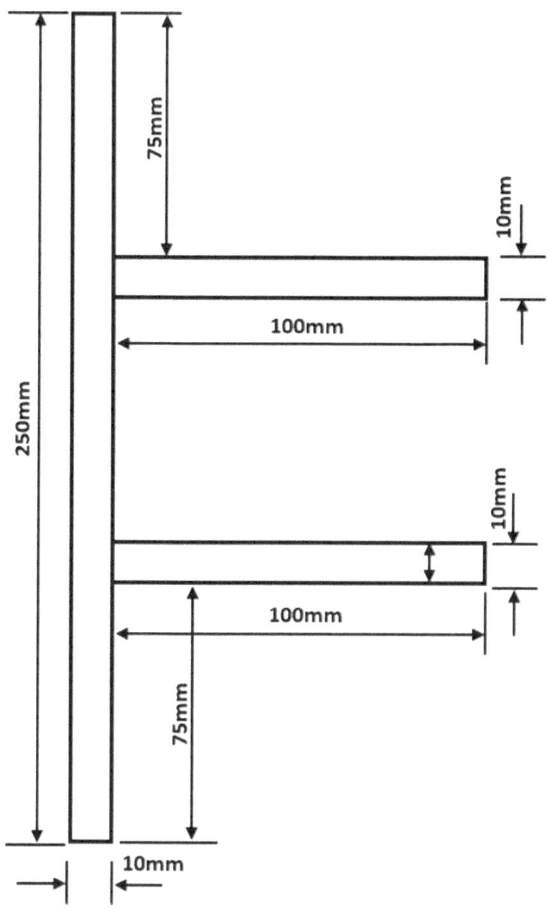


**Problem definition:** Cooling Fins are extended surfaces used to remove heat. If the inner side of the vertical wall is at 100 °C and the outer sides (including fins) convect to an ambient of 30 °C, how effective is the following fin design? Note: fin effectiveness is defined as the ratio of heat transferred with fin to that without fin. Do this problem as two dimension as well as two-dimensional axisymmetric with the inner wall located at R = 75 mm. (All dimensions are in mm). The fins along with wall are made of Al Alloy with thermal conductivity of 146.5 W/mK.

### 13.3.1 Pre-processor

Create geometric coordinates (0.075,0,0), (0.085,0,0), (0.075,0.25,0), (0.085,0.25,0), (0.085,0.175,0), (0.185,0.175,0), (0.185,0165,0), (0.085,0.165,0) (0.185,0.0.85,0), (0.185,0.085,0), (0.185,0.075,0), (0.085,0. 075,0).

Commands : POINT, ADD
Menu : Geometry → Key point → Create → Add
Parameters : (To be filled by the user)

| **Point Data** | 0:0:0 |
|---|---|

After entering, the coordinate data click 'Apply' or press 'Cntrl + Enter'. Likewise, each coordinate is created using the above command options.

**Create a line**

Commands : CURVE, LINE
Menu : Geometry → Curve → Create → Line

| **End points** | Use Mouse to pick the points |
|---|---|

**Create surface using four curves**

Menu : Geometry → Surface → Create → On Geometry→ Bounded
Command : SURFACE, BOUNDED
Parameters : (To be filled by the user)

| **Boundary Curves** | Use Mouse to select the curves |
|---|---|

**Create elements using quadrilateral elements**

Commands : MESH, QUAD
Menu : Mesh → MeshGen→ QUAD
Parameters : (To be filled by the user)

| | |
|---|---|
| **Surface** | 1 |
| **Element Size** | 0.005 |
| **Method** | Mapped |
| **Type** | 4-Node |
| **Divisions** | |
| **Bias** | |

**Note**:

Right click/left click mouse point to alter sub-divisions and Bias.

**Create material property**

Command : MATERIAL, HTISOTROPIC
Menu : Property → Material → Thermal→ Isotropic
Parameters : (To be filled by the user)

| | |
|---|---|
| **Elements** | ALL[0] |
| **Thermal Conductivity** | 146.5 |
| **Density** | |
| **Specific Heat** | |
| **Label** | |

**Create the Heat Transfer Boundary Conditions**

Command : HTCONV, ADD
Menu : Load/BC → Thermal → Convection
Parameters : (To be filled by the user)

| **Entity Type** | Edge |
|---|---|
| **Entity Edge** | Use Mouse to select the Element edges/faces |
| **Film coefficient** | 5 |
| **Ambient temperature** | 30 |
| **Label** | |

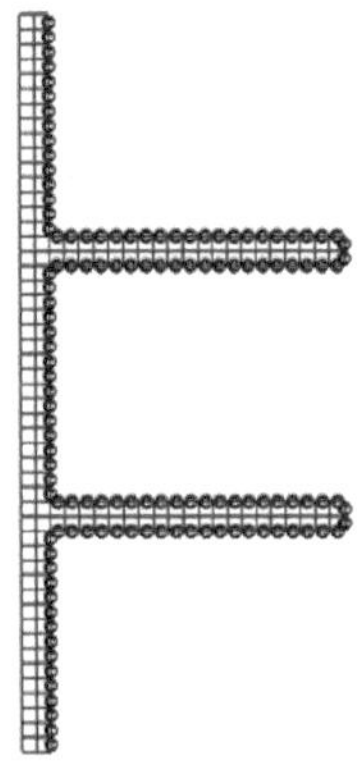

Command : HTTAMP, ADD
Menu : Load/BC → Thermal → Temperature
Parameters : (To be filled by the user)

| **Node IDs** | Use Mouse to select the Nodes |
|---|---|
| **Temperature** | 100 |
| **Label** | |

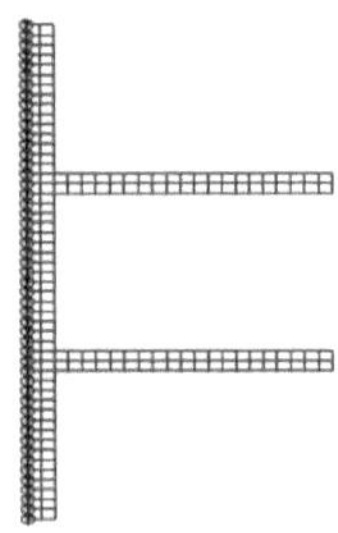

### *13.3.2 Solution*

**Set analysis type**

| | | |
|---|---|---|
| Commands | : | ANTYPE, ADD |
| Menu | : | Analysis → Analysis Type |
| Parameters | : | (To be filled by the user) |

| Analysis Type | HT-Steady State |
|---|---|

**Save the job and submit the job into FEAST[SMT]**

<u>**Note**</u>:

For solving user can click  icon in right end of tool bar.

### *13.3.3 Post-processing*

**Temperature Contour**

| | | |
|---|---|---|
| Command | : | POST, CONTOUR |
| Menu | : | Post → Contour |

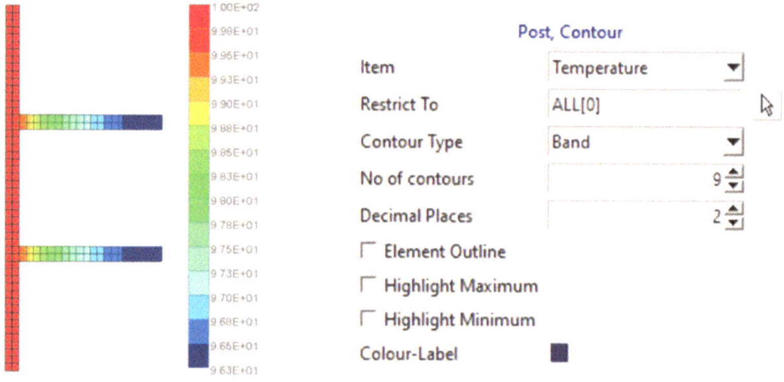

## 13.4 Exercise Problems

A composite slab made of 10 mm Aluminum Alloy AA2014 and 10 mm Stainless SS304L is exposed to thermal environments as shown in Figure. Determine the steady state temperature distribution within the slab. Use standard properties of AA2015 and SS304L.

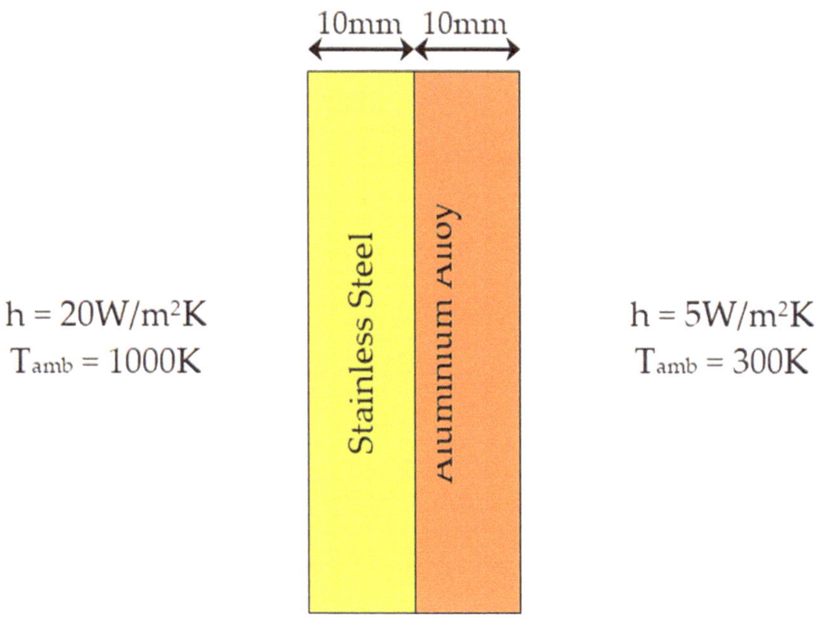


A 2 in. diameter stainless steel pipe of 1 mm thickness carrying hot water at 350 K is insulated with a porous insulator of thickness 10 mm having thermal conductivity of 0.1 W/mK. Determine the temperature distribution within this composite system if the thermal environments are as shown in Figure.

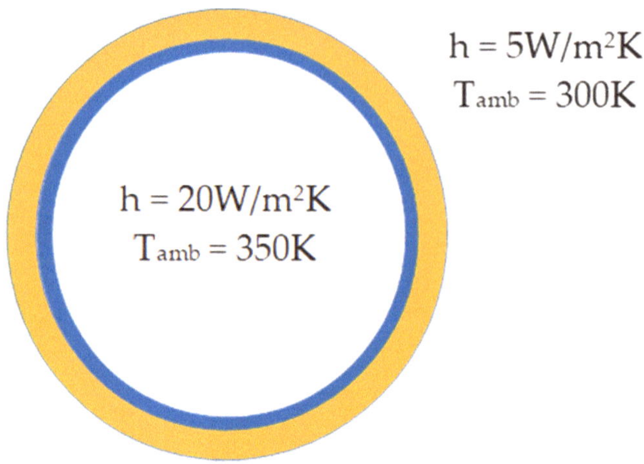


A unit cube made of Al Alloy AA2014 is at T = 500 K at time t = 0 s. Determine the temperature distribution of the cube at t = 1000 s if it is having free convection (h = 5 W/m$^2$K) on all sides to an ambient of 300 K. Do the same problem if the cube is made up of Stainless steel SS304L.

How many degree of freedoms are there for a heat transfer problem when compared to a static analysis problem?

What conditions are required to make a transient conduction problem to be well posed? What about steady state conduction problem?

What thermo-physical properties are required to solve a steady state conduction problem? What about a transient conduction problem?

How does FEM formulation of heat transfer account for a boundary in which no boundary condition is explicitly defined?

Transient problems in FEM are usually solved by implicit formulation. Can we use explicit formulation without the requirement of inverting the mass matrix M? (Hint: Can we use lumped mass matrix?).

Identify the types of non-linearity that can be present in a heat transfer problem. How do they differ from the non-linearity types in a static analysis problem?

Why is it preferred to work in Kelvin scale in heat transfer? (Hint: Radiation).

# Chapter 14
# DynamicAnalysis Using FEAST$^{SMT}$ Software

T. J. Raj Thilak

> *I never teach my pupils. I only attempt to provide the conditions in which they can learn.*
> *—Albert Einstein*

## 14.1 Free Vibration Analysis of a C-section Beam

A C-section beam having dimensions as shown in Fig. 14.1 is made of 20 layers of 0.1 mm thick carbon fiber reinforced composite and is considered for carrying out free–free vibration analysis. Since it is an orthotropic material, all the properties along the fiber and across the fiber are given as input. Since the thickness of the structure is very small compared to the other dimensions of the structure, composite shell element is used to idealize the structure.

Material Properties

$E_L$—266000 MPa; $E_T$—5952 MPa; $E_N$—5952 MPa;

$\upsilon_{LT}$—0.346; $\upsilon_{LN}$—0; $\upsilon_{TN}$—0;

$G_{LT}$—4413 MPa; $G_{LN}$—4413 MPa; $G_{TN}$—4413 MPa;

$\rho$—1700 kg/m$^3$

Lamina Angle Sequence—$[0/90/0/90/45/-45/45/-45/0/90]_S$.

T. J. R. Thilak (✉)
FSDD/SDAG/STR, VSSC, VeliThiruvananthapuram, Kerala 695022, India
e-mail: tj_rajthilak@vssc.gov.in; rajthilaks@gmail.com



S. U. Nair and S. Somanath (eds.), *Introduction to Finite Element Analysis*,
https://doi.org/10.1007/978-981-19-7989-7_14

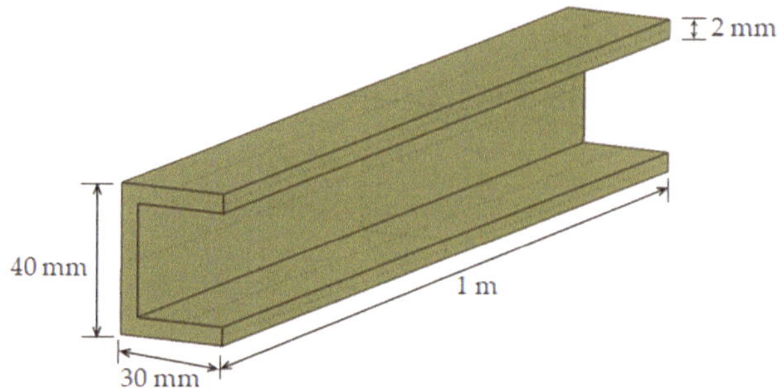


**Fig. 14.1** C-section beam

### *14.1.1 Pre-processor*

**Create geometric coordinates (0:0:0), (0:0.04:0), (0.03:0.04:0), (0.03:0:0)**

Commands : POINT, ADD
Menu : Geometry → Key point → Create → Add
Parameters : (To be filled by the user)

| **Point Data** | 0:0:0 |
|---|---|

After entering the coordinate data click 'Apply' or press 'Ctrl+Enter'. Likewise, each coordinate is created using the above command options.

**Create a line**

Commands : CURVE, LINE
Menu : Geometry → Curve → Create → Line

| **End points** | Use Mouse to pick the points |
|---|---|

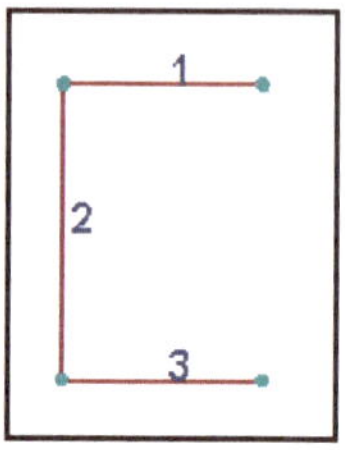

### Extrude surface using curves

Command : SURFACE, EXTRUDE
Menu : Geometry → Surface → Create → Extrude
Parameters : (To be filled by the user)

| **List of Curve** | Use Mouse to select the curves |
|---|---|
| **Extrusion Vector** | 0:0:-1 |
| **Twist** | 0 |
| **Scale** | 0 |
| **Anchor** | 0 |

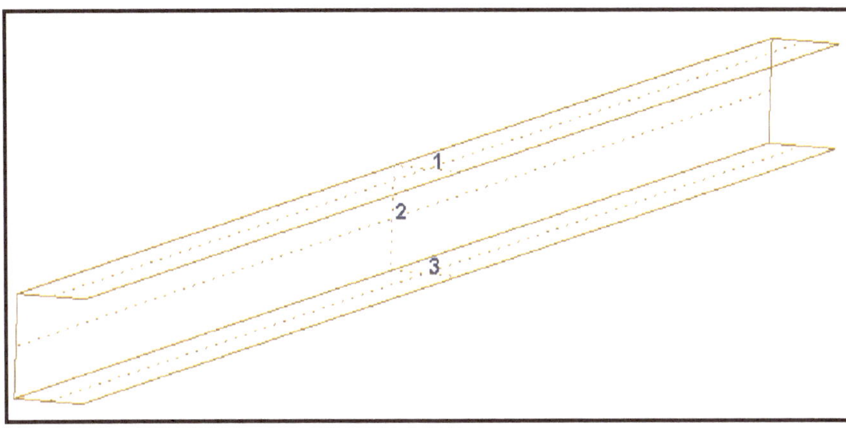

**Create elements using quadrilateral elements**

Commands : MESH, QUAD
Menu : Mesh → MeshGen→ QUAD
Parameters : (To be filled by the user)

| **Surface** | 1T3 |
|---|---|
| **Element Size** | 0.01 |
| **Method** | Mapped |
| **Type** | 4-Node |
| **Divisions** | |
| **Bias** | |

**Note**:

Right click/left click mouse point to alter sub-divisions and Bias.

Initially the geometry was created in meters. If the user wants to change the unit to millimeter, then the user can scale the model by 1000 to convert all the dimensions from meter to millimeter.

**Modify the Node Scale**

Commands : NODE, SCALE
Menu : Mesh → Node→ Transform→ Scale
Parameters : (To be filled by the user)

| **List of Nodes** | ALL |
|---|---|
| **Scale** | 1000:1000:1000 |
| **Anchor** | 0 |
| **Copies** | 0 |

**Note**:

**Create thickness for layers**

Commands : THICKNESS, ADD
Menu : Property → Physical → Thickness
Parameters : (To be filled by the user)

| **Elements** | |
|---|---|
| **Thickness** | 0.1 |
| **Label** | |

**Note**:

Do not specify the element IDs.

**Create orthotropic material property for layers**

Command : MATERIAL, ORTHOTROPIC
Menu : Property → Material→ Structural→ Orthotropic
Parameters : (To be filled by the user)

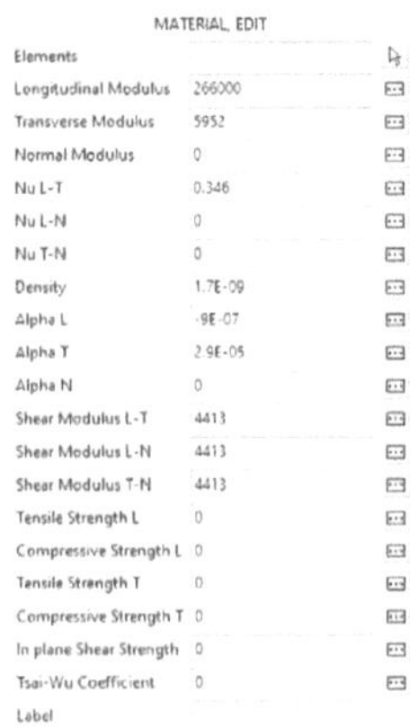


**Note**:

Do not specify the element IDs.

**Create material angle**

Command : MATANGLE, ADD
Menu : Property → Physical → Material angle
Parameters: (To be filled by the user)

| **Angle** | 0 |
|---|---|
| **Reference Direction** | Element edges |
| **LCS** | |
| **Label** | |

Create another set of material angles with value 45/−45/90.

Verify whether two sets of material angle values were defined in 'Edit' option.

**Create layup details**

Command : MATERIAL, LAYERED
Menu : Property→ Material→ Structural→ Layered
Parameters : (To be filled by the user)

| **Elements** | ALL[0] |
|---|---|
| **Layup Data** | 1/1/1/1/4/1/1/1/1/1/4/1 |
| **Label** | |

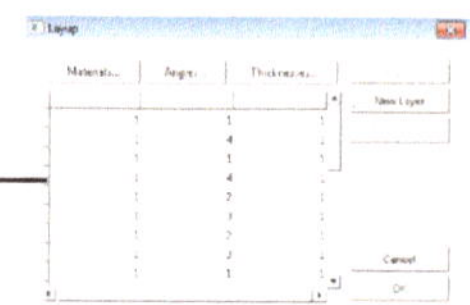

**Note**:

Layered material is also considered as a material property in the FEAST$^{SMT}$ software. The layup details have to be supplied from the bottom layer to the top layer. The bottom layer to top layer direction is same as the direction of the element normal which can be identified by the following command or the menu.

Command : ELEMENT, NORMAL
Menu : Mesh→ Element→ Verify→ Normal

## *14.1.2 Solution*

**Set analysis type**

Commands : ANTYPE, ADD
Menu : Analysis → Analysis Type
Parameters : (To be filled by the user)

| Analysis Type | Free Vibration |
|---|---|

**Set free vibration general data**

Command : FREEVIBGEN,ADD
Menu : Analysis → Free Vibration → General
Parameters : (To be filled by the user)

| **Mode Extraction** | Number of modes |
|---|---|
| **Number of modes** | 20 |
| **Mass Option** | Coupled |
| **Effective Mass** | No |

**Note**:

There are three options available in FEAST$^{SMT}$ software for constructing the mass matrix.

(i) Lumped Mass Matrix
(ii) Consistent Mass Matrix
(iii) Coupled Mass Matrix.

**Lumped Mass Matrix**: This option calculates the total mass of the element by calculating the volume of the element and multiplying it by density. The total mass is divided by the number of nodes in the element and the resulting mass is lumped at the nodes in all the translation DOF. It gives the lower bound of the estimated frequency.

**Consistent Mass Matrix**: This option calculates the mass matrix consistent with the calculation of the stiffness matrix, i.e. it uses shape functions of the element to calculate the mass matrix. The mass matrix thus calculated is a full symmetric matrix at the element level. It gives the upper bound of the estimated frequency.

**Coupled Mass Matrix**: This option calculates the mass matrix as an average value from both the options mentioned above.

**Save the job and submit the job into FEAST$^{SMT}$**

**Note**:

For solving user can click **!** icon in right end of tool bar.

### 14.1.3 Post-processing

**View Results Frequency**

Command : POST, TABLEVIEW
Menu : Post → View Table
Parameters : (To be filled by the user)

| Item | Frequencies |
|---|---|

**Deformed shape**

Command : POST, Deflection
Menu : Post → Deflection

| Item | Mode Shape |
|---|---|
| Mode | 9 - 278.517 |
| Scale Factor | 1 |

| Mode | Frequency(Hz) |
|---|---|
| Rigid body modes | |
| 7 | 90.2917 |
| 8 | 199.683 |
| 9 | 278.517 |
| 10 | 466.519 |

__Note__:

If in a dynamic analysis, the load is not specified then the resulting dynamic analysis is a free vibration analysis. If the support conditions are also not mentioned that is the structure is free in the air without any support, then the resulting dynamic analysis is free-free vibration analysis. Since the support conditions are not mentioned the structure can rigidly move in all six directions. Therefore, the first six modes in this analysis will be rigid body modes with zero frequency and the first elastic mode of the structure will be the seventh mode. This property of the free–free vibration analysis can be used for any model to check hanging elements in the structure, provided density is assigned for all the elements in the structure.

**Mode Shape Contour**

| | | |
|---|---|---|
| Command | : | POST, CONTOUR |
| Menu | : | Post → Contour |

Figure 14.2 shows the first bending mode of the C-section beam. Frequencies along with the corresponding mode shape and generalized mass define the dynamic characteristics of the structure. These dynamic characteristics provide an insight into the different shapes in which the structure will vibrate based on the excitation frequency and its location. In this example, if the structure is excited at this frequency at a location away from the zero displacement in the mode shape, the structure will form a shape shown in Fig. 14.2. The amplitude of the deformation depends on the amplitude of the force applied at that frequency.

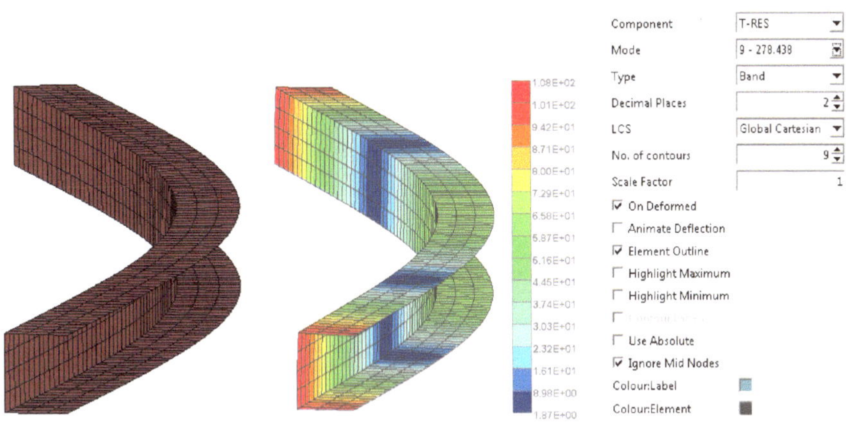


**Fig. 14.2** Elastic mode of a C-section beam

## 14.2 Frequency Response Analysis of a Bracket

Electronic packages are usually mounted on brackets in real-life systems. The packages are tested using base excitation given at all frequencies in the range of interest to check integrity of electronic package under operating environments. The applicable dynamic analysis for this problem is base excitation in frequency response. A typical aluminum bracket with dimensions as shown in Fig. 14.3 and thickness 3 mm is used for analysis. An electronic package is mounted on top of the bracket via four M4 bolts and the CG of the package is located 50 mm from the top surface of the bracket. Since the frequencies of the package are not considered, the package is not modeled explicitly. Instead the mass of the package is simulated using mass element placed

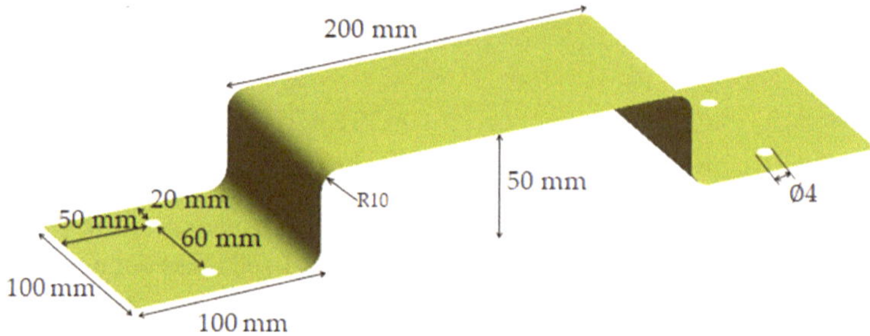


**Fig. 14.3** Electronic package mounted on a bracket

at the CG location and this mass element is connected to the bracket via rigid links. The bracket is mounted on the system using four numbers of M4 bolts. Therefore base excitation is given in these bolt locations. The objective of this analysis is to find acceleration response at CG of the electronic package for 1 g acceleration at all frequencies in the range 0.1–5 Hz. Since the bracket is symmetric in one plane, only one-half is modeled and the other half is mirrored to create the entire model. Since the bracket is very thin compared to other dimensions, shell element is used to discretize the model.

### *14.2.1 Pre-processor*

**Create geometric coordinates (0:0:0), (100:0:0), (100:50:0), (200:50:0)**

Commands : POINT, ADD
Menu : Geometry → Key point → Create → Add
Parameters : (To be filled by the user)

| **Point Data** | 0:0:0 |
|---|---|

After entering the coordinate data click 'Apply' or press 'Cntrl+Enter'. Likewise each coordinate is created using the above command options.

**Create a line**

Commands : CURVE, LINE
Menu : Geometry → Curve → Create → Line
Parameters : (To be filled by the user)

| End points | Use Mouse to pick the points |
|---|---|

Create lines joining key points (P1/P2), (P2/P3) and (P3/P4).

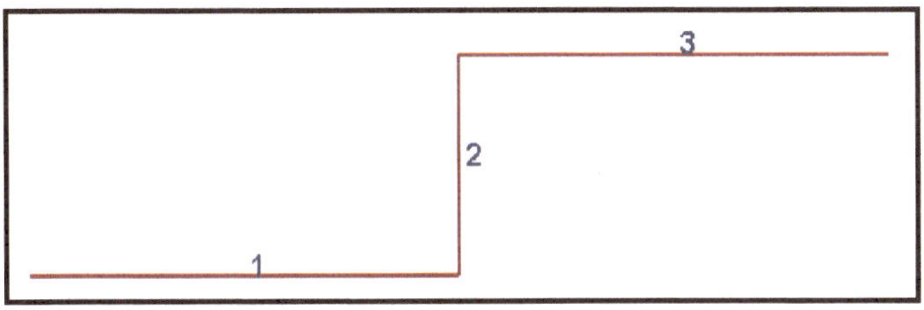


**Create surface on using curves**

Command : SURFACE, EXTRUDE
Menu : Geometry → Surface → Create → Extrude
Parameters : (To be filled by the user)

| List of curves | Use Mouse to select the curves |
|---|---|
| Extrusion Vector | 0:0:-100 |
| Twist | |
| Scale | |
| Anchor | |

### Create Filet

Command : SURFACE, FILLET
Menu : Geometry → Surface → Create → On Geometry → Fillet
Parameters : (To be filled by the user)

| Surface | Use Mouse to select the surfaces |
|---|---|
| Radius | 10 |

Create filet Connecting surfaces 1/2 and 2/3.

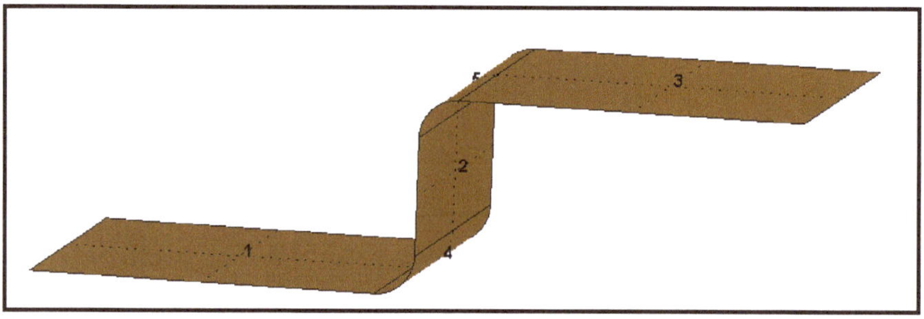


### Create elements using quadrilateral elements

Commands : MESH, QUAD
Menu : Mesh → MeshGen → QUAD
Parameters : (To be filled by the user)

| Surface | 1T5 |
|---|---|
| Element Size | 5 |
| Method | Mapped |
| Type | 4-Node |
| Divisions | |
| Bias | |

**Note**:

Right click/left click mouse point to alter sub-divisions and Bias.

Make sure that the element edge length should be 5 and each edge should contain elements equal to (edge length/5). So that 5 elements will be available for discretizing filets.

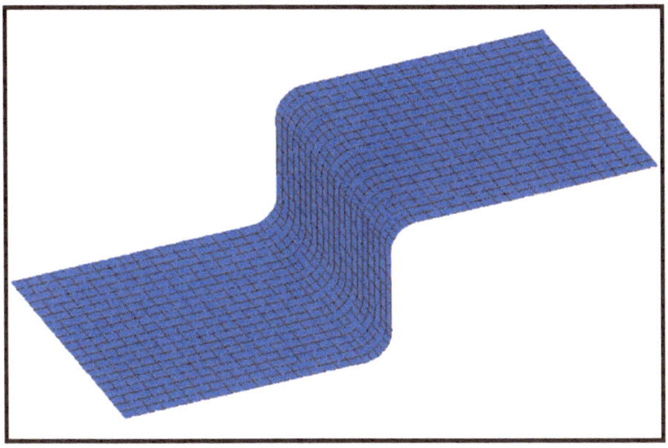

**Create Key points—(50:0:−20), (50:0:−80)**

Command : POINT, ADD
Menu : Geometry → Keypoint → Create → Add
Parameters : (To be filled by the user)

**Create cutout**

Command : MESH, CUTOUT
Menu : Mesh → MeshGen → Cutout
Parameters : (To be filled by the user)

| | |
|---|---|
| **Domain Elements** | Use mouse to select the elements |
| **Center** | P5 |
| **Radius** | 4 |

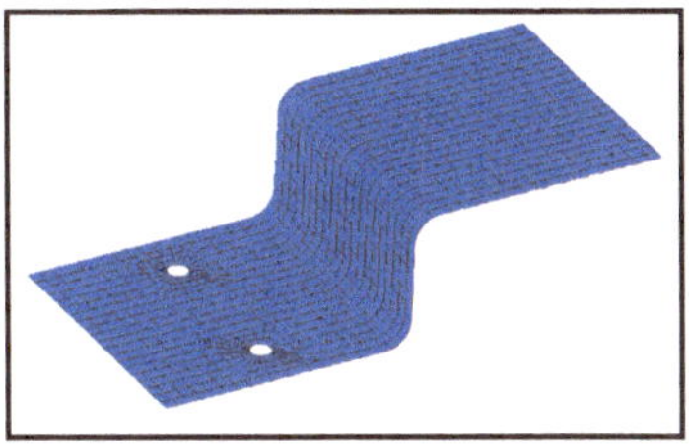

Similarly create a cut out using key point at P6.

**Create mesh for entire structure by mirroring the existing half**

Command : ELEMENT, MIRROR
Menu : Mesh → Element → Transform→ Mirror
Parameters : (To be filled by the user)

| List of elements | ALL |
|---|---|
| **Plane** | X Axis |
| **Location** | P4 |

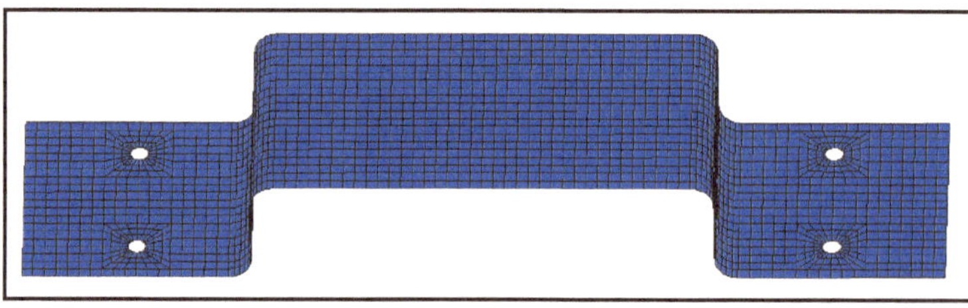

**Note**:

While mirroring the mesh, the second half of the mesh is an independent structure and the software creates duplicate nodes at the plane of symmetry. The second half of the structure should be connected to the first half by deleting the duplicate nodes at the intersection. Otherwise, the two halves behave like two independent structure. Merge option in the software is used to delete the duplicate nodes. If the distance between two nodes is within the 'Tolerance' specified, then it merges two nodes into a single node.

**Merge duplicate nodes**

Command : NODE, MERGE
Menu : Mesh → Node → Modify → Merge
Parameters : (To be filled by the user)

| Nodes | ALL |
|---|---|
| **Tolerance** | 1E-04 |

**Create nodes specifying coordinate—(50,0,−20), (50,0,−80), (350,0,−20), (350,0,−80), (200,100,−50)**

Commands : NODE, ADD
Menu : Mesh→ Node→ Create→ Add
Parameters : (To be filled by the user)

| Coordinate | 50/0/-20 |
|---|---|
| Type | Cartesian |
| ID | |

After entering the coordinate data click 'Apply' or press 'Cntrl+Enter'. Likewise each coordinate is created using the above command options.

Note:

These nodes are required to create rigid links which can be used to simulate the bolts and to create the mass element that simulates the electronic package. The mass element is connected to the bracket through four rigid links at the mounting location of the electronic package.

**Create a rigid link in cut out Location**

Command : RLINK, ADD
Menu : Load/BC→ Structural→ Rigid Link
Parameters : (To be filled by the user)

| Slave nodes | 1204T1227 |
|---|---|
| Master node | 2549 |
| Slave DOFs | 123456 |
| Master DOF | Any |
| LCS ID | |
| Label | |

| Slave nodes | 465/731/2165/2271 |
|---|---|
| Master Node | 2553 |
| Slave DOFs | 123456 |
| Master DOF | Any |
| LCS | |
| Label | |

Similarly, rigid links are created at all bolt locations and mass locations.

**Specify Displacement constraints**

Commands : DISPBC, ADD
Menu : Load/BC→ Structural→ DispBC
Parameters : (To be filled by the user)

| Node IDs | 2549T2552 |
|---|---|
| BC Value | /0/0/0/0/0 |
| LCS ID | 0 |
| Set ID | 1 |

**Note**:

Since the base is to be excited for 1 g in X-direction, all the DOF except $U_x$ is fixed at the bolt location. This displacement constraint simulates the shaking of the structure in X-direction with 1 g acceleration.

**Create thickness**

Commands : THICKNESS, ADD
Menu : Property → Physical → Thickness

| Elements | ALL[0] |
|---|---|
| Thickness | 3 |
| Label | |

**Specify material properties**

Commands : MATERIAL, ISOTROPIC
Menu : Property → Material → Structural → Isotropic
Parameters : (To be filled by the user)

| Elements | All[0] |
|---|---|
| Young's Modulus | 70000 |
| Nu | 0.3 |
| Density | 2.8E-9 |
| Alpha | 0 |
| Label | |

**Lump the mass of the Electronic Package using Mass Element**

Commands : MASS, ADD
Menu : Load/BC→ Structural → Mass
Parameters : (To be filled by the user)

| **Nodes** | 2553 |
|---|---|
| **MX** | 50 |
| **MY** | 50 |
| **MZ** | 50 |
| **IXX** | 0 |
| **IYY** | 0 |
| **IZZ** | 0 |
| **IXY** | 0 |
| **IXZ** | 0 |
| **IYX** | 0 |
| **Label** | |

### *14.2.2 Solver*

**Set analysis type**

Commands : ANTYPE, ADD
Menu : Analysis → Analysis Type
Parameters : (To be filled by the user)

| **Analysis Type** | Frequency Response |
|---|---|

**Set frequency response data**

Command : FREQRESPGEN,ADD
Menu : Analysis → Frequency Response → General
Parameters : (To be filled by the user)

| | |
|---|---|
| **Response Extraction** | Auto |
| **Start Frequency** | 0.1 |
| **End Frequency** | 5 |
| **Finer Increment** | 0.01 |
| **Coarser Increment** | 0.1 |
| **Number of modes** | 5 |
| **Mass Option** | Lumped |
| **Stress Option** | No |
| **Node List** | 465/731/2165/2271/2549T2553 |

**Set Damping data**

Command : DAMPING, ADD
Menu : Analysis → Damping
Parameters : (To be filled by the user)

| | |
|---|---|
| **Damping factors** | 0.1/0.02/5/0.02 |

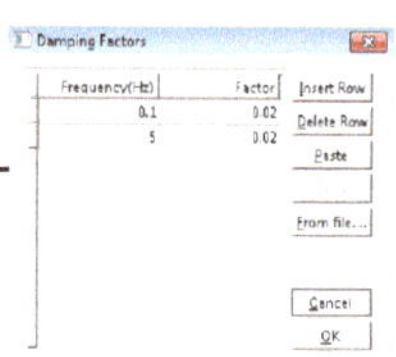


<u>**Note**</u>:

If the damping ratio is constant for all the modes, the value can be given for any two distinct frequencies and the damping ratio is assumed to be constant for all the frequencies. If it is not uniform, the value for in-between frequencies will be linearly interpolated from the extreme values.

**Add BaseExcitation Force**

Command : BASEEXCITATION, ADD
Menu : Load/BC→ Structural→ Base excitation
Parameters : (To be filled by the user)

| | |
|---|---|
| **Nodes** | 2549T2552 |
| **Magnitude** | 9810 |
| **Component** | Ux |
| **LCS** | 0 |
| **Label** | |

**Save the job and submit the job into FEAST$^{SMT}$**

**Note**:

For solving user can click ! icon in right end of tool bar.

### *14.2.3 Post-processing*

**View Results Frequency**

Command : POST, VIEWTABLE
Menu : Post→ View Table
Parameter : (To be filled by the user)

| | |
|---|---|
| Item | Frequencies |

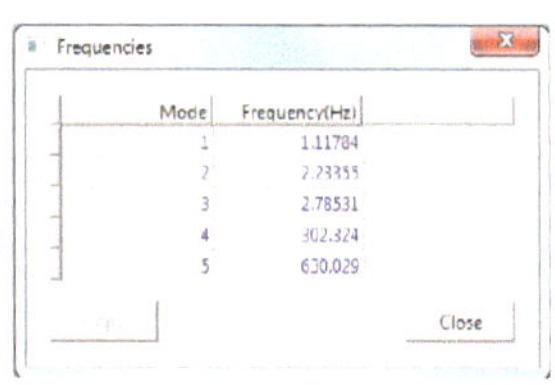


**View History plot**

Command : POST,HISTORYPLOT
Menu : Post→ History Plot

| | |
|---|---|
| Item | Acceleration |
| Component | T-RES |
| Complex As | Modulus |

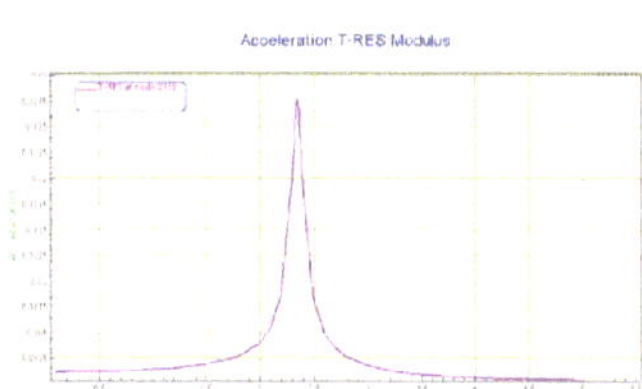

**Note**:

The result shows that the magnitude of the acceleration magnifies 14.7 times that of the input given at the base. Therefore, the electronic package has to be designed rugged enough to withstand such a large amplification.

## 14.3 Transient Analysis of a Bridge

Transient response analysis is performed on structures to monitor the response of the structure when there is a sudden application of load. Figure 14.4 shows a schematic of bridge whose response has to be monitored for the movement of a heavy truck. The bridge is made of concrete and the mass of the truck is 25 tons. It is assumed that the mass of the truck does not alter the dynamics of the bridge. The bottom of the bridge is constrained in all the DOF to simulate the foundation. The vehicle movement on the bridge is simulated by applying load due to vehicle at different points at different instants. Since the slabs and columns of the bridge are very thick with respect to the other dimensions of the bridge, brick element is used to discretize the bridge. To exhibit the features of the FEAST$^{SMT}$ software, the model is created without creating any geometries. The model is constructed by creating nodes and then elements. The distance between the front and back tires and right and left tires is assumed as 1 m. It is also assumed that the truck is moving at a speed of 4.5 km/h.

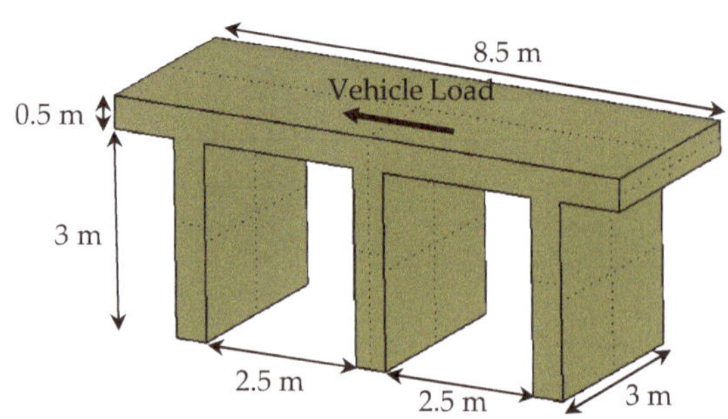


**Fig. 14.4** Schematic of a bridge

## 14.4 Pre-processor

### Create nodes specifying coordinate—(−0.25,0,1.5)

Commands : NODE, ADD
Menu : Mesh → Node → Create → Add
Parameters : (To be filled by the user)

| **Coordinate** | -0.25/0/1.5 |
|---|---|
| **Type** | Cartesian |
| **ID** | |

### Create one-dimensional element using Node extrude

Command : ELEMENT, NEXTRUDE
Menu : Mesh → Element → Create → NExtrude
Parameters : (To be filled by the user)

| **Nodes** | 1 |
|---|---|
| **Extrusion Vector** | 0.5:0:0 |
| **Twist** | |
| **Scale** | |
| **Anchor** | |
| **Segments** | 2 |
| **Type** | 2-Node |

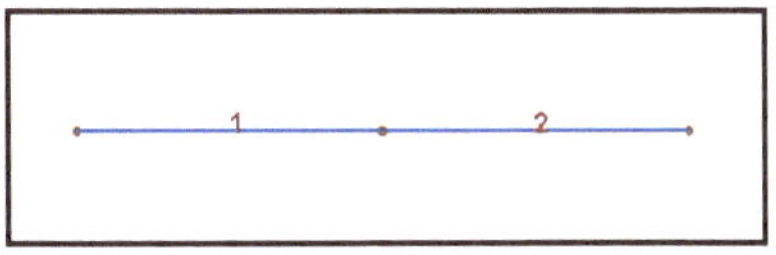


### Extrude one-dimensional elements to create two-dimensional elements

Command : ELEMENT, EXTRUDE
Menu : Mesh → Element → Create → Extrude
Parameters : (To be filled by the user)

| **List of Elements** | ALL[0] |
|---|---|
| **Extrusion Vector** | 0:0:-3 |
| **Twist** | 0 |
| **Scale** | 0 |
| **Anchor** | 0 |
| **Segments** | 12 |

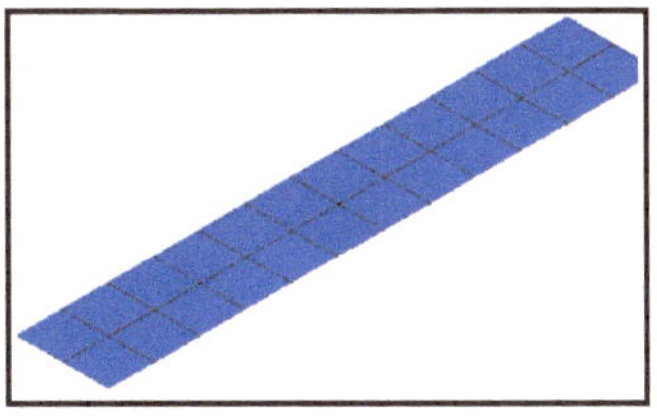

<u>**Note**</u>:

The 'Delete Source' in the parameter box is switched ON to delete the existing lower dimension elements after creating the higher dimension elements.

**Extrude two-dimensional elements to create three-dimensional elements**

Command : ELEMENT, EXTRUDE
Menu : Mesh→ Element→ Create → Extrude
Parameters : (To be filled by the user)

| **List of Elements** | ALL[0] |
|---|---|
| **Extrusion Vector** | 0:3.5:0 |
| **Twist** | 0 |
| **Scale** | 0 |
| **Anchor** | 0 |
| **Segments** | 14 |

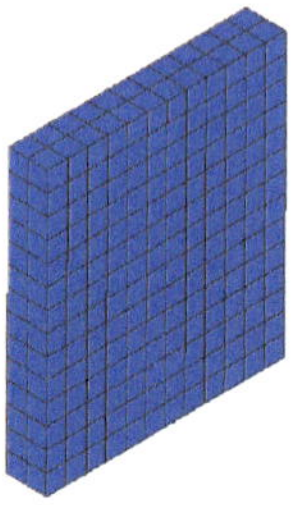

**Create two-dimensional elements using on Face option**

Command : ELEMENT, ONFACE
Menu : Mesh → Element → Create→ OnFace
Parameters : (To be filled by the user)

| **Faces** | Use Mouse to select the element face |
|---|---|

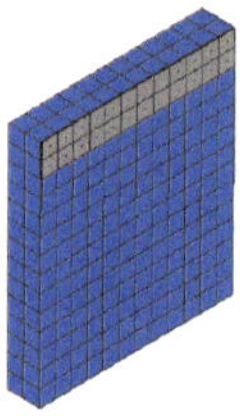

<u>**Note**</u>:

Repeat the above step to create two-dimensional elements at other columns.

**Extrude the shell element**

Command : ELEMENT, EXTRUDE
Menu : Mesh → Element → Create → Extrude
Parameters : (To be filled by the user)

| **List of Elements** | Use Mouse to select the shell element |
|---|---|
| **Extrusion Vector** | 4:0:0 |
| **Twist** | 0 |
| **Scale** | 0 |
| **Anchor** | 0 |
| **Segments** | 16 |

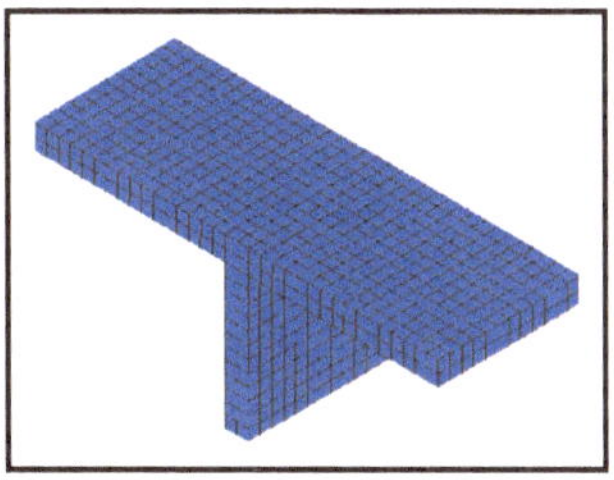

**Note**:

Repeat the above step to create three-dimensional elements from the existing two-dimensional elements.

**Copy and translate the elements**

Command : ELEMENT, TRANSLATE
Menu : Mesh → Element → Transform → Translate
Parameters : (To be filled by the user)

| **List of elements** | Use Mouse to select the shell element |
|---|---|
| **Translation** | 3:0:0 |
| **Copies** | 1 |

**Note**:

Copies "0" represents, software moves the element.

**Note**:

Repeat the above step to create other two slabs.

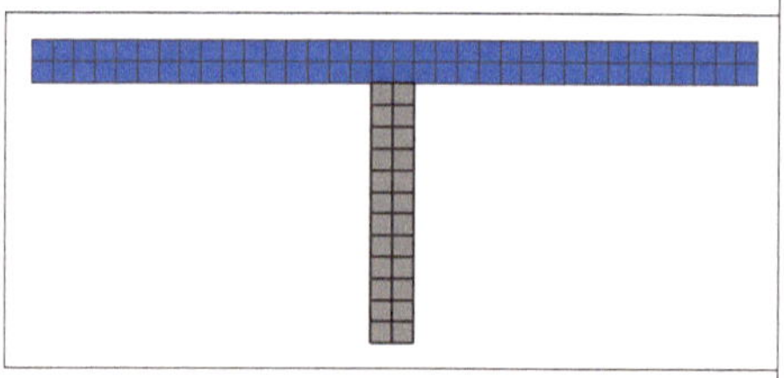
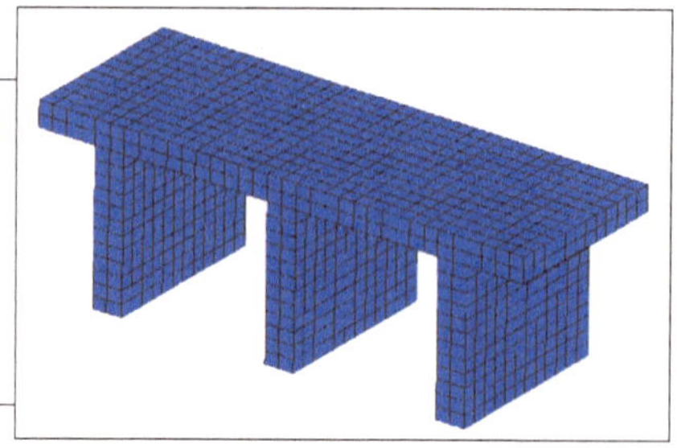

**Merge duplicate nodes**

Command : NODE, MERGE
Menu : Mesh → Node → Modify → Merge
Parameters :( To be filled by the user)

| **Nodes** | ALL[0] |
|---|---|
| **Tolerance** | 1E-04 |

**Specify material properties**

Commands : MATERIAL, ISOTROPIC
Menu : Property → Material → Structural → Isotropic
Parameters : (To be filled by the user)

| **Elements** | All[0] |
|---|---|
| **Young's Modulus** | 2.73861E+10 |
| **Nu** | 0.3 |
| **Density** | 2400 |
| **Alpha** | 0 |
| **Label** | |

## Specify Displacement constraints

Commands : DISPBC, ADD
Menu : Load/BC→ Structural → DispBC
Parameters : (To be filled by the user)

| **Nodes** | 2549T2552 |
|---|---|
| **DispBC** | 0/0/0/// |
| **LCS** | 0 |
| **Label** | |

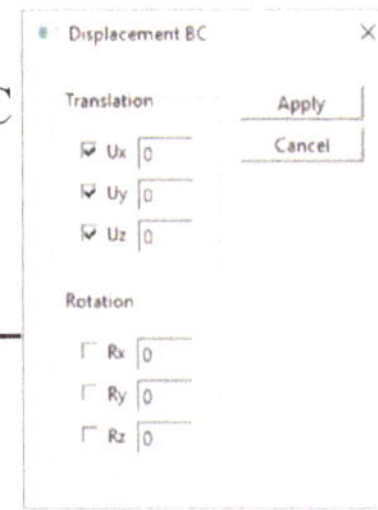


## Create the Function using Real Table

Commands : FUNCTION, TABSCALAR
Menu : Property→Function→ Scalar Table
Parameters : (To be filled by the user)

| **Table Data** | 0/0/0.1/245250/1/245250/1.1/0 |
|---|---|

Same way to create five more real table functions.

| 1 | 0 | 2.0 | 0 | 3.0 | 0 | 4.0 | 0 | 5.0 | 0 |
|---|---|---|---|---|---|---|---|---|---|
| 1.1 | 245250 | 2.1 | 245250 | 3.1 | 245250 | 4.1 | 245250 | 5.1 | 245250 |
| 2.0 | 245250 | 3.0 | 245250 | 4.0 | 245250 | 5.0 | 245250 | 6.0 | 245250 |
| 2.1 | 0 | 3.1 | 0 | 4.1 | 0 | 5.1 | 0 | 6.1 | 0 |

## Specify loads applied

Commands : POINTLOAD,ADD
Menu : Load/BC→ Structural→ Point Load
Parameters : (To be filled by the user)

| **Nodes** | 964/968/1120/1124 |
|---|---|
| **Magnitude** | F1 |
| **Component** | FY |
| **LCS** | 1 |
| **Label** | |

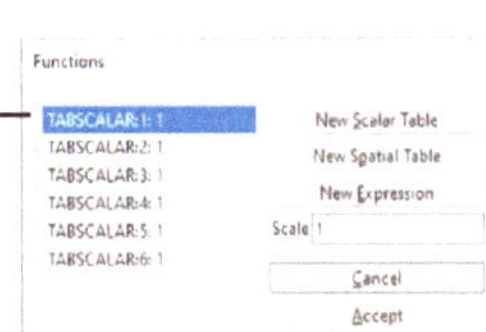

Note:

Repeat the above step to create five more Force data using the above real function data at different locations.

## 14.4.1 Solver

**Set analysis type**

Commands : ANTYPE, ADD
Menu : Analysis → Analysis Type
Parameters : (To be filled by the user)

| **Analysis Type** | Transient Response |
|---|---|

**Set transient response data**

Command : TRANSRESPGEN,ADD
Menu : Analysis → Transient Response→ General
Parameters : (To be filled by the user)

| **Mode Extraction** | 1.0/100.0 |
|---|---|
| **Number of modes** | 20 |
| **No. of time steps** | 1000 |
| **Start time** | 0 |
| **End time** | 7 |
| **Mass type** | Lumped |
| **Stress output** | YES |
| **Nodes** | ALL[0] |

**Set damping data**

Command : DAMPING, ADD
Menu : Analysis → Damping
Parameters : (To be filled by the user)

| **Damping factors** | 0.1/0.05/5/0.02 |
|---|---|

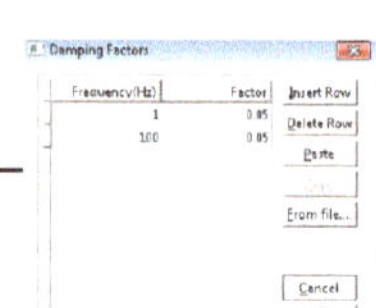

**Save the job and submit the job into FEAST$^{SMT}$**

<u>**Note**</u>:

For solving user can click  icon in right end of tool bar.

## *14.4.2 Post-processing*

**View Results Frequency**

Command : POST, TABLEVIEW
Menu : Post → View Table
Parameters : (To be filled by the user)

| Item | Frequencies |
|---|---|

**View History plot**

Command : POST,HISTORYPLOT
Menu : Post → History Plot

| Item | Displacement |
|---|---|
| Component | RES |
| Nodes | Use Mouse to pick the Nodes |

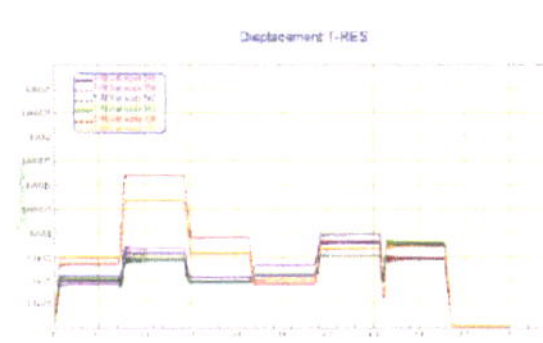

**View Stress plot**

Command : POST,HISTORYPLOT
Menu : Post → History Plot

| Item | Stress |
|---|---|
| Surface | Top |
| Component | von Mises |
| Nodes | Use Mouse to pick the Nodes |

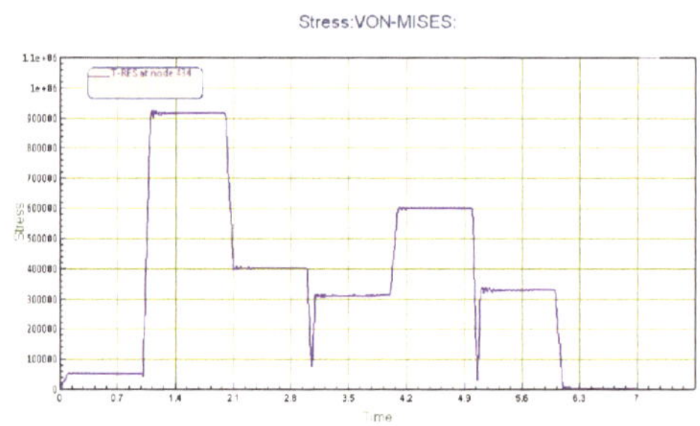


The maximum displacement occurs at the slab region and is approximately 0.18 mm and the maximum stress occurs at the bottom portion of the middle column and the value is approximately 1 MPa which is safe for the bridge.

## 14.5 Exercise Problems

14.5.1 Is it possible to find the frequency of a simple pendulum using finite element method? Justify your answer.

14.5.2 How to model a weight lifting bar with weight attached at both the ends? Assume the bar to be initially straight.

14.5.3 What is the significance of rigid body modes in an unconstrained structure?

14.5.4 A diver weighing 70 kg jumps from an aluminum springboard to the swimming pool. If the modal damping of the springboard is 2% for all the modes, find the time for the springboard to come at rest. The length of the springboard is 2 m and its cross section is 0.02 m × 0.3 m. The springboard is hinged at one end and a fulcrum support is provided at a distance of 0.25 m from the hinged end. Try discretizing the springboard initially with beam elements and then with plate element. Comment on the results.

14.5.5 Repeat the above problem by considering only first two modes and then with first ten modes. Comment on the results.

14.5.6 A casing containing electronic components weighing 50 kg has to be tested for sine vibrations on a vibration shaker. The test article is mounted onto the shaker through a cylindrical fixture as shown in the figure. The input sine frequency varies from 1 to 100 Hz with an acceleration of 1 g. Find response at center of gravity of the casing due to base excitation.

**Hint**: Assume the entire mass of the test article is concentrated at the C.G and model it as a lumped mass connecting to the fixture through rigid links.

14.5.7 How to find the fundamental frequencies of a square plate by modeling only one fourth or one-half of the plate?

14.5.8 Usually for a structure with refined mesh, the number of frequencies is equal to the number of independent degrees of freedom. However, the user is interested in first few fundamental modes. Even for performing a response analysis, first few modes are only considered. Justify.

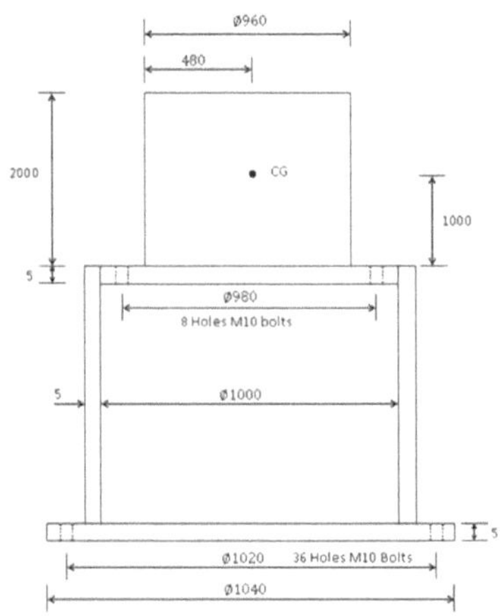


14.5.9 What is the nature of convergence for a free vibration analysis? Is it possible to comment about the convergence of a mode shape or frequency just by looking at the mode shape of it? Explain it

14.5.10 A user has modeled a structure in SI units and got the frequencies. While transferring the model to a different agency, the agency requested him to send the model in with length units alone converted to "mm". What are all the other unit changes other than length dimension, the user has to introduce in his model to get the same frequency?

# Chapter 15
# Buckling Analysis Using FEAST$^{SMT}$ Software

**T. J. Raj Thilak**

> *You cannot teach a man anything; you can only help him discover it in himself.*
> *–Galileo*

## 15.1 Buckling Analysis of a Cylinder with Stiffeners

Buckling is an instability phenomenon caused mainly due to compressive loading. Linear buckling is an eigenvalue problem in which the eigenvalues are the critical load factors and corresponding eigenvectors are buckling mode shapes. The problem under consideration is a stiffened cylindrical shell with bottom and top flanges as shown in Fig. 15.1.

The thickness of the structure is 5 mm. The cylinder is reinforced with 36 equally spaced square stiffeners with cross section 20 mm $\times$ 20 mm. The cylinder is bolted at the bottom flange and a compressive load of 100 kN is applied at the top flange. Since the cylinder is very thin compared to the height and diameter, it is modeled using shell element. The stiffeners are modeled using beam element since its cross section is very small compared to its length.

T. J. R. Thilak (✉)
FSDD/SDAG/STR, VSSC, Veli, Thiruvananthapuram, Kerala 695022, India
e-mail: tj_rajthilak@vssc.gov.in; rajthilaks@gmail.com

 
S. U. Nair and S. Somanath (eds.), *Introduction to Finite Element Analysis*,
https://doi.org/10.1007/978-981-19-7989-7_15

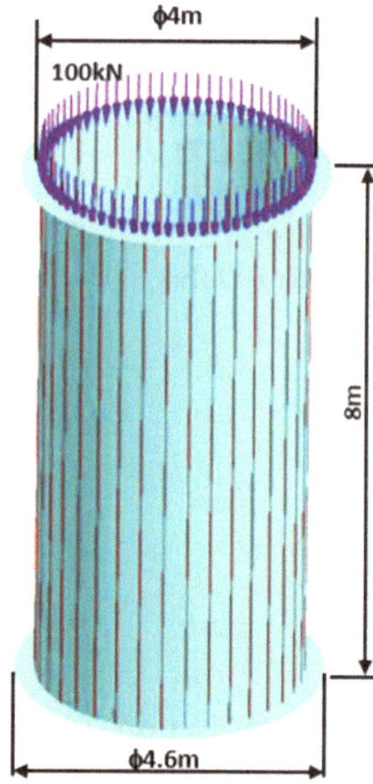


**Fig. 15.1** Cylinder with compressive loading

## 15.2 Pre-processor

**Create geometric coordinates (2:0:0), (2:8:0)**

Commands : POINT, ADD
Menu : Geometry → Key point → Create → Add
Parameters : (To be filled by the user)

| **Point Data** | 0:0:0 |
|---|---|

After entering the coordinate data click 'Apply' or press 'Cntrl+Enter'. Likewise, each coordinate is created using the above command options.

**Create a line**

Commands : CURVE, LINE
Menu : Geometry → Curve → Create → Line

| **End points** | Use Mouse to pick the points |
|---|---|

**Meshing using beam elements**

Commands : MESH, BAR
Menu : Mesh→ MeshGen→ Bar
Parameters : (To be filled by the user)

| | |
|---|---|
| **Curve** | 1 |
| **Element Size** | 0.2 |
| **Type** | 2-node |
| **Divisions** | 40 |
| **Bias** | 1.0 |

Note:

Right click/left click mouse point to alter sub-divisions and Bias.

**Sweep the beam element**

Commands : ELEMENT, SWEEP
Menu : Mesh → Element→Create→ Sweep
Parameters : (To be filled by the user)

| | |
|---|---|
| **List of Elements** | ALL[0] |
| **Angle** | 360 |
| **Axis** | Y Axis |
| **Location** | 0:0:0 |
| **Twist** | |
| **Scale** | |
| **Anchor** | |
| **Segments** | 72 |

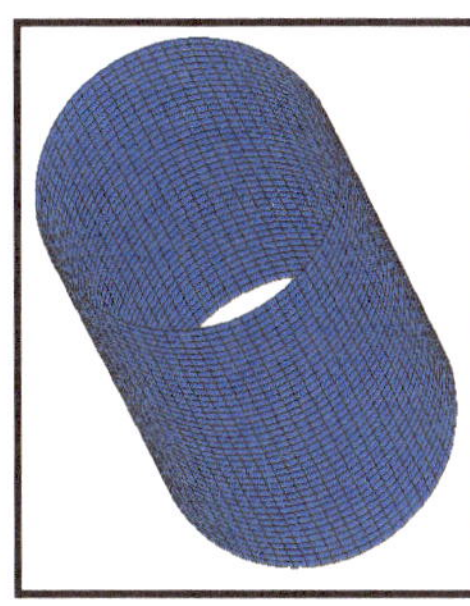

### Create Flange using Edge offset

Commands : EDGEMESH, OFFSET
Menu : Mesh → FE Mesh → Using Edges → Offset
Parameters : (To be filled by the user)

| Element edges | Select element boundary edges |
|---|---|
| Offset value | 0.3 |
| Segments | 2 |
| Reference Point | |
| Output ID | |
| Node output ID | |

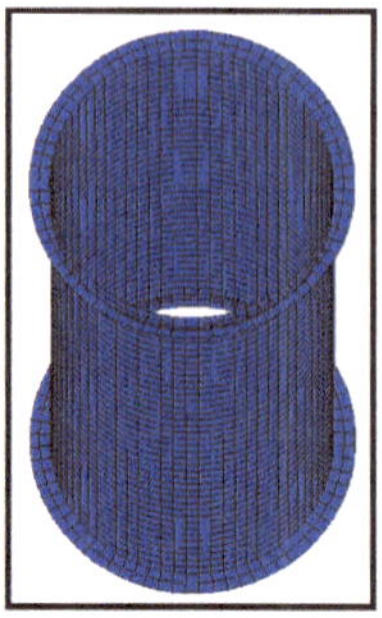

### Create Stiffeners

Commands : EDGEMESH, STIFFENER
Menu : Mesh → FE Mesh → Using Edges → Stiffener
Parameters : (To be filled by the user)

| Element edges | Use mouse to select the Element edges |
|---|---|
| Material ID | 1 |
| Property ID | 1 |

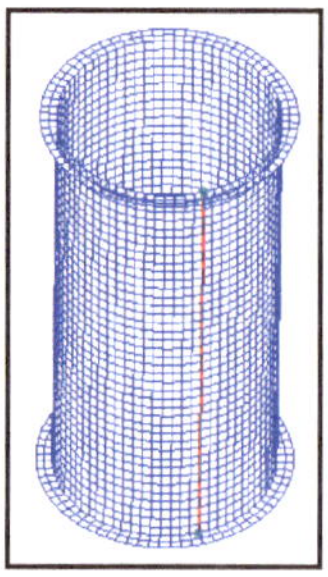

**Create more stiffeners using Rotate element option**

Commands : ELEMENT, ROTATE
Menu : Mesh→ Element→Transform→Rotate
Parameters : (To be filled by the user)

| **List of Elements** | Use mouse to select the elements |
|---|---|
| **Angle** | 10 |
| **Axis** | Y Axis |
| **Location** | 0:0:0 |
| **Copies** | 35 |

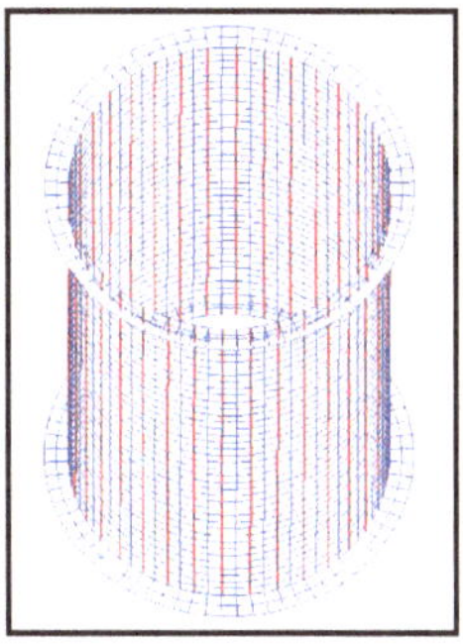

**Create thickness**

Commands : THICKNESS, ADD
Menu : Property → Physical → Thickness
Parameters : (To be filled by the user)

| **Elements** | Select 2-D element |
|---|---|
| **Thickness** | 0.005 |
| **Label** | |

**Specify Beam Properties**

Command : BEAMPROP, ADD
Menu : Property → Physical → Beam Properties → CrossSection
Parameters : (To be filled by the user)

| Elements | All[0] |
|---|---|
| Cross section | RECT/0.02/0.02 |
| Centroid offsets | 0/0 |
| Label | |

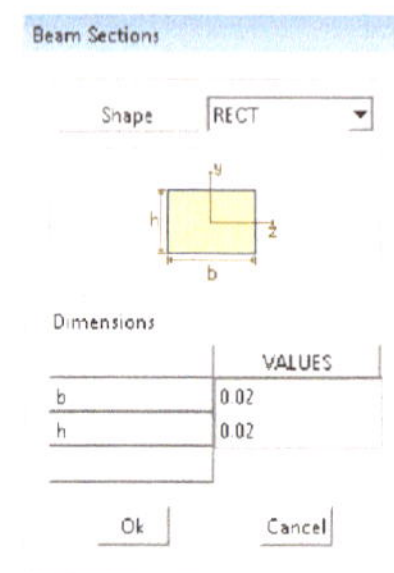


**Create material property**

Command : MATERIAL, ISOTROPIC
Menu : Property → Material → Structural → Isotropic
Parameters : (To be filled by the user)

| Elements | ALL[0] |
|---|---|
| Young's Modulus | 2.1E+11 |
| Nu | 0.3 |
| Density | 7800 |
| Alpha | 0 |
| Label | |

**Create one Node—(0, 4, 0)**

Commands : NODE, ADD
Menu : Mesh → Node → Create → Add
Parameters : (To be filled by the user)

| Coordinate | 0:4:0 |
|---|---|
| Coordinate Type | Cartesian |
| ID | |

**Specify Reference Node for beam**

Commands : BEAMLCS, ADD
Menu : Property → Physical → Beam Properties → BeamLCS
Parameters : (To be filled by the user)

| | |
|---|---|
| **Elements** | Use Mouse to select beam element |
| **Type** | Towards a node |
| **Node** | 4820 |

**Merge duplicate nodes**

Command : NODE, MERGE
Menu : Mesh → Node → Modify → Merge
Parameters : (To be filled by the user)

| | |
|---|---|
| **Nodes** | ALL |
| **Tolerance** | 1E-04 |

**Specify Displacement constraints**

Commands : DISPBC, ADD
Menu : Load/BC → Structural → DispBC
Parameters : (To be filled by the user)

| | |
|---|---|
| **Nodes** | Select the Bottom Flange Center Nodes(2961/2962/2969T3383B6) |
| **DispBC** | 0/0/0/0/0/0 |
| **LCS** | 0 |
| **Label** | |

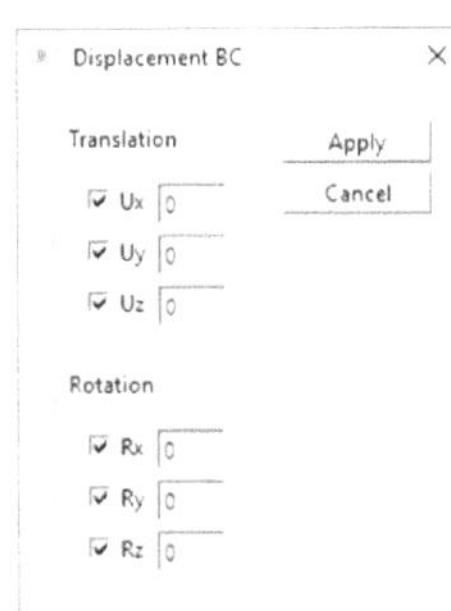

**Specify Point loads**

Commands : POINTLOAD,ADD
Menu : Load/BC → Structural → Point Load
Parameters : (To be filled by the user)

| | |
|---|---|
| **Nodes** | Cylinder top nodes (1/4/84T2913B41) |
| **Magnitude** | -100 |
| **Component** | Fy |
| **LCS** | 1 |
| **Label** | |

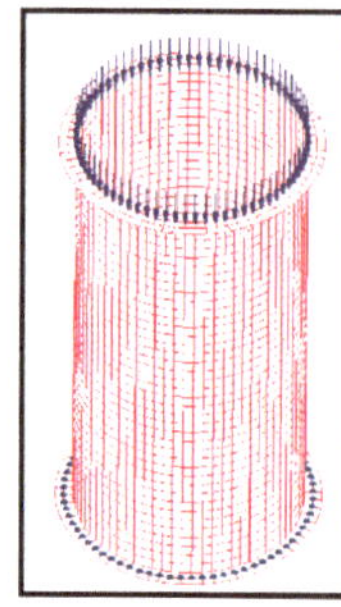

## 15.3 Solution

**Set analysis type**

Commands : ANTYPE, ADD
Menu : Analysis → Analysis Type
Parameters : (To be filled by the user)

| | |
|---|---|
| **Analysis Types** | Buckling |

**Set buckling data**

Commands : BUCKLINGGEN, ADD
Menu : Analysis → Buckling → General
Parameters : (To be filled by the user)

| | |
|---|---|
| **Number of modes** | 5 |
| **Stress output** | Yes |

**Save the job and submit the job into FEAST$^{SMT}$**

<u>**Note**</u>:

For solving user can click ! icon in right end of tool bar.

## 15.4 Result Viewing Using Post-processing

**View Results Frequency**

Command : POST, TABLEVIEW
Menu : Post → View Table
Parameters : (To be filled by the user)

| **Item** | Critical Load Factor |
|---|---|

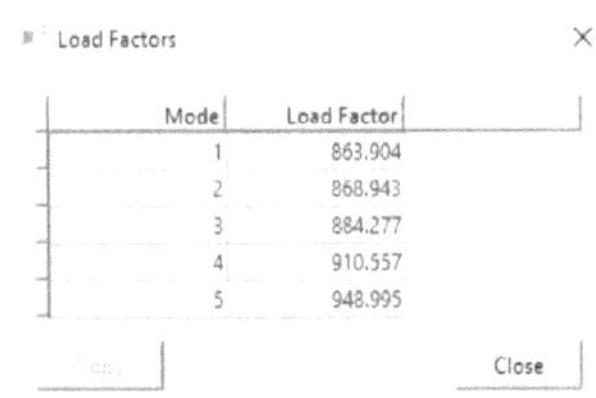
Load Factors

| Mode | Load Factor |
|---|---|
| 1 | 863.904 |
| 2 | 868.943 |
| 3 | 884.277 |
| 4 | 910.557 |
| 5 | 948.995 |

Close

**Deformed shape**

Command : POST, DEFLECTION
Menu : Post → Deflection

| Item | Buckling Mode |
|---|---|
| Modes | 1 -863.904 |
| Scale Factor | 1 |

**Mode Shape Contour**

| Command | : | POST, CONTOUR |
|---|---|---|
| Menu | : | Post → Contour |

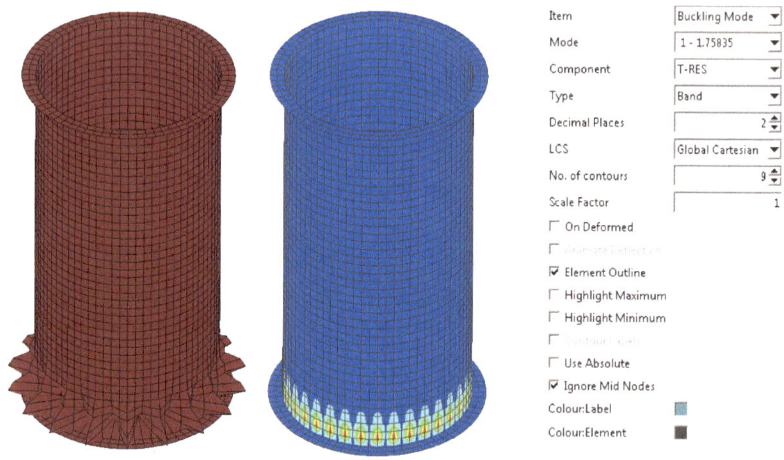


## 15.5 Exercise Problems

15.5.1 Take the dimension of a diet Pepsi can. How many cans can be stacked vertically in a truck without crushing the bottom most can?

15.5.2 What is the importance of corrugation in a water bottle? Justify your answer by doing a buckling analysis of the water bottle with and without the corrugation.

15.5.3 Find the critical load for the configuration shown below

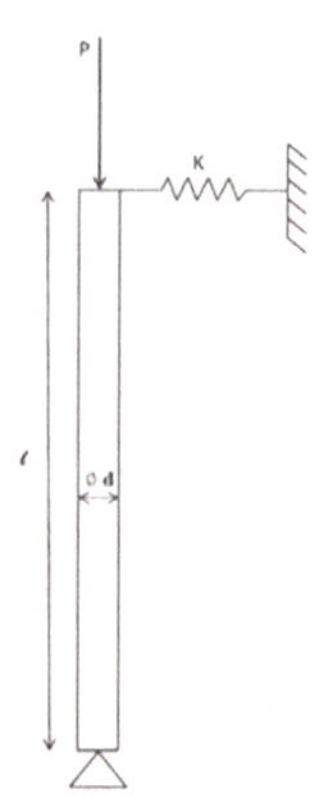


l = 10 m

d = 0.5 m

P = 1000 N

K = 100 N/m

15.5.4 What is the meaning of negative buckling load factor in a buckling analysis?

15.5.5 Can a structure have zero buckling load factor? Explain.

15.5.6 Find the buckling mode of the truss structure shown below. The structure is made of aluminum with area of cross section 0.1 $m^2$.

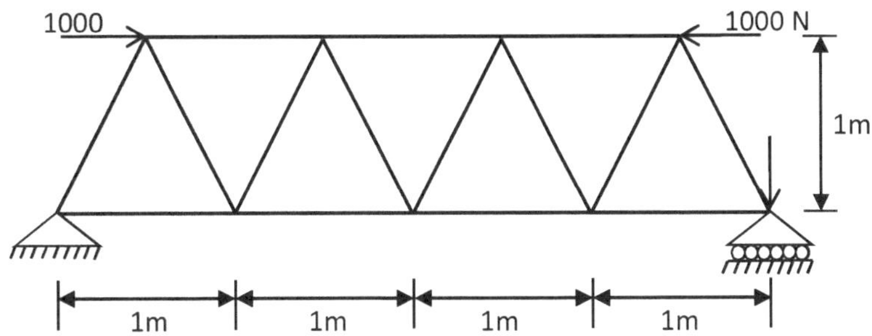


15.5.7 Find the buckling mode of the steel beam as shown below subjected to a thermal load of 200 C.

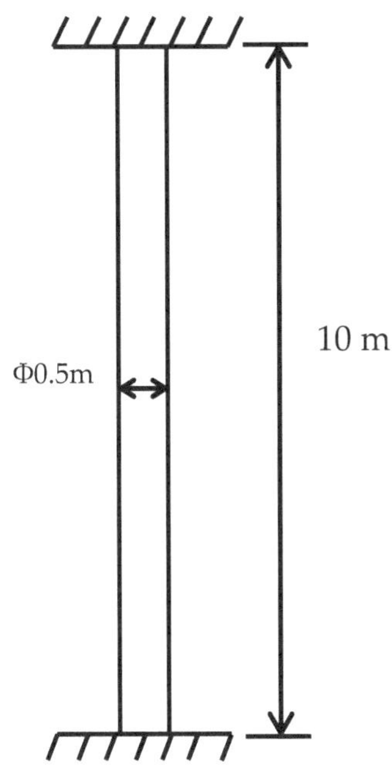

# Index




S. U. Nair and S. Somanath (eds.), *Introduction to Finite Element Analysis*, https://doi.org/10.1007/978-981-19-7989-7